STUDENT SOLUTIONS MANUAL

MARK SCHERVISH
Carnegie Mellon University

PROBABILITY AND STATISTICS

FOURTH EDITION

Morris DeGroot
Carnegie Mellon University

Mark Schervish
Carnegie Mellon University

Addison-Wesley
is an imprint of

Reproduced by Pearson Addison-Wesley from electronic files supplied by the author.

Publishing as Pearson Addison-Wesley, 75 Arlington Street, Boston, MA 02116.

ISBN-13: 978-0-321-71598-2
ISBN-10: 0-321-71598-5

www.pearsonhighered.com

Contents

Preface

This manual contains solutions to the odd-numbered exercises in *Probability and Statistics*, 4th edition, by Morris DeGroot and myself. I have preserved many of the solutions to the exercises that existed in the 3rd edition. Certainly errors have been introduced, and I will post any errors brought to my attention on my web page `http://www.stat.cmu.edu/ mark/` along with errors in the text itself. Feel free to send me comments.

Some of the solutions make use of the results in earlier exercises. If a result is proven in an earlier exercise, its proof is not repeated, even if the earlier exercise had an even number so that its solution is not in this manual.

Mark J. Schervish

Chapter 1

Introduction to Probability

1.4 Set Theory

1. Assume that $x \in B^c$. We need to show that $x \in A^c$. We shall show this indirectly. Assume, to the contrary, that $x \in A$. Then $x \in B$ because $A \subset B$. This contradicts $x \in B^c$. Hence $x \in A$ is false and $x \in A^c$.

3. To prove the first result, let $x \in (A \cup B)^c$. This means that x is not in $A \cup B$. In other words, x is neither in A nor in B. Hence $x \in A^c$ and $x \in B^c$. So $x \in A^c \cap B^c$. This proves that $(A \cup B)^c \subset A^c \cap B^c$. Next, suppose that $x \in A^c \cap B^c$. Then $x \in A^c$ and $x \in B^c$. So x is neither in A nor in B, so it can't be in $A \cup B$. Hence $x \in (A \cup B)^c$. This shows that $A^c \cap B^c \subset (A \cup B)^c$. The second result follows from the first by applying the first result to A^c and B^c and then taking complements of both sides.

5. To prove the first result, let $x \in (\cup_i A_i)^c$. This means that x is not in $\cup_i A_i$. In other words, for every $i \in I$, x is not in A_i. Hence for every $i \in I$, $x \in A_i^c$. So $x \in \cap_i A_i^c$. This proves that $(\cup_i A_i)^c \subset \cap_i A_i^c$. Next, suppose that $x \in \cap_i A_i^c$. Then $x \in A_i^c$ for every $i \in I$. So for every $i \in I$, x is not in A_i. So x can't be in $\cup_i A_i$. Hence $x \in (\cup_i A_i)^c$. This shows that $\cap_i A_i^c \subset (\cup_i A_i)^c$. The second result follows from the first by applying the first result to A_i^c for $i \in I$ and then taking complements of both sides.

7. (a) These are the points not in A, hence they must be either below 1 or above 5. That is $A^c = \{x : x < 1 \text{ or } x > 5\}$.

   (b) These are the points in either A or B or both. So they must be between 1 and 5 or between 3 and 7. That is, $A \cup B = \{x : 1 \le x \le 7\}$.

   (c) These are the points in B but not in C. That is $BC^c = \{x : 3 < x \le 7\}$. (Note that $B \subset C^c$.)

   (d) These are the points in none of the three sets, namely $A^c B^c C^c = \{x : 0 < x < 1 \text{ or } x > 7\}$.

   (e) These are the points in the answer to part (b) and in C. There are no such values and $(A \cup B)C = \emptyset$.

9. (a) For each n, $B_n = B_{n+1} \cup A_n$, hence $B_n \supset B_{n+1}$ for all n. For each n, $C_{n+1} \cap A_n = C_n$, so $C_n \subset C_{n+1}$.

   (b) Suppose that $x \in \cap_{n=1}^{\infty} B_n$. Then $x \in B_n$ for all n. That is, $x \in \cup_{i=n}^{\infty} A_i$ for all n. For $n = 1$, there exists $i \ge n$ such that $x \in A_i$. Assume to the contrary that there are at most finitely many i such that $x \in A_i$. Let m be the largest such i. For $n = m + 1$, we know that there is $i \ge n$ such that $x \in A_i$. This contradicts m being the largest i such that $x \in A_i$. Hence, x is in infinitely many A_i. For the other direction, assume that x is in infinitely many A_i. Then, for every n, there is a value of $j > n$ such that $x \in A_j$, hence $x \in \cup_{i=n}^{\infty} A_i = B_n$ for every n and $x \in \cap_{n=1}^{\infty} B_n$.

(c) Suppose that $x \in \cup_{n=1}^{\infty} C_n$. That is, there exists n such that $x \in C_n = \cap_{i=n}^{\infty} A_i$, so $x \in A_i$ for all $i \geq n$. So, there at most finitely many i (a subset of $1, \ldots, n-1$) such that $x \notin A_i$. Finally, suppose that $x \in A_i$ for all but finitely many i. Let k be the last i such that $x \notin A_i$. Then $x \in A_i$ for all $i \geq k+1$, hence $x \in \cap_{i=k+1}^{\infty} A_i = C_{k+1}$. Hence $x \in \cup_{n=1}^{\infty} C_n$.

11. (a) All of the events mentioned can be determined by knowing the voltages of the two subcells. Hence the following set can serve as a sample space

$$S = \{(x, y) : 0 \leq x \leq 5 \text{ and } 0 \leq y \leq 5\},$$

where the first coordinate is the voltage of the first subcell and the second coordinate is the voltage of the second subcell. Any more complicated set from which these two voltages can be determined could serve as the sample space, so long as each outcome could at least hypothetically be learned.

(b) The power cell is functional if and only if the sum of the voltages is at least 6. Hence, $A = \{(x, y) \in S : x + y \geq 6\}$. It is clear that $B = \{(x, y) \in S : x = y\}$ and $C = \{(x, y) \in S : x > y\}$. The powercell is not functional if and only if the sum of the voltages is less than 6. It needs less than one volt to be functional if and only if the sum of the voltages is greater than 5. The intersection of these two is the event $D = \{(x, y) \in S : 5 < x + y < 6\}$. The restriction "$\in S$" that appears in each of these descriptions guarantees that the set is a subset of S. One could leave off this restriction and add the two restrictions $0 \leq x \leq 5$ and $0 \leq y \leq 5$ to each set.

(c) The description can be worded as "the power cell is not functional, and needs at least one more volt to be functional, and both subcells have the same voltage." This is the intersection of A^c, D^c, and B. That is, $A^c \cap D^c \cap B$. The part of D^c in which $x + y \geq 6$ is not part of this set because of the intersection with A^c.

(d) We need the intersection of A^c (not functional) with C^c (second subcell at least as big as first) and with B^c (subcells are not the same). In particular, $C^c \cap B^c$ is the event that the second subcell is strictly higher than the first. So, the event is $A^c \cap B^c \cap C^c$.

1.5 The Definition of Probability

1. Define the following events:

$$\begin{aligned} A &= \{\text{the selected ball is red}\}, \\ B &= \{\text{the selected ball is white}\}, \\ C &= \{\text{the selected ball is either blue, yellow, or green}\}. \end{aligned}$$

We are asked to find $\Pr(C)$. The three events A, B, and C are disjoint and $A \cup B \cup C = S$. So $1 = \Pr(A) + \Pr(B) + \Pr(C)$. We are told that $\Pr(A) = 1/5$ and $\Pr(B) = 2/5$. It follows that $\Pr(C) = 2/5$.

3. (a) If A and B are disjoint then $B \subset A^c$ and $BA^c = B$, so $\Pr(BA^c) = \Pr(B) = 1/2$.

(b) If $A \subset B$, then $B = A \cup (BA^c)$ with A and BA^c disjoint. So $\Pr(B) = \Pr(A) + \Pr(BA^c)$. That is, $1/2 = 1/3 + \Pr(BA^c)$, so $\Pr(BA^c) = 1/6$.

(c) According to Theorem 1.4.11, $B = (BA) \cup (BA^c)$. Also, BA and BA^c are disjoint so, $\Pr(B) = \Pr(BA) + \Pr(BA^c)$. That is, $1/2 = 1/8 + \Pr(BA^c)$, so $\Pr(BA^c) = 3/8$.

5. Using the same notation as in Exercise 4, we now want $\Pr(E_1^c \cap E_2^c)$. According to Theorems 1.4.9 and 1.5.3, this equals $1 - \Pr(E_1 \cup E_2) = 0.4$.

7. Rearranging terms in Eq. (1.5.1) of the text, we get

$$\Pr(A \cap B) = \Pr(A) + \Pr(B) - \Pr(A \cup B) = 0.4 + 0.7 - \Pr(A \cup B) = 1.1 - \Pr(A \cup B).$$

So $\Pr(A \cap B)$ is largest when $\Pr(A \cup B)$ is smallest and vice-versa. The smallest possible value for $\Pr(A \cup B)$ occurs when one of the events is a subset of the other. In the present exercise this could only happen if $A \subset B$, in which case $\Pr(A \cup B) = \Pr(B) = 0.7$, and $\Pr(A \cap B) = 0.4$. The largest possible value of $\Pr(A \cup B)$ occurs when either A and B are disjoint or when $A \cup B = S$. The former is not possible since the probabilities are too large, but the latter is possible. In this case $\Pr(A \cup B) = 1$ and $\Pr(A \cap B) = 0.1$.

9. The required probability is

$$\begin{aligned}\Pr(A \cap B^C) + \Pr(A^C B) &= [\Pr(A) - \Pr(A \cap B)] + [\Pr(B) - \Pr(A \cap B)] \\ &= \Pr(A) + \Pr(B) - 2\Pr(A \cap B).\end{aligned}$$

11. (a) The set of points for which $(x - 1/2)^2 + (y - 1/2)^2 < 1/4$ is the interior of a circle that is contained in the unit square. (Its center is $(1/2, 1/2)$ and its radius is $1/2$.) The area of this circle is $\pi/4$, so the area of the remaining region (what we want) is $1 - \pi/4$.

(b) We need the area of the region between the two lines $y = 1/2 - x$ and $y = 3/2 - x$. The remaining area is the union of two right triangles with base and height both equal to $1/2$. Each triangle has area $1/8$, so the region between the two lines has area $1 - 2/8 = 3/4$.

(c) We can use calculus to do this. We want the area under the curve $y = 1 - x^2$ between $x = 0$ and $x = 1$. This equals

$$\int_0^1 (1 - x^2)dx = x - \left.\frac{x^3}{3}\right|_{x=0}^{1} = \frac{2}{3}.$$

(d) The area of a line is 0, so the probability of a line segment is 0.

13. We know from Exercise 12 that

$$\Pr\left(\bigcup_{i=1}^{n} A_i\right) = \sum_{i=1}^{n} \Pr(B_i).$$

Furthermore, from the definition of the events $B_1, \ldots, B_n$ it is seen that $B_i \subset A_i$ for $i = 1, \ldots, n$. Therefore, by Theorem 1.5.4, $\Pr(B_i) \leq \Pr(A_i)$ for $i = 1, \ldots, n$. It now follows that

$$\Pr\left(\bigcup_{i=1}^{n} A_i\right) \leq \sum_{i=1}^{n} \Pr(A_i).$$

(Of course, if the events $A_1, \ldots, A_n$ are disjoint, there is equality in this relation.)

For the second part, apply the first part with A_i replaced by A_i^c for $i = 1, \ldots, n$. We get

$$\Pr\left(\bigcup A_i^c\right) \leq \sum_{i=1}^{n} \Pr(A_i^c). \tag{S.1.1}$$

Exercise 5 in Sec. 1.4 says that the left side of (S.1.1) is $\Pr\left([\bigcap A_i]^c\right)$. Theorem 1.5.3 says that this last probability is $1 - \Pr\left(\bigcap A_i\right)$. Hence, we can rewrite (S.1.1) as

$$1 - \Pr\left(\bigcap A_i\right) \leq \sum_{i=1}^{n} \Pr(A_i^c).$$

Finally take one minus both sides of the above inequality (which reverses the inequality) and produces the desired result.

1.6 Finite Sample Spaces

1. The safe way to obtain the answer at this stage of our development is to count that 18 of the 36 outcomes in the sample space yield an odd sum. Another way to solve the problem is to note that regardless of what number appears on the first die, there are three numbers on the second die that will yield an odd sum and three numbers that will yield an even sum. Either way the probability is 1/2.

3. The only differences greater than or equal to 3 that are available are 3, 4 and 5. These large difference only occur for the six outcomes in the upper right and the six outcomes in the lower left of the array in Example 1.6.5 of the text. So the probability we want is $1 - 12/36 = 2/3$.

5. The probability of being in an odd-numbered grade is $2x + x + x = 4x = 4/7$.

7. The possible genotypes of the offspring are aa and Aa, since one parent will definitely contribute an a, while the other can contribute either A or a. Since the parent who is Aa contributes each possible allele with probability 1/2 each, the probabilities of the two possible offspring are each 1/2 as well.

1.7 Counting Methods

1. Each pair of starting day and leap year/no leap year designation determines a calendar, and each calendar correspond to exactly one such pair. Since there are seven days and two designations, there are a total of $7 \times 2 = 14$ different calendars.

3. This is a simple matter of permutations of five distinct items, so there are $5! = 120$ ways.

5. Let the sample space consist of all four-tuples of dice rolls. There are $6^4 = 1296$ possible outcomes. The outcomes with all four rolls different consist of all of the permutations of six items taken four at a time. There are $P_{6,4} = 360$ of these outcomes. So the probability we want is $360/1296 = 5/18$.

7. There are 20^{12} possible outcomes in the sample space. If the 12 balls are to be thrown into different boxes, the first ball can be thrown into any one of the 20 boxes, the second ball can then be thrown into any one of the other 19 boxes, etc. Thus, there are $20 \cdot 19 \cdot 18 \cdots 9$ possible outcomes in the event. So the probability is $20!/[8!20^{12}]$.

9. There are $6!$ possible arrangements in which the six runners can finish the race. If the three runners from team A finish in the first three positions, there are $3!$ arrangements of these three runners among these three positions and there are also $3!$ arrangements of the three runners from team B among the last three positions. Therefore, there are $3! \times 3!$ arrangements in which the runners from team A finish in the first three positions and the runners from team B finish in the last three positions. Thus, the probability is $(3!3!)/6! = 1/20$.

11. In terms of factorials, $P_{n,k} = n!/[k!(n-k)!]$. Since we are assuming that n and $n = k$ are large, we can use Stirling's formula to approximate both of them. The approximation to $n!$ is $(2\pi)^{1/2}n^{n+1/2}e^{-n}$, and the approximation to $(n-k)!$ is $(2\pi)^{1/2}(n-k)^{n-k+1/2}e^{-n+k}$. The approximation to the ratio is the ratio of the approximations because the ratio of each approximation to its corresponding factorial converges to 1. That is,

$$\frac{n!}{k!(n-k)!} \approx \frac{(2\pi)^{1/2}n^{n+1/2}e^{-n}}{k!(2\pi)^{1/2}(n-k)^{n-k+1/2}e^{-n+k}} = \frac{e^{-k}n^k}{k!}\left(1-\frac{k}{n}\right)^{-n-k-1/2}.$$

Further simplification is available if one assumes that k is small compared to n, that is $k/n \approx 0$. In this case, the last factor is approximately e^k, and the whole approximation simplifies to $n^k/k!$. This makes sense because, if $n/(n-k)$ is essentially 1, then the product of the k largest factors in $n!$ is essentially n^k.

1.8 Combinatorial Methods

1. We have to assign 10 houses to one pollster, and the other pollster will get to canvas the other 10 houses. Hence, the number of assignments is the number of combinations of 20 items taken 10 at a time,

$$\binom{20}{10} = 184{,}756.$$

3. Since $93 = 63 + 30$, the two numbers are the same.

5. The number is $\dfrac{4251!}{(97!4154!)} = \dbinom{4251}{97}$, an integer.

7. There are $\binom{n}{k}$ possible sets of k seats to be occupied, and they are all equally likely. There are $n-k+1$ sets of k adjacent seats, so the probability we want is

$$\frac{n-k+1}{\binom{n}{k}} = \frac{(n-k+1)!k!}{n!}.$$

9. This problem is slightly tricky. The total number of ways of choosing the n seats that will be occupied by the n people is $\binom{2n}{n}$. Offhand, it would seem that there are only two ways of choosing these seats so that no two adjacent seats are occupied, namely:

X0X0...0 and 0X0X...0X

Upon further consideration, however, $n-1$ more ways can be found, namely:

X00X0X...0X, X0X00X0X...0X, etc.

Therefore, the total number of ways of choosing the seats so that no two adjacent seats are occupied is $n+1$. The probability is $(n+1)/\binom{2n}{n}$.

11. This exercise is similar to Exercise 10. Let the sample space consist of all subsets (unordered) of 12 out of the 100 people in the group. There are $\binom{100}{12}$ such subsets. The number of subsets that contain A and B is the number of subsets of size 10 out of the other 98 people, $\binom{98}{10}$, so the probability we want is

$$\frac{\binom{98}{10}}{\binom{100}{12}} = \frac{12 \times 11}{100 \times 99} = 0.01333.$$

13. This exercise is similar to Exercise 12. Here, we want four designated bulbs to be in the same group. The probability is

$$\frac{\binom{20}{6} + \binom{20}{10}}{\binom{24}{10}} = 0.1140.$$

15. (a) If we express 2^n as $(1+1)^n$ and expand $(1+1)^n$ by the binomial theorem, we obtain the desired result.

    (b) If we express 0 as $(1-1)^n$ and expand $(1-1)^n$ by the binomial theorem, we obtain the desired result.

17. Call the four players A, B, C, and D. The number of ways of choosing the positions in the deck that will be occupied by the four aces is $\binom{52}{4}$. Since player A will receive 13 cards, the number of ways of choosing the positions in the deck for the four aces so that all of them will be received by player A is $\binom{13}{4}$. Similarly, since player B will receive 13 other cards, the number of ways of choosing the positions for the four aces so that all of them will be received by player B is $\binom{13}{4}$. A similar result is true for each of the other players. Therefore, the total number of ways of choosing the positions in the deck for the four aces so that all of them will be received by the same player is $4\binom{13}{4}$. Thus, the final probability is $4\binom{13}{4} / \binom{52}{4}$.

19. From the description of what counts as a collection of customer choices, we see that each collection consists of a tuple $(m_1, \ldots, m_n)$, where m_i is the number of customers who choose item i for $i = 1, \ldots, n$. Each m_i must be between 0 and k and $m_1 + \cdots + m_n = k$. Each such tuple is equivalent to a sequence of $n + k - 1$ 0's and 1's as follows. The first m_1 terms are 0 followed by a 1. The next m_2 terms are 0 followed by a 1, and so on up to m_{n-1} 0's followed by a 1 and finally m_n 0's. Since $m_1 + \cdots + m_n = k$ and since we are putting exactly $n - 1$ 1's into the sequence, each such sequence has exactly $n + k - 1$ terms. Also, it is clear that each such sequence corresponds to exactly one tuple of customer choices. The numbers of 0's between successive 1's give the numbers of customers who choose that item, and

the 1's indicate where we switch from one item to the next. So, the number of combinations of choices is the number of such sequences: $\binom{n+k-1}{k}$.

21. We are asked for the number of unordered samples with replacement, as constructed in Exercise 19. Here, $n = 365$, so there are $\binom{365+k}{k}$ different unordered sets of k birthdays chosen with replacement from $1, \dots, 365$.

1.9 Multinomial Coefficients

1. We have three types of elements that need to be assigned to 21 houses so that exactly seven of each type are assigned. The number of ways to do this is the multinomial coefficient

$$\binom{21}{7,7,7} = 399{,}072{,}960.$$

3. We need to divide the 300 members of the organization into three subsets: the 5 in one committee, the 8 in the second committee, and the 287 in neither committee. There are $\binom{300}{5,8,287}$ ways to do this.

5. There are $\binom{n}{n_1, n_2, n_3, n_4, n_5, n_6}$ many ways to arrange n_j j's (for $j = 1, \dots, 6$) among the n rolls. The number of possible equally likely rolls is 6^n. So, the probability we want is $\frac{1}{6^n}\binom{n}{n_1, n_2, n_3, n_4, n_5, n_6}$.

7. There are $\binom{25}{10,8,7}$ ways of distributing the 25 cards to the three players. There are $\binom{12}{6,2,4}$ ways of distributing the 12 red cards to the players so that each receives the designated number of red cards. There are then $\binom{13}{4,6,3}$ ways of distributing the other 13 cards to the players, so that each receives the designated total number of cards. The product of these last two numbers of ways is, therefore, the number of ways of distributing the 25 cards to the players so that each receives the designated number of red cards and the designated total number of cards. So, the final probability is $\binom{12}{6,2,4}\binom{13}{4,6,3}\Big/\binom{25}{10,8,7}$.

9. There are $\binom{52}{13,13,13,13}$ ways of distributing the cards to the four players. Call these four players A, B, C, and D. There is only one way of distributing the cards so that player A receives all red cards, player B receives all yellow cards, player C receives all blue cards, and player D receives all green cards. However, there are 4! ways of assigning the four colors to the four players and therefore there are 4! ways of distributing the cards so that each player receives 13 cards of the same color. So, the probability we need is

$$\frac{4!}{\binom{52}{13,13,13,13}} = \frac{4!(13!)^4}{52!} \approx 4.474 \times 10^{-28}.$$

11. We shall use induction. Since we have already proven the binomial theorem, we know that the conclusion to the multinomial theorem is true for every n if $k = 2$. We shall use induction again, but this time using k instead of n. For $k = 2$, we already know the result is true. Suppose that the result is true for all $k \leq k_0$ and for all n. For $k = k_0 + 1$ and arbitrary n we must show that

$$(x_1 + \cdots + x_{k_0+1})^n = \sum \binom{n}{n_1, \ldots, n_{k_0+1}} x_1^{n_1} \cdots x_{k_0+1}^{n_{k_0+1}}, \tag{S.1.2}$$

where the summation is over all $n_1, \ldots, n_{k_0+1}$ such that $n_1 + \cdots + n_{k_0+1} = n$. Let $y_i = x_i$ for $i = 1, \ldots, k_0 - 1$ and let $y_{k_0} = x_{k_0} + x_{k_0+1}$. We then have

$$(x_1 + \cdots + x_{k_0+1})^n = (y_1 + \cdots + y_{k_0})^n.$$

Since we have assumed that the theorem is true for $k = k_0$, we know that

$$(y_1 + \cdots + y_{k_0})^n = \sum \binom{n}{m_1, \ldots, m_{k_0}} y_1^{m_1} \cdots y_{k_0}^{m_{k_0}}, \tag{S.1.3}$$

where the summation is over all $m_1, \ldots, m_{k_0}$ such that $m_1 + \cdots + m_{k_0} = n$. On the right side of (S.1.3), substitute $x_{k_0} + x_{k_0+1}$ for y_{k_0} and apply the binomial theorem to obtain

$$\sum \binom{n}{m_1, \ldots, m_{k_0}} y_1^{m_1} \cdots y_{k_0-1}^{m_{k_0-1}} \sum_{i=0}^{m_{k_0}} \binom{m_{k_0}}{i} x_{k_0}^{i} x_{k_0+1}^{m_{k_0}-i}. \tag{S.1.4}$$

In (S.1.4), let $n_i = m_i$ for $i = 1, \ldots, k_0 - 1$, let $n_{k_0} = i$, and let $n_{k_0+1} = m_{k_0} - i$. Then, in the summation in (S.1.4), $n_1 + \cdots + n_{k_0+1} = n$ if and only if $m_1 + \cdots + m_{k_0} = n$. Also, note that

$$\binom{n}{m_1, \ldots, m_{k_0}} \binom{m_{k_0}}{i} = \binom{n}{n_1, \ldots, n_{k_0+1}}.$$

So, (S.1.4) becomes

$$\sum \binom{n}{n_1, \ldots, n_{k_0+1}} x_1^{n_1} \cdots x_{k_0+1}^{n_{k_0+1}},$$

where this last sum is over all $n_1, \ldots, n_{k_0+1}$ such that $n_1 + \cdots + n_{k_0+1} = n$.

1.10 The Probability of a Union of Events

1. Let A_i be the event that person i receives exactly two aces for $i = 1, 2, 3$. We want $\Pr(\cup_{i=1}^{3} A_i)$. We shall apply Theorem 1.10.1 directly. Let the sample space consist of all permutations of the 52 cards where the first five cards are dealt to person 1, the second five to person 2, and the third five to person 3. A permutation of 52 cards that leads to the occurrence of event A_i can be constructed as follows. First, choose which of person i's five locations will receive the two aces. There are $C_{5,2}$ ways to do this. Next, for each such choice, choose the two aces that will go in these locations, distinguishing the order in which they are placed. There are $P_{4,2}$ ways to do this. Next, for each of the preceding choices, choose the locations for the other two aces from among the 47 locations that are not dealt to person i, distinguishing order. There are $P_{47,2}$ ways to do this. Finally, for each of the preceding choices, choose

a permutation of the remaining 48 cards among the remaining 48 locations. There are 48! ways to do this. Since there are 52! equally likely permutations in the sample space, we have

$$\Pr(A_i) = \frac{C_{5,2}P_{4,2}P_{47,2}48!}{52!} = \frac{5!4!47!48!}{2!3!2!45!52!} \approx 0.0399.$$

Careful examination of the expression for $\Pr(A_i)$ reveals that it can also be expressed as

$$\Pr(A_i) = \frac{\binom{4}{2}\binom{48}{3}}{\binom{52}{5}}.$$

This expression corresponds to a different, but equally correct, way of describing the sample space in terms of equally likely outcomes. In particular, the sample space would consist of the different possible five-card sets that person i could receive without regard to order.

Next, compute $\Pr(A_iA_j)$ for $i \neq j$. There are still $C_{5,2}$ ways to choose the locations for person i's aces amongst the five cards and for each such choice, there are $P_{4,2}$ ways to choose the two aces in order. For each of the preceding choices, there are $C_{5,2}$ ways to choose the locations for person j's aces and 2 ways to order the remaining two aces amongst the two locations. For each combination of the preceding choices, there are 48! ways to arrange the remaining 48 cards in the 48 unassigned locations. Then, $\Pr(A_iA_j)$ is

$$\Pr(A_iA_j) = \frac{2C_{5,2}^2P_{4,2}48!}{52!} = \frac{2(5!)^2 4!48!}{(2!)^3(3!)^2 52!} \approx 3.694 \times 10^{-4}.$$

Once again, we can rewrite the expression for $\Pr(A_iA_j)$ as

$$\Pr(A_iA_j) = \frac{\binom{4}{2}\binom{48}{3,3,42}}{\binom{52}{5,5,42}}.$$

This corresponds to treating the sample space as the set of all pairs of five-card subsets.

Next, notice that it is impossible for all three players to receive two aces, so $\Pr(A_1A_2A_3) = 0$. Applying Theorem 1.10.1, we obtain

$$\Pr\left(\cup_{i=1}^3 A_i\right) = 3 \times 0.0399 - 3 \times 3.694 \times 10^{-4} = 0.1186.$$

3. As seen from Fig. S.1.1, the required percentage is $P_1 + P_2 + P_3$. From the given values, we have, in percentages,

$$\begin{aligned}
P_7 &= 5,\\
P_4 &= 20 - P_7 = 15,\\
P_5 &= 20 - P_7 = 15,\\
P_6 &= 10 - P_7 = 5,\\
P_1 &= 60 - P_4 - P_6 - P_7 = 35,\\
P_2 &= 40 - P_4 - P_5 - P_7 = 5,\\
P_3 &= 30 - P_5 - P_6 - P_7 = 5.
\end{aligned}$$

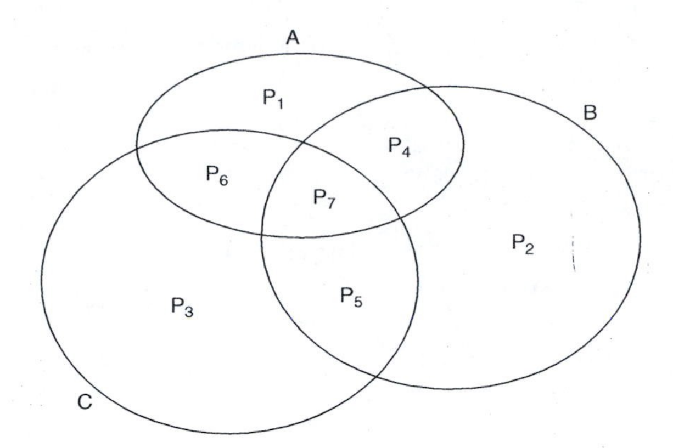

Figure S.1.1: Figure for Exercise 3 of Sec. 1.10.

Therefore, $P_1 + P_2 + P_3 = 45$.

5. Determine first the probability that at least one guest will receive the proper hat. This probability is the value p_n specified in the matching problem, with $n = 4$, namely

$$p_4 = 1 - \frac{1}{2} + \frac{1}{6} - \frac{1}{24} = \frac{5}{8}.$$

So, the probability that no guest receives the proper hat is $1 - 5/8 = 3/8$.

7. Let A_1 denote the event that no student from the freshman class is selected, and let A_2, A_3, and A_4 denote the corresponding events for the sophomore, junior, and senior classes, respectively. The probability that at least one student will be selected from each of the four classes is equal to $1 - \Pr(A_1 \cup A_2 \cup A_3 \cup A_4)$. We shall evaluate $\Pr(A_1 \cup A_2 \cup A_3 \cup A_4)$ by applying Theorem 1.10.2. The event A_1 will occur if and only if the 15 selected students are sophomores, juniors, or seniors. Since there are 90 such students out of a total of 100 students, we have $\Pr(A_1) = \binom{90}{15} \Big/ \binom{100}{15}$. The values of $\Pr(A_i)$ for $i = 2, 3, 4$ can be obtained in a similar fashion. Next, the event A_1A_2 will occur if and only if the 15 selected students are juniors or seniors. Since there are a total of 70 juniors and seniors, we have $\Pr(A_1A_2) = \binom{70}{15} \Big/ \binom{100}{15}$. The probability of each of the six events of the form A_iA_j for $i < j$ can be obtained in this way. Next the event $A_1A_2A_3$ will occur if and only if all 15 selected students are seniors. Therefore, $\Pr(A_1A_2A_3) = \binom{40}{15} \Big/ \binom{100}{15}$. The probabilities of the events $A_1A_2A_4$ and $A_1A_3A_4$ can also be obtained in this way. It should be noted, however, that $\Pr(A_2A_3A_4) = 0$ since it is impossible that all 15 selected students will be freshmen. Finally, the event $A_1A_2A_3A_4$ is also obviously impossible, so $\Pr(A_1A_2A_3A_4) = 0$. So, the probability we want is

$$1 - \left[\frac{\binom{90}{15}}{\binom{100}{15}} + \frac{\binom{80}{15}}{\binom{100}{15}} + \frac{\binom{70}{15}}{\binom{100}{15}} + \frac{\binom{60}{15}}{\binom{100}{15}} - \frac{\binom{70}{15}}{\binom{100}{15}} - \frac{\binom{60}{15}}{\binom{100}{15}} - \frac{\binom{50}{15}}{\binom{100}{15}} - \frac{\binom{50}{15}}{\binom{100}{15}} - \frac{\binom{40}{15}}{\binom{100}{15}} - \frac{\binom{30}{15}}{\binom{100}{15}} + \frac{\binom{40}{15}}{\binom{100}{15}} + \frac{\binom{30}{15}}{\binom{100}{15}} + \frac{\binom{20}{15}}{\binom{100}{15}} \right].$$

9. Let $p_n = 1 - q_n$. As discussed in the text, $p_{10} < p_{300} < 0.63212 < p_{53} < p_{21}$. Since p_n is smallest for $n = 10$, then q_n is largest for $n = 10$.

11. Consider choosing 5 envelopes at random into which the 5 red letters will be placed. If there are exactly r red envelopes among the five selected envelopes ($r = 0, 1, \ldots, 5$), then exactly $x = 2r$ envelopes will contain a card with a matching color. Hence, the only possible values of x are 0, 2, 4..., 10. Thus, for $x = 0, 2, \ldots, 10$ and $r = x/2$, the desired probability is the probability that there are exactly r red envelopes among the five selected envelopes, which is $\dfrac{\binom{5}{r}\binom{5}{5-r}}{\binom{10}{5}}$.

13. We know that

$$\bigcap_{i=1}^{\infty} A_i = \left(\bigcup_{i=1}^{\infty} A_i^c\right)^c.$$

Hence,

$$\Pr\left(\bigcap_{i=1}^{\infty} A_i\right) = 1 - \Pr\left(\bigcup_{i=1}^{\infty} A_i^c\right).$$

However, since $A_1 \supset A_2 \supset \ldots$, then $A_1^c \subset A_2^c \subset \ldots$. Therefore, by Exercise 12,

$$\Pr\left(\bigcup_{i=1}^{\infty} A_i^c\right) = \lim_{n\to\infty} \Pr(A_n^c) = \lim_{n\to\infty} [1 - \Pr(A_n)] = 1 - \lim_{n\to\infty} Pr(A_n).$$

It now follows that

$$\Pr\left(\bigcap_{i=1}^{\infty} A_i\right) = \lim_{n\to\infty} \Pr(A_n).$$

1.12 Supplementary Exercises

1. No, since both A and B might occur.

3. $\dfrac{\binom{250}{18}\cdot\binom{100}{12}}{\binom{350}{30}}$.

5. The region where total utility demand is at least 215 is shaded in Fig. S.1.2. The area of the shaded region is

$$\frac{1}{2} \times 135 \times 135 = 9112.5$$

The probability is then $9112.5/29204 = 0.3120$.

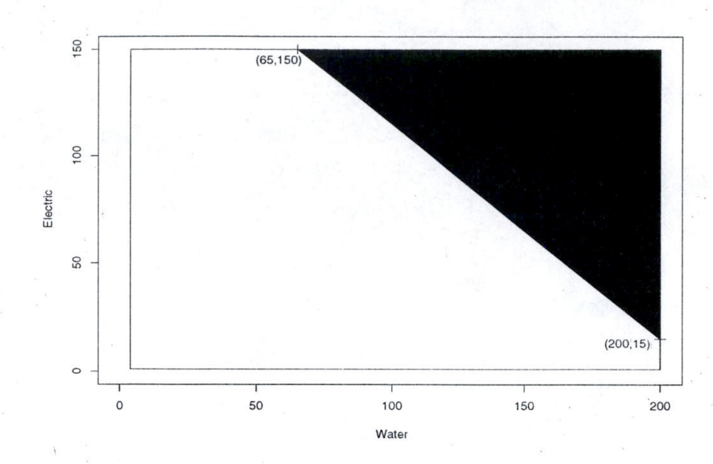

Figure S.1.2: Region where total utility demand is at least 215 in Exercise 5 of Sec. 1.12.

7. The presence of the blue balls is irrelevant in this problem, since whenever a blue ball is drawn it is ignored. Hence, the answer is the same as in part (a) of Exercise 6.

9. There are $\binom{10}{5}$ ways of choosing the five envelopes into which the red cards will be placed. There are $\binom{7}{j}\binom{3}{5-j}$ ways of choosing exactly j red envelopes and $5-j$ green envelopes. Therefore the probability that exactly j red envelopes will contain red cards is

$$\binom{7}{j}\binom{3}{5-j}\Big/\binom{10}{5} \qquad \text{for } j = 2, 3, 4, 5.$$

But if j red envelopes contain red cards, then $j-2$ green envelopes must also contain green cards. Hence, this is also the probability of exactly $k = j + (j-2) = 2j-2$ matches.

11. We can use Fig. S.1.1 by relabeling the events A, B, and C in the figure as A_1, A_2, and A_3 respectively. It is now easy to see that the probability that exactly one of the three events occurs is $p_1 + p_2 + p_3$. Also,

$$\begin{aligned}\Pr(A_1) &= p_1 + p_4 + p_6 + p_7, \\ \Pr(A_1 \cap A_2) &= p_4 + p_7, \text{ etc.}\end{aligned}$$

By breaking down each probability in the given expression in this way, we obtain the desired result.

13. (a) In order for the winning combination to have no consecutive numbers, between every pair of numbers in the winning combination there must be at least one number not in the winning combination. That is, there must be at least $k-1$ numbers not in the winning combination to be in between the pairs of numbers in the winning combination. Since there are k numbers in the winning combination, there must be at least $k + k - 1 = 2k - 1$ numbers available in order for it to be possible to have no consecutive numbers in the winning combination. So, n must be at least $2k-1$ to allow consecutive numbers.

(b) Let $i_1, \ldots, i_k$ and $j_1, \ldots, j_k$ be as described in the problem. For one direction, suppose that $i_1, \ldots, i_k$ contains at least one pair of consecutive integers, say $i_{a+1} = i_a + 1$. Then

$$j_{a+1} = i_{a+1} - a = i_a + 1 - a = i_a - (a-1) = j_a.$$

So, $j_1, \ldots, j_k$ contains repeats. For the other direction, suppose that $j_1, \ldots, j_k$ contains repeats, say $j_{a+1} = j_a$. Then

$$i_{a+1} = j_{a+1} + a = j_a + a = i_a + 1.$$

So $i_1, \ldots, i_k$ contains a pair of consecutive numbers.

(c) Since $i_1 < i_2 < \cdots < i_k$, we know that $i_a + 1 \le i_{a+1}$, so that $j_a = i_a - a + 1 \le i_{a+1} - a = j_{a+1}$ for each $a = 1, \ldots, k-1$. Since $i_k \le n$, $j_k = i_k - k + 1 \le n - k + 1$. The set of all $(j_1, \ldots, j_k)$ with $1 \le j_1 < \cdots < j_k \le n - k + 1$ is just the number of combinations of $n - k + 1$ items taken k at a time, that is $\binom{n-k+1}{k}$.

(d) By part (b), there are no pairs of consecutive integers in the winning combination $(i_1, \ldots, i_k)$ if and only if $(j_1, \ldots, j_k)$ has no repeats. The total number of winning combinations is $\binom{n}{k}$. In part (c), we computed the number of winning combinations with no repeats among $(j_1, \ldots, j_k)$ to be $\binom{n-k+1}{k}$. So, the probability of no consecutive integers is

$$\frac{\binom{n-k+1}{k}}{\binom{n}{k}} = \frac{(n-k)!(n-k+1)!}{n!(n-2k+1)!}.$$

(e) The probability of at least one pair of consecutive integers is one minus the answer to part (d).

Chapter 2

Conditional Probability

2.1 The Definition of Conditional Probability

1. If $A \subset B$, then $A \cap B = A$ and $\Pr(A \cap B) = \Pr(A)$. So $\Pr(A|B) = \Pr(A)/\Pr(B)$.

3. Since $A \cap S = A$ and $\Pr(S) = 1$, it follows that $\Pr(A \mid S) = \Pr(A)$.

5. Let R_i be the event that a red ball is drawn on the ith draw, and let B_i be the event that a blue ball is drawn on the ith draw for $i = 1, \ldots, 4$. Then

$$\begin{aligned}
\Pr(R_1) &= \frac{r}{r+b}, \\
\Pr(R_2 \mid R_1) &= \frac{r+k}{r+b+k}, \\
\Pr(R_3 \mid R_1 \cap R_2) &= \frac{r+2k}{r+b+2k}, \\
\Pr(B_4 \mid R_1 \cap R_2 \cap R_3) &= \frac{b}{r+b+3k}.
\end{aligned}$$

The desired probability is the product of these four probabilities, namely

$$\frac{r(r+k)(r+2k)b}{(r+b)(r+b+k)(r+b+2k)(r+b+3k)}.$$

7. We know that $\Pr(A) = 0.6$ and $\Pr(A \cap B) = 0.2$. Therefore, $\Pr(B \mid A) = \dfrac{0.2}{0.6} = \dfrac{1}{3}$.

9. (a) If card A has been selected, each of the other four cards is equally likely to be the other selected card. Since three of these four cards are red, the required probability is $3/4$.

   (b) We know, without being told, that at least one red card must be selected, so this information does not affect the probabilities of any events. We have

$$\Pr(\text{both cards red}) = \Pr(R_1)\Pr(R_2 \mid R_1) = \frac{4}{5} \cdot \frac{3}{4} = \frac{3}{5}.$$

11. This is the conditional version of Theorem 1.5.3. From the definition of conditional probability, we have

$$\Pr(A^c|B) = \frac{\Pr(A^c \cap B)}{\Pr(B)},$$

$$1-\Pr(A|B) = 1-\frac{\Pr(A\cap B)}{\Pr(B)}, \\ = \frac{\Pr(B)-\Pr(A\cap B)}{\Pr(B)}. \quad \text{(S.2.1)}$$

According to Theorem 1.5.6 (switching the names A and B), $\Pr(B) - \Pr(A\cap B) = \Pr(A^c \cap B)$. Combining this with (S.2.1) yields $1-\Pr(A|B) = \Pr(A^c|B)$.

13. Let A_1 denote the event that the selected coin has a head on each side, let A_2 denote the event that it has a tail on each side, let A_3 denote the event that it is fair, and let B denote the event that a head in obtained. Then

$$\Pr(A_1) = \frac{3}{9}, \quad \Pr(A_2)=\frac{4}{9}, \quad \Pr(A_3)=\frac{2}{9},$$
$$\Pr(B \mid A_1) = 1, \quad \Pr(B\mid A_2)=0, \quad \Pr(B\mid A_3)=\frac{1}{2}.$$

Hence,

$$\Pr(B)=\sum_{i=1}^{3}\Pr(A_i)\Pr(B\mid A_i)=\frac{4}{9}.$$

15. The analysis is similar to that given in the previous exercise, and the probability is 0.47.

17. Clearly, we must assume that $\Pr(B_j \cap C) > 0$ for all j, otherwise (2.1.5) is undefined. By applying the definition of conditional probability to each term, the right side of (2.1.5) can be rewritten as

$$\sum_{i=1}^{k}\frac{\Pr(B_j\cap C)}{\Pr(C)}\frac{\Pr(A\cap B_j\cap C)}{\Pr(B_j\cap C)} = \frac{1}{\Pr(C)}\sum_{i=1}^{k}\Pr(A\cap B_j\cap C).$$

According to the law of total probability, the last sum above is $\Pr(A\cap C)$, hence the ratio is $\Pr(A|C)$.

2.2 Independent Events

1. If $\Pr(B) < 1$, then $\Pr(B^c) = 1-\Pr(B) > 0$. We then compute

$$\begin{aligned}\Pr(A^c|B^c) &= \frac{\Pr(A^c\cap B^c)}{\Pr(B^c)} \\ &= \frac{1-\Pr(A\cup B)}{1-\Pr(B)} \\ &= \frac{1-\Pr(A)-\Pr(B)+\Pr(A\cap B)}{1-\Pr(B)} \\ &= \frac{1-\Pr(A)-\Pr(B)+\Pr(A)\Pr(B)}{1-\Pr(B)} \\ &= \frac{[1-\Pr(A)][1-\Pr(B)]}{1-\Pr(B)} \\ &= 1-\Pr(A) = \Pr(A^c).\end{aligned}$$

3. Since the event $A\cap B$ is a subset of the event A, and $\Pr(A) = 0$, it follows that $\Pr(A\cap B) = 0$. Hence, $\Pr(A\cap B) = 0 = \Pr(A)\Pr(B)$.

5. The probability that both systems will malfunction is $(0.001)^2 = 10^{-6}$. The probability that at least one of the systems will function is therefore $1 - 10^{-6}$.

7. Let E_1 be the event that A is in class, and let E_2 be the event that B is in class. Let C be the event that at least one of the students is in class. That is, $C = E_1 \cup E_2$.

   (a) We want $\Pr(C)$. We shall use Theorem 1.5.7 to compute the probability. Since E_1 and E_2 are independent, we have $\Pr(E_1 \cap E_2) = \Pr(E_1)\Pr(E_2)$. Hence

$$\Pr(C) = \Pr(E_1) + \Pr(E_2) - \Pr(E_1 \cap E_2) = 0.8 + 0.6 - 0.8 \times 0.6 = 0.92.$$

   (b) We want $\Pr(E_1|C)$. We computed $\Pr(C) = 0.92$ in part (a). Since $E_1 \subset C$, $\Pr(E_1 \cap C) = \Pr(E_1) = 0.8$. So, $\Pr(E_1|C) = 0.8/0.92 = 0.8696$.

9. The probability that exactly n tosses will be required on a given performance is $1/2^n$. Therefore, the probability that exactly n tosses will be required on all three performances is $(1/2^n)^3 = 1/8^n$. The probability that the same number of tosses will be required on all three performances is $\sum_{n=1}^{\infty} \frac{1}{8^n} = \frac{1}{7}$.

11. (a) We must determine the probability that at least two of the four oldest children will have blue eyes. The probability p_j that exactly j of these four children will have blue eyes is

$$p_j = \binom{4}{j}\left(\frac{1}{4}\right)^j\left(\frac{3}{4}\right)^{4-j}.$$

   The desired probability is therefore $p_2 + p_3 + p_4$.

   (b) The two different types of information provided in Exercise 10 and part (a) are similar to the two different types of information provided in part (a) and part (b) of Exercise 9 of Sec. 2.1.

13. The probability of obtaining a particular sequence of ten particles in which one particle penetrates the shield and nine particles do not is $(0.01)(0.99)^9$. Since there are 10 such sequences in the sample space, the desired probability is $10(0.01)(0.99)^9$.

15. If n particles are emitted, the probability that at least one particle will penetrate the shield is $1-(0.99)^n$. In order for this value to be at least 0.8 we must have

$$\begin{aligned} 1 - (0.99)^n &\geq 0.8 \\ (0.99)^n &\leq 0.2 \\ n\log(0.99) &\leq \log(0.2). \end{aligned}$$

Since $\log(0.99)$ is negative, this final relation is equivalent to the relation

$$n \geq \frac{\log(0.2)}{\log(0.99)} \approx 160.1.$$

So 161 or more particles are needed.

17. In order for the target to be hit for the first time on the third throw of boy A, all five of the following independent events must occur: (1) A misses on his first throw, (2) B misses on his first throw, (3) A misses on his second throw, (4) B misses on his second throw, (5) A hits on his third throw. The probability of all five events occurring is $\frac{2}{3} \cdot \frac{3}{4} \cdot \frac{2}{3} \cdot \frac{3}{4} \cdot \frac{1}{3} = \frac{1}{12}$.

19. Let A_1 denote the event that no red balls are selected, let A_2 denote the event that no white balls are selected, and let A_3 denote the event that no blue balls are selected. We must determine the value of $\Pr(A_1 \cup A_2 \cup A_3)$. We shall apply Theorem 1.10.1. The event A_1 will occur if and only if all ten selected balls are white or blue. Since there is probability 0.8 that any given selected ball will be white or blue, we have $\Pr(A_1) = (0.8)^{10}$. Similarly, $\Pr(A_2) = (0.7)^{10}$ and $\Pr(A_3) = (0.5)^{10}$. The event $A_1 \cap A_2$ will occur if and only if all ten selected balls are blue. Therefore $\Pr(A_1 \cap A_2) = (0.5)^{10}$. Similarly, $\Pr(A_2 \cap A_3) = (0.2)^{10}$ and $\Pr(A_1 \cap A_3) = (0.3)^{10}$. Finally, the event $A_1 \cap A_2 \cap A_3$ cannot possibly occur, so $\Pr(A_1 \cap A_2 \cap A_3) = 0$. So, the desired probability is

$$(0.8)^{10} + (0.7)^{10} + (0.5)^{10} - (0.5)^{10} - (0.2)^{10} - (0.3)^{10} \approx 0.1356.$$

21. For the "only if" direction, we need to prove that if $A_1, \ldots, A_k$ are independent then

$$\Pr(A_{i_1} \cap \cdots \cap A_{i_m} | A_{j_1} \cap \cdots \cap A_{j_\ell}) = \Pr(A_{i_1} \cap \cdots \cap A_{i_m}),$$

for all disjoint subsets $\{i_1, \ldots, i_m\}$ and $\{j_1, \ldots, j_\ell\}$ of $\{1, \ldots, k\}$. If $A_1, \ldots, A_k$ are independent, then

$$\Pr(A_{i_1} \cap \cdots \cap A_{i_m} \cap A_{j_1} \cap \cdots \cap A_{j_\ell}) = \Pr(A_{i_1} \cap \cdots \cap A_{i_m}) \Pr(A_{j_1} \cap \cdots \cap A_{j_\ell}),$$

hence it follows that

$$\Pr(A_{i_1} \cap \cdots \cap A_{i_m} | A_{j_1} \cap \cdots \cap A_{j_\ell}) = \frac{\Pr(A_{i_1} \cap \cdots \cap A_{i_m} \cap A_{j_1} \cap \cdots \cap A_{j_\ell})}{\Pr(A_{j_1} \cap \cdots \cap A_{j_\ell})} = \Pr(A_{i_1} \cap \cdots \cap A_{i_m}).$$

For the "if" direction, assume that $\Pr(A_{i_1} \cap \cdots \cap A_{i_m} | A_{j_1} \cap \cdots \cap A_{j_\ell}) = \Pr(A_{i_1} \cap \cdots \cap A_{i_m})$ for all disjoint subsets $\{i_1, \ldots, i_m\}$ and $\{j_1, \ldots, j_\ell\}$ of $\{1, \ldots, k\}$. We must prove that $A_1, \ldots, A_k$ are independent. That is, we must prove that for every subset $\{s_1, \ldots, s_n\}$ of $\{1, \ldots, k\}$, $\Pr(A_{s_1} \cap \cdots \cap A_{s_n}) = \Pr(A_{s_1}) \cdots \Pr(A_{s_n})$. We shall do this by induction on n. For $n = 1$, we have that $\Pr(A_{s_1}) = \Pr(A_{s_1})$ for each subset $\{s_1\}$ of $\{1, \ldots, k\}$. Now, assume that for all $n \le n_0$ and for all subsets $\{s_1, \ldots, s_n\}$ of $\{1, \ldots, k\}$ it is true that $\Pr(A_{s_1} \cap \cdots \cap A_{s_n}) = \Pr(A_{s_1}) \cdots \Pr(A_{s_n})$. We need to prove that for every subset $\{t_1, \ldots, t_{n_0+1}\}$ of $\{1, \ldots, k\}$

$$\Pr(A_{t_1} \cap \cdots \cap A_{t_{n_0+1}}) = \Pr(A_{t_1}) \cdots \Pr(A_{t_{n_0+1}}). \tag{S.2.2}$$

It is clear that

$$\Pr(A_{t_1} \cap \cdots \cap A_{t_{n_0+1}}) = \Pr(A_{t_1} \cap \cdots \cap A_{t_{n_0}} | A_{t_{n_0+1}}) \Pr(A_{t_{n_0+1}}). \tag{S.2.3}$$

We have assumed that $\Pr(A_{t_1} \cap \cdots \cap A_{t_{n_0}} | A_{t_{n_0+1}}) = \Pr(A_{t_1} \cap \cdots \cap A_{t_{n_0}})$ for all disjoint subsets $\{t_1, \ldots, t_{n_0}\}$ and $\{t_{n_0+1}\}$ of $\{1, \ldots, k\}$. Since the right side of this last equation is the probability of the intersection of only n_0 events, then we know that

$$\Pr(A_{t_1} \cap \cdots \cap A_{t_{n_0}}) = \Pr(A_{t_1}) \cdots \Pr(A_{t_{n_0}}).$$

Combining this with Eq. (S.2.3) implies that (S.2.2) holds.

23. (a) Conditional on B the events $A_1, \ldots, A_{11}$ are independent with probability 0.8 each. The conditional probability that a particular collection of eight programs out of the 11 will compile is $0.8^8 0.2^3 = 0.001342$. There are $\binom{11}{8} = 165$ different such collections of eight programs out of the 11, so the probability of exactly eight programs will compile is $165 \times 0.001342 = 0.2215$.

(b) Conditional on B^c the events $A_1, \ldots, A_{11}$ are independent with probability 0.4 each. The conditional probability that a particular collection of eight programs out of the 11 will compile is $0.4^8 0.6^3 = 0.0001416$. There are $\binom{11}{8} = 165$ different such collections of eight programs out of the 11, so the probability of exactly eight programs will compile is $165 \times 0.0001416 = 0.02335$.

2.3 Bayes' Theorem

1. It must be true that $\sum_{i=1}^{k} \Pr(B_i) = 1$ and $\sum_{i=1}^{k} \Pr(B_i \mid A) = 1$. However, if $\Pr(B_1 \mid A) < \Pr(B_1)$ and $\Pr(B_i \mid A) \leq \Pr(B_i)$ for $i = 2, \ldots, k$, we would have $\sum_{i=1}^{k} \Pr(B_i \mid A) < \sum_{i=1}^{k} \Pr(B_i)$, a contradiction. Therefore, it must be true that $\Pr(B_i \mid A) > \Pr(B_i)$ for at least one value of i ($i = 2, \ldots, k$).

3. Let C denote the event that the selected item is nondefective. Then

$$\Pr(A_2 \mid C) = \frac{(0.3)(0.98)}{(0.2)(0.99) + (0.3)(0.98) + (0.5)(0.97)} = 0.301.$$

Commentary: It should be noted that if the selected item is observed to be defective, the probability that the item was produced by machine M_2 is decreased from the prior value of 0.3 to the posterior value of 0.26. However, if the selected item is observed to be nondefective, this probability changes very little, from a prior value of 0.3 to a posterior value of 0.301. In this example, therefore, obtaining a defective is more informative than obtaining a nondefective, but it is much more probable that a nondefective will be obtained.

5. The desired probability Pr(Lib.|NoVote) can be calculated as follows:

$$\frac{\Pr(\text{Lib.}) \Pr(\text{NoVote}|\text{Lib.})}{\Pr(\text{Cons.}) \Pr(\text{NoVote}|\text{Cons.}) + \Pr(\text{Lib.}) \Pr(\text{NoVote}|\text{Lib.}) + \Pr(\text{Ind.}) \Pr(\text{NoVote}|\text{Ind.})}$$
$$= \frac{(0.5)(0.18)}{(0.3)(0.35) + (0.5)(0.18) + (0.2)(0.50)} = \frac{18}{59}.$$

7. (a) Let π_i denote the posterior probability that coin i was selected. The prior probability of each coin is 1/5. Therefore

$$\pi_i = \frac{\frac{1}{5} p_i}{\sum_{j=1}^{5} \frac{1}{5} p_j} \quad \text{for} \quad i = 1, \ldots, 5.$$

The five values are $\pi_1 = 0$, $\pi_2 = 0.1$, $\pi_3 = 0.2$, $\pi_4 = 0.3$, and $\pi_5 = 0.4$.

(b) The probability of obtaining another head is equal to

$$\sum_{i=1}^{5} \Pr(\text{Coin } i) \Pr(\text{Head} \mid \text{ Coin } i) = \sum_{i=1}^{5} \pi_i p_i = \frac{3}{4}.$$

(c) The posterior probability π_i of coin i would now be

$$\pi_i = \frac{\frac{1}{5}(1-p_i)}{\sum_{j=1}^{5}\frac{1}{5}(1-p_j)} \quad \text{for } i=1,\ldots,5.$$

Thus, $\pi_1 = 0.4, \pi_2 = 0.3, \pi_3 = 0.2, \pi_4 = 0.1$, and $\pi_5 = 0$. The probability of obtaining a head on the next toss is therefore $\sum_{i=1}^{5}\pi_i p_i = \frac{1}{4}$.

9. We shall continue to use the notation from the solution to Exercise 14 in Sec. 2.1. Let C be the event that exactly one out of seven observed parts is defective. We are asked to find $\Pr(B_j|C)$ for $j = 1, 2, 3$. We need $\Pr(C|B_j)$ for each j. Let A_i be the event that the ith part is defective. For all i, $\Pr(A_i|B_1) = 0.02$, $\Pr(A_i|B_2) = 0.1$, and $\Pr(A_i|B_3) = 0.3$. Since the seven parts are conditionally independent given each state of the machine, the probability of each possible sequence of seven parts with one defective is $\Pr(A_i|B_j)[1 - \Pr(A_i|B_j)]^6$. There are seven distinct such sequences, so

$$\begin{aligned}
\Pr(C|B_1) &= 7 \times 0.02 \times 0.98^6 = 0.1240,\\
\Pr(C|B_2) &= 7 \times 0.1 \times 0.9^6 = 0.3720,\\
\Pr(C|B_3) &= 7 \times 0.3 \times 0.7^6 = 0.2471.
\end{aligned}$$

The expression in the denominator of Bayes' theorem is

$$\Pr(C) = 0.8 \times 0.1240 + 0.1 \times 0.3720 + 0.1 \times 0.2471 = 0.1611.$$

Bayes' theorem now says

$$\begin{aligned}
\Pr(B_1|C) &= \frac{0.8 \times 0.1240}{0.1611} = 0.6157,\\
\Pr(B_2|C) &= \frac{0.1 \times 0.3720}{0.1611} = 0.2309,\\
\Pr(B_3|C) &= \frac{0.1 \times 0.2471}{0.1611} = 0.1534.
\end{aligned}$$

11. This time, we want $\Pr(B_4|E^c)$. We know that $\Pr(E^c) = 1 - \Pr(E) = 1/4$ and $\Pr(E^c|B_4) = 1 - \Pr(E|B_4) = 1/4$. This means that E^c and B_4 are independent so that $\Pr(B_4|E^c) = \Pr(B_4) = 1/4$.

13. (a) Let B_1 be the event that the coin is fair, and let B_2 be the event that the coin has two heads. Let H_i be the event that we obtain a head on the ith toss for $i = 1, 2, 3, 4$. We shall apply Bayes' theorem conditional on H_1H_2.

$$\begin{aligned}
&\Pr(B_1|H_1 \cap H_2 \cap H_3)\\
&= \frac{\Pr(B_1|H_1 \cap H_2)\Pr(H_3|B_1 \cap H_1 \cap H_2)}{\Pr(B_1|H_1 \cap H_2)\Pr(H_3|B_1 \cap H_1 \cap H_2) + \Pr(B_2|H_1 \cap H_2)\Pr(H_3|B_2 \cap H_1 \cap H_2)}\\
&= \frac{(1/5) \times (1/2)}{(1/5) \times (1/2) + (4/5) \times 1} = \frac{1}{9}.
\end{aligned}$$

(b) If the coin ever shows a tail, it can't have two heads. Hence the posterior probability of B_1 becomes 1 after we observe a tail.

15. The law of total probability tells us how to compute $\Pr(E_1)$.

$$\Pr(E_1) = \sum_{i=1}^{11} \Pr(B_i)\frac{i-1}{10}.$$

Using the numbers in Example 2.3.8 for $\Pr(B_i)$ we obtain 0.274. This is smaller than the value 0.5 computed in Example 2.3.7 because the prior probabilities in Example 2.3.8 are much higher for the B_i with low values of i, particularly $i = 2, 3, 4$, and they are much smaller for those B_i with large values of i. Since $\Pr(E_1)$ is a weighted average of the values $(i-1)/10$ with the weights being $\Pr(B_i)$ for $i = 1, \ldots 11$, the more weight we give to small values of $(i-1)/10$, the smaller the weighted average will be.

2.4 The Gambler's Ruin Problem

1. Clearly a_i in Eq. (2.4.9) is an increasing function of i. Hence, if $a_{98} < 1/2$, then $a_i < 1/2$ for all $i \le 98$. For $i = 98$, Eq. (2.4.9) yields almost exactly 4/9, which is less that 1/2.

3. If the initial fortune of gambler A is i dollars, then for conditions (a), (b), and (c), the initial fortune of gambler B is $i/2$ dollars. Hence, $k = 3i/2$. If we let $r = (1-p)/p > 1$, then it follows from Eq. (2.4.8) that the probability that A will win under conditions (a), (b), or (c) is

$$\frac{r^i - 1}{r^{3i/2} - 1} = \frac{1 - (1/r_i)}{r^{i/2} - (1/r_i)}.$$

If i and j are positive integers with $i < j$, it now follows that

$$\frac{1 - (1/r_j)}{r^{j/2} - (1/r_j)} < \frac{1 - (1/r_j)}{r^{i/2} - (1/r_j)} < \frac{1 - (1/r_i)}{r^{i/2} - (1/r_i)}.$$

Thus the larger the initial fortune of gambler A is, the smaller is his probability of winning. Therefore, he has the largest probability of winning under condition (a).

5. In this exercise, $p = 1/2$ and $k = i + 2$. Therefore $a_i = i/(i+2)$. In order to make $a_i \ge 0.99$, we must have $i \ge 198$.

7. In this exercise $p = 1/3$ and $k = i + 2$. Therefore, by Eq. (2.4.9)

$$a_i = \frac{2^i - 1}{2^{i+2} - 1} = \frac{1 - (1/2^i)}{4 - (1/2^i)}.$$

But for every number x $(0 < x < 1)$, we have $\dfrac{1-x}{4-x} < \dfrac{1}{4}$. Hence, $a_i < 1/4$ for every positive integer i.

9. This problem can be expressed as a gambler's ruin problem. We consider the initial fortune of gambler A to be five dollars and the initial fortune of gambler B to be ten dollars. Gambler A wins one dollar from gambler B each time that box B is selected, and gambler B wins one dollar from gambler A each time that box A is selected. Since $i = 5, k = 15$, and $p = 1/2$, it follows from Eq. (2.4.6) that the probability that gambler A will win (and box B will become empty) is 1/3. Therefore, the probability that box A will become empty first is 2/3.

2.5 Supplementary Exercises

1. Let $\Pr(D) = p > 0$. Then

$$\begin{aligned}\Pr(A) &= p\Pr(A \mid D) + (1-p)\Pr(A \mid D^c) \\ &\geq p\Pr(B \mid D) + (1-p)\Pr(B \mid D^c) = \Pr(B).\end{aligned}$$

3. Since $\Pr(A \mid B) = \dfrac{\Pr(A \cap B)}{\Pr(B)}$ and $\Pr(B \mid A) = \dfrac{\Pr(A \cap B)}{\Pr(A)}$, we have $\dfrac{\Pr(A \cap B)}{(1/5)} + \dfrac{\Pr(A \cap B)}{(1/3)} = \dfrac{2}{3}$. Hence $\Pr(A \cap B) = 1/12$, and $\Pr(A^c \cup B^c) = 1 - \Pr(A \cap B) = 11/12$.

5. The probability of obtaining the number 6 exactly three times in ten rolls is $a = \binom{10}{3}\left(\frac{1}{6}\right)^3\left(\frac{5}{6}\right)^7$. Hence, the probability of obtaining the number 6 on the first three rolls and on none of the subsequent rolls is $b = \left(\frac{1}{6}\right)^3\left(\frac{5}{6}\right)^7$. Hence, the required probability is $\dfrac{b}{a} = 1/\binom{10}{3}$.

7. The three events are always independent under the stated conditions. The proof is a straightforward generalization of the proof of Exercise 2 in Sec. 2.2.

9. Let $\Pr(A) = p$. Then $\Pr(A \cap B) = \Pr(A \cap B \cap C) = 0$, $\Pr(A \cap C) = 4p^2$, $\Pr(B \cap C) = 8p^2$. Therefore, by Theorem 1.10.1, $5p = p + 2p + 4p - [0 + 4p^2 + 8p^2] + 0$, and $p = 1/6$.

11. $1 - \Pr(\text{losing 50 times}) = 1 - \left(\frac{49}{50}\right)^{50}$.

13. Let A, B, and C stand for the events that each of the students is in class on a particular day.

    (a) We want $\Pr(A \cup B \cup C)$. We can use Theorem 1.10.1. Independence makes it easy to compute the probabilities of the various intersections.

$$\Pr(A \cup B \cup C) = 0.3 + 0.5 + 0.8 - [0.3 \times 0.5 + 0.3 \times 0.8 + 0.5 \times 0.8] + 0.3 \times 0.5 \times 0.8 = 0.93.$$

    (b) Once again, use independence to calculate probabilities of intersections.

$$\begin{aligned}&\Pr(A \cap B^c \cap C^c) + \Pr(A^c \cap B \cap C^c) + \Pr(A^c \cap B^c \cap C) \\ &= (0.3)(0.5)(0.2) + (0.7)(0.5)(0.2) + (0.7)(0.5)(0.8) = 0.38.\end{aligned}$$

15. $\Pr(\text{Each box contains one red ball}) = \dfrac{3!}{3^3} = \dfrac{2}{9} = \Pr(\text{Each box contains one white ball})$.

    So $\Pr(\text{Each box contains both colors}) = \left(\frac{2}{9}\right)^2$.

17. $\Pr(U + V = j)$ is as follows, for $j = 0, 1, \ldots, 18$:

j	Prob.	j	Prob.
0	0.01	10	0.09
1	0.02	11	0.08
2	0.03	12	0.07
3	0.04	13	0.06
4	0.05	14	0.05
5	0.06	15	0.04
6	0.07	16	0.03
7	0.08	17	0.02
8	0.09	18	0.01
9	0.10		

Thus

$$\begin{aligned}\Pr(U+V=W+X) &= \sum_{j=0}^{18} \Pr(U+V=j)\Pr(W+X=j) \\ &= (0.01)^2+(0.02)^2+\cdots+(0.01)^2 = 0.067\,.\end{aligned}$$

19. Let E_i be the event that A and B are both selected for committee i ($i = 1, 2, 3$) and let $\Pr(E_i) = p_i$. Then

$$p_1 = \frac{\binom{6}{1}}{\binom{8}{3}} \approx 0.1071, \quad p_2 = \frac{\binom{6}{2}}{\binom{8}{4}} \approx 0.2143, \quad p_3 = \frac{\binom{6}{3}}{\binom{8}{5}} \approx 0.3571.$$

Since $E_1, E_2,$ *and* E_3 are independent, it follows from Theorem 1.10.1 that the required probability is

$$\begin{aligned}\Pr(E_1 \cup E_2 \cup E_3) &= p_1+p_2+p_3-p_1p_2-p_2p_3-p_1p_3+p_1p_2p_3 \\ &\approx 0.5490.\end{aligned}$$

21. A will win if he wins on his first toss (probability 1/2) or if all three players miss on their first tosses (probability 1/8) and then A subsequently wins. Hence,

$$\Pr(A \text{ wins}) = \frac{1}{2} + \frac{1}{8}\Pr(A \text{ wins}),$$

and $\Pr(A \text{ wins}) = 4/7$.

Similarly, B will win if A misses on his first toss and B wins on his first toss, or if all three players miss on their first tosses and then B subsequently wins. Hence,

$$\Pr(B \text{ wins}) = \frac{1}{4} + \frac{1}{8}\Pr(B \text{ wins}),$$

and $\Pr(B \text{ wins}) = 2/7$. Thus, $\Pr(C \text{ wins}) = 1 - 4/7 - 2/7 = 1/7$.

23. Let A be the event that the person you meet is a statistician, and let B be the event that he is shy. Then

$$\Pr(A \mid B) = \frac{(0.8)(0.1)}{(0.8)(0.1)+(0.15)(0.9)} = 0.372.$$

25. (a) $$\Pr(\text{Defective} \mid \text{Removed}) = \frac{(0.9)(0.3)}{(0.9)(0.3)+(0.2)(0.7)} = \frac{27}{41} = 0.659.$$

(b) $$\Pr(\text{Defective} \mid \text{Not Removed}) = \frac{(0.1)(0.3)}{(0.1)(0.3)+(0.8)(0.7)} = \frac{3}{59} = 0.051.$$

27. Let A denote the event that the family has at least one boy, and B the event that it has at least one girl. Then

$$\begin{aligned}\Pr(B) &= 1-(1/2)^n,\\ \Pr(A\cap B) &= 1-\Pr(\text{All girls})-\Pr(\text{All boys}) = 1-(1/2)^n-(1/2)^n.\end{aligned}$$

Hence,

$$\Pr(A \mid B) = \frac{\Pr(A\cap B)}{\Pr(B)} = \frac{1-(1/2)^{n-1}}{1-(1/2)^n}$$

29. (a) Let X denote the number of aces selected.
Then

$$\begin{aligned}\Pr(X=i) &= \frac{\binom{4}{i}\binom{48}{13-i}}{\binom{52}{13}}, \quad i=0,1,2,3,4.\\ \Pr(X\ge 2 \mid X\ge 1) &= \frac{1-\Pr(X=0)-\Pr(X=1)}{1-\Pr(X=0)}\\ &\approx \frac{1-0.3038-0.4388}{1-0.3038} = 0.3697.\end{aligned}$$

(b) Let A denote the event that the ace of hearts and no other aces are obtained, and let H denote the event that the ace of hearts is obtained.
Then

$$\Pr(A) = \frac{\binom{48}{12}}{\binom{52}{13}} \approx 0.1097, \quad \Pr(H) = \frac{13}{52} = 0.25.$$

The required probability is

$$\frac{\Pr(H)-\Pr(A)}{\Pr(H)} \approx \frac{0.25-0.1097}{0.25} = 0.5612.$$

31. The probability that two specified letters will be placed in the correct envelopes is $1[n(n-1)]$. The probability that none of the other $n-2$ letters will then be placed in the correct envelopes is q_{n-2}. Therefore, the probability that only the two specified letters, and no other letters, will be placed in the correct envelopes is $\frac{1}{n(n-1)}q_{n-2}$. It follows that the probability that exactly two of the n letters will be placed in the correct envelopes, without specifying which pair will be correctly placed, is $\binom{n}{2}\frac{1}{n(n-1)}q_{n-2} = \frac{1}{2}q_{n-2}$.

33. By Exercise 3 of Sec. 1.10, the probability that a family subscribes to exactly one of the three newspapers is 0.45. As can be seen from the solution to that exercise, the probability that a family subscribes only to newspaper A is 0.35. Hence, the required probability is $35/45 = 7/9$.

35. The second situation, with stakes of two dollars, is equivalent to the situation in which A and B have initial fortunes of 25 dollars and bet one dollar on each play. In the notation of Sec. 2.4, we have $i = 50$ and $k = 100$ in the first situation and $i = 25$ and $k = 50$ in the second situation. Hence, if $p = 1/2$, it follows from Eq. (2.4.6) that gambler A has the same probability 1/2 of ruining gambler B in either situation. If $p \neq 1/2$, then it follows from Eq. (2.4.9) that the probabilities α_1 and α_2 of winning in the two situations equal the values

$$\begin{aligned} \alpha_1 &= \frac{([1-p]/p)^{50} - 1}{([1-p]/p)^{100} - 1} = \frac{1}{([1-p]/p)^{50} + 1}, \\ \alpha_2 &= \frac{([1-p]/p)^{25} - 1}{([1-p]/p)^{50} - 1} = \frac{1}{([1-p]/p)^{25} + 1}. \end{aligned}$$

Hence, if $p < 1/2$, then $([1-p]/p) > 1$ and $\alpha_2 > \alpha_1$. If $p > 1/2$, then $([1-p]/p) < 1$ and $\alpha_1 > \alpha_2$.

Chapter 3

Random Variables and Distributions

3.1 Random Variables and Discrete Distributions

1. Each of the 11 integers from 10 to 20 has the same probability of being the value of X. Six of the 11 integers are even, so the probability that X is even is $6/11$.

3. By looking over the 36 possible outcomes enumerated in Example 1.6.5, we find that $X = 0$ for 6 outcomes, $X = 1$ for 10 outcomes, $X = 2$ for 8 outcomes, $X = 3$ for 6 outcomes, $X = 4$ for 4 outcomes, and $X = 5$ for 2 outcomes. Hence, the p.f. $f(x)$ is as follows:

x	0	1	2	3	4	5
$f(x)$	3/18	5/18	4/18	3/18	2/18	1/18

5. For $x = 2, 3, 4, 5$, the probability of obtaining exactly x red balls is $\binom{7}{x}\binom{3}{5-x}/\binom{10}{5}$.

7. Suppose that a machine produces a defective item with probability 0.7 and produces a nondefective item with probability 0.3. If X denotes the number of defective items that are obtained when 8 items are inspected, then the random variable X will have the binomial distribution with parameters $n = 8$ and $p = 0.7$. By the same reasoning, however, if Y denotes the number of nondefective items that are obtained, then Y will have the binomial distribution with parameters $n = 8$ and $p = 0.3$. Furthermore, $Y = 8 - X$. Therefore, $X \geq 5$ if and only if $Y \leq 3$ and it follows that $\Pr(X \geq 5) = \Pr(Y \leq 3)$. Probabilities for the binomial distribution with $n = 8$ and $p = 0.3$ are given in the table in the back of the book. The value of $\Pr(Y \leq 3)$ will be the sum of the entries for $k = 0, 1, 2$, and 3.

9. We need $\sum_{x=0}^{\infty} f(x) = 1$, which means that $c = 1/\sum_{x=0}^{\infty} 2^{-x}$. The last sum is known from Calculus to equal $1/(1 - 1/2) = 2$, so $c = 1/2$.

11. In order for the specified function to be a p.f., it must be the case that $\sum_{x=1}^{\infty} \frac{c}{x} = 1$ or equivalently $\sum_{x=1}^{\infty} \frac{1}{x} = \frac{1}{c}$. But $\sum_{x=1}^{\infty} \frac{1}{x} = \infty$, so there cannot be such a constant c.

3.2 Continuous Distributions

1. We compute $\Pr(X \le 8/27)$ by integrating the p.d.f. from 0 to 8/27.

$$\Pr\left(X \le \frac{8}{27}\right) = \int_0^{8/27} \frac{2}{3} x^{-1/3} dx = x^{2/3}\Big|_0^{8/27} = \frac{4}{9}.$$

3. The p.d.f. has the appearance of Fig. S.3.1.

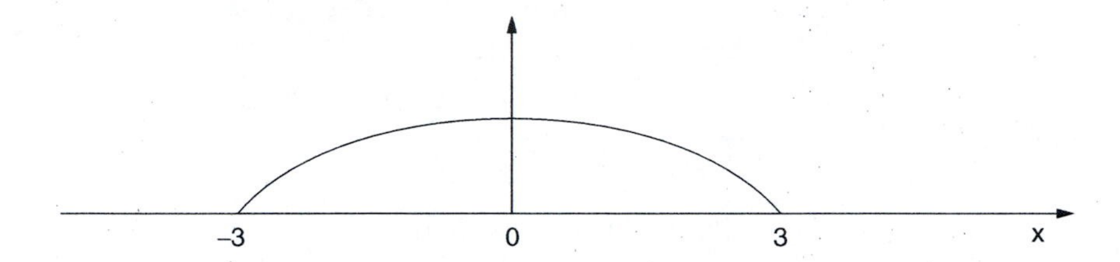

Figure S.3.1: Figure for Exercise 3 of Sec. 3.2.

(a) $\Pr(X < 0) = \frac{1}{36}\int_{-3}^{0} (9 - x^2)dx = 0.5.$

(b) $\Pr(-1 < X < 1) = \frac{1}{36}\int_{-1}^{1} (9 - x^2)dx = 0.4815.$

(c) $\Pr(X > 2) = \frac{1}{36}\int_{2}^{3} (9 - x^2)dx = 0.07407.$

The answer in part (a) could also be obtained directly from the fact that the p.d.f. is symmetric about the point $x = 0$. Therefore, the probability to the left of $x = 0$ and the probability to the right of $x = 0$ must each be equal to 1/2.

5. (a) $\int_0^t \frac{1}{8} x\, dx = 1/4$, or $t^2/16 = 1/4$. Hence, $t = 2$.

(b) $\int_t^4 (x/8)\, dx = 1/2$, or $1 - t^2/16 = 1/2$. Hence, $t = \sqrt{8}$.

7. Since the uniform distribution extends over an interval of length 10 units, the value of the p.d.f. must be 1/10 throughout the interval. Hence,

$$\int_0^7 f(x)\, dx = \frac{7}{10}.$$

9. Since $\int_0^\infty 1/(1+x)\, dx = \infty$, there is no constant c such that $\int_0^\infty f(x)\, dx = 1$.

11. Since $\int_0^1 (1/x)dx = \infty$, there is no constant c such that $\int_0^1 f(x)\, dx = 1$.

13. We find $\Pr(X < 20) = \int_0^{20} cx dx = 200c$. Setting this equal to 0.9 yields $c = 0.0045$.

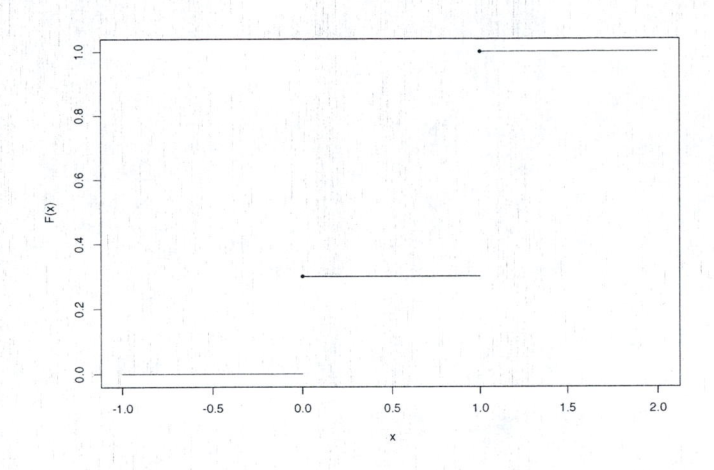

Figure S.3.2: C.d.f. of X in Exercise 1 of Sec. 3.3.

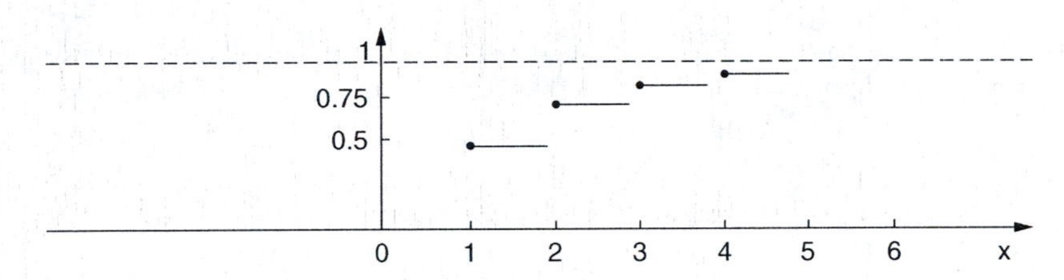

Figure S.3.3: C.d.f. for Exercise 3 of Sec. 3.3.

3.3 The Cumulative Distribution Function

1. The c.d.f. $F(x)$ of X is 0 for $x < 0$. It jumps to $0.3 = \Pr(X = 0)$ at $x = 0$, and it jumps to 1 and stays there at $x = 1$. The c.d.f. is sketched in Fig. S.3.2.

3. Here $\Pr(X = n) = 1/2^n$ for $n = 1, 2, \ldots$. Therefore, the c.d.f. must have the appearance of Fig. S.3.3.

5. $$f(x) = \frac{dF(x)}{dx} = \begin{cases} 0 & \text{for } x < 0, \\ \frac{2}{9}x & \text{for } 0 < x < 3, \\ 0 & \text{for } x > 3. \end{cases}$$

The value of $f(x)$ at $x = 0$ and $x = 3$ is irrelevant. This p.d.f. has the appearance of Fig. S.3.4.

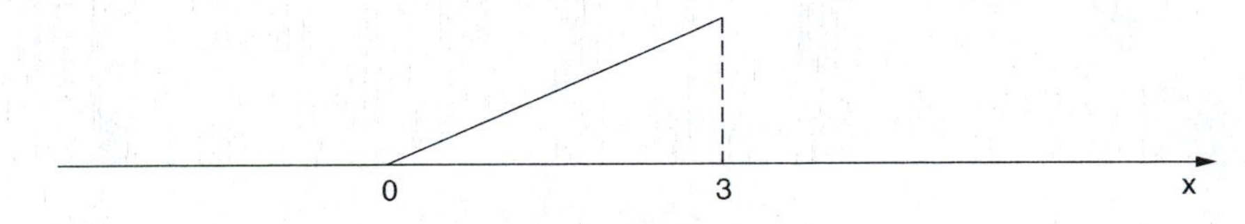

Figure S.3.4: Figure for Exercise 5 of Sec. 3.3.

7. The c.d.f. equals 0 for $x < -2$ and it equals 1 for $x > 8$. For $-2 \le x \le 8$, the c.d.f. equals

$$F(x) = \int_{-2}^{x} \frac{dy}{10} = \frac{x+2}{10}.$$

The c.d.f. has the appearance of Fig. S.3.5.

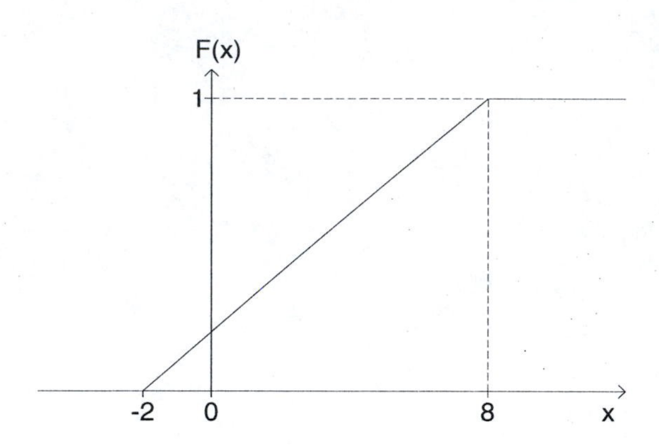

Figure S.3.5: Figure for Exercise 7 of Sec. 3.3.

9. $\Pr(Y = 0) = \Pr(X \leq 1) = 1/5$ and $\Pr(Y = 5) = \Pr(X \geq 3) = 2/5$. Also, Y is distributed uniformly between $Y = 1$ and $Y = 3$, with a total probability of 2/5. Therefore, over this interval $F(y)$ will be linear with a total increase of 2/5. The c.d.f. is plotted in Fig. S.3.6.

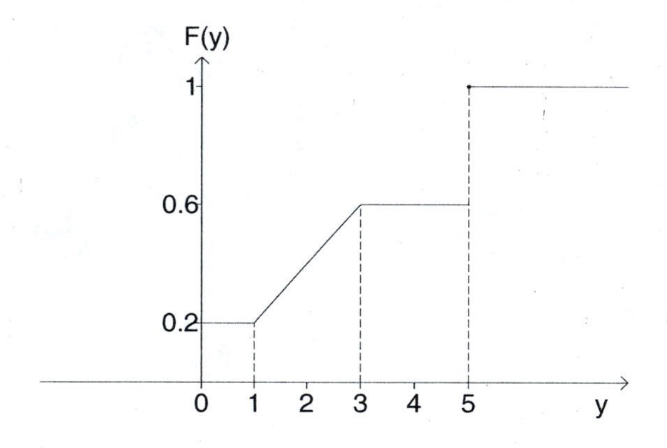

Figure S.3.6: C.d.f. for Exercise 9 of Sec. 3.3.

11. As in Exercise 10, we set $F(x) = p$ and solve for x.

$$\frac{1}{9}x^2 = p; \quad x^2 = 9p; \quad x = 3p^{1/2}.$$

The quantile function of X is $F^{-1}(p) = 3p^{1/2}$.

13. VaR at probability level 0.95 is the negative of the 0.05 quantile. Using the result from Example 3.3.8, the 0.05 quantile of the uniform distribution on the interval $[-12, 24]$ is $0.05 \times 24 - 0.95 \times 12 = -10.2$. So, VaR at probability level 0.95 is 10.2

15. Since $f(x) = 0$ for $x \le 0$ and for $x \ge 1$, the c.d.f. $F(x)$ will be flat (0) for $x \le 0$ and flat (1) for $x \ge 1$. Between 0 and 1, we compute $F(x)$ by integrating the p.d.f. For $0 < x < 1$,

$$F(x) = \int_0^x 2y dy = x^2.$$

The requested plot is in Fig. S.3.7.

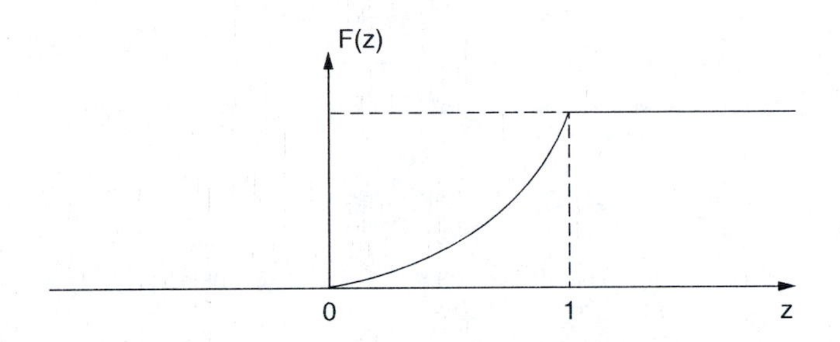

Figure S.3.7: C.d.f. for Exercise 15 of Sec. 3.3.

17. (a) Let $0 < p_1 < p_2 < 1$. Define $A_i = \{x : F(x) \ge p_i\}$ for $i = 1, 2$. Since $p_1 < p_2$ and F is nondecreasing, it follows that $A_2 \subset A_1$. Hence, the smallest number in A_1 (which equals $F^{-1}(p_1)$ by definition) is no greater than the smallest number in A_2 (which equals $F^{-1}(p_2)$ by definition. That is, $F^{-1}(p_1) \le F^{-1}(p_2)$, and the quantile function is nondecreasing.

    (b) Let $x_0 = \lim_{\substack{p \to 0 \\ p > 0}} F^{-1}(p)$. We are asked to prove that x_0 is the greatest lower bound of the set $C = \{c : F(c) > 0\}$. First, we show that no $x > x_0$ is a lower bound on C. Let $x > x_0$ and $x_1 = (x + x_0)/2$. Then $x_0 < x_1 < x$. Because $F^{-1}(p)$ is nondecreasing, it follows that there exists $p > 0$ such that $F^{-1}(p) < x_1$, which in turn implies that $p \le F(x_1)$, and $F(x_1) > 0$. Hence $x_1 \in C$, and x is not a lower bound on C. Next, we prove that x_0 is a lower bound on C. Let $x \in C$. We need only prove that $x_0 \le x$. Because $F^{-1}(p)$ is nondecreasing, we must have $\lim_{\substack{p \to 0 \\ p > 0}} F^{-1}(p) \le F^{-1}(q)$ for all $q > 0$. Hence, $x_0 \le F^{-1}(p)$ for all $q > 0$. Because $x \in C$, we have $F(x) > 0$. Let $q = F(x)$ so that $q > 0$. Then $x_0 \le F^{-1}(q) \le x$. The proof that x_1 is the least upper bound on the set of all d such that $F(d) < 1$ is very similar.

    (c) Let $0 < p < 1$. Because F^{-1} is nondecreasing, $F^{-1}(p^-)$ is the least upper bound on the set $C = \{F^{-1}(q) : q < p\}$. We need to show that $F^{-1}(p)$ is also that least upper bound. Clearly, $F^{-1}(p)$ is an upper bound, because F^{-1} is nondecreasing and $p > q$ for all $q < p$. To see that $F^{-1}(p)$ is the least upper bound, let y be an upper bound. We need to show $F^{-1}(p) \le y$. By definition, $F^{-1}(p)$ is the greatest lower bound on the set $D = \{x : F(x) \ge p\}$. Because y is an upper bound on C, it follows that $F^{-1}(q) \le y$ for all $q < p$. Hence, $F(y) \ge q$ for all $q < p$. Because F is nondecreasing, we have $F(y) \ge p$, hence $y \in D$, and $F^{-1}(p) \le y$.

19. First, we show that $F^{-1}(F(x)) \le x$. By definition $F^{-1}(F(x))$ is the smallest y such that $F(y) \ge F(x)$. Clearly $F(x) \ge F(x)$, hence $F^{-1}(F(x)) \le x$. Next, we show that, if $p > F(x)$, then $F^{-1}(p) > x$. Let $p > F(x)$. By Exercise 17, we know that $F^{-1}(p) \ge x$. By definition, $F^{-1}(p)$ is the greatest lower bound on the set $C_p = \{y : F(y) \ge p\}$. All $y \in C_p$ satisfy $F(y) > (p + F(x))/2$. Since F is continuous from the right, $F(F^{-1}(p)) \ge (p + F(x))/2$. But $F(x) < (p + F(x))/2$, so $x \ne F^{-1}(p)$, hence $F^{-1}(p) > x$.

3.4 Bivariate Distributions

1. (a) Let the constant value of the p.d.f. on the rectangle be c. The area of the rectangle is 2. So, the integral of the p.d.f. is $2c = 1$, hence $c = 1/2$.

   (b) $\Pr(X \geq Y)$ is the integral of the p.d.f. over that part of the rectangle where $x \geq y$. This region is shaded in Fig. S.3.8. The region is a trapezoid with area $1 \times (1+2)/2 = 1.5$. The integral of the

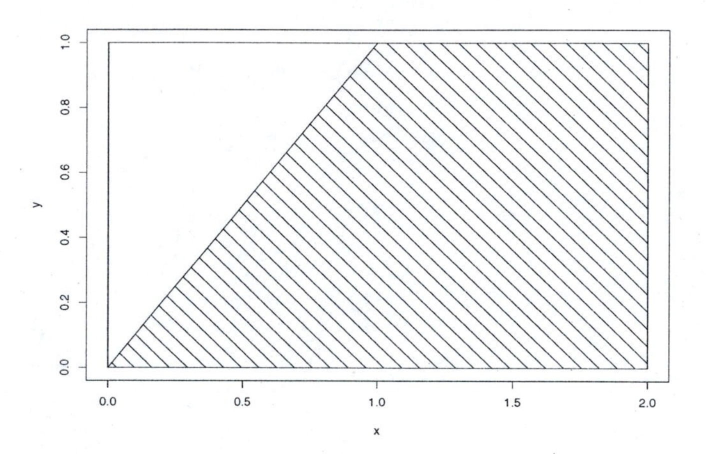

Figure S.3.8: Region where $x \geq y$ in Exercise 1b of Sec. 3.4.

   constant 1/2 over this region is then $0.75 = \Pr(X \geq Y)$.

3. (a) If we sum $f(x, y)$ over the 25 possible pairs of values (x, y), we obtain $40c$. Since this sum must be equal to 1, it follows that $c = 1/40$.

   (b) $f(0, -2) = (1/40) \cdot 2 = 1/20$.

   (c) $\Pr(X = 1) = \sum_{y=-2}^{2} f(1, y) = 7/40$.

   (d) The answer is found by summing $f(x, y)$ over the following pairs: $(-2, -2)$, $(-2, -1)$, $(-1, -2)$, $(-1, -1)$, $(-1, 0)$, $(0, -1)$, (0, 0), (0, 1), (1, 0), (1, 1), (1, 2), (2, 1), and (2, 2). The sum is 0.7.

5. (a) By sketching the curve $y = 1 - x^2$, we find that $y \leq 1 - x^2$ for all points on or below this curve. Also, $y \geq 0$ for all points on or above the x-axis. Therefore, $0 \leq y \leq 1 - x^2$ only for points in the shaded region in Fig. S.3.9.

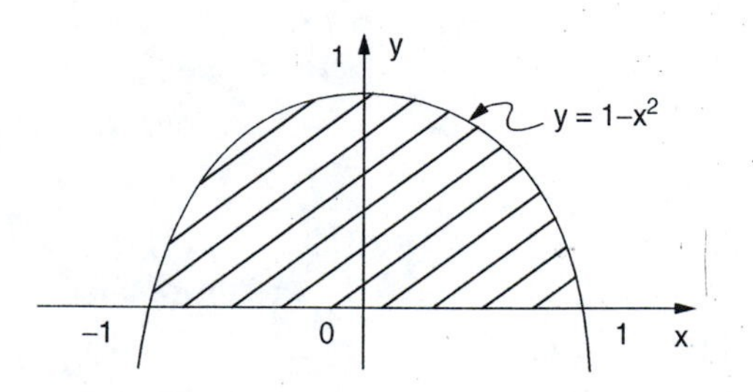

Figure S.3.9: Figure for Exercise 5a of Sec. 3.4.

Hence,

$$\int_{-\infty}^{\infty}\int_{-\infty}^{\infty} f(x,y)\,dx\,dy = \int_{-1}^{1}\int_{0}^{1-x^2} c\,(x^2+y)\,dy\,dx = \frac{4}{5}c\cdot$$

Therefore, $c = 5/4$.

(b) Integration is done over the shaded region in Fig. S.3.10.

$$\Pr\left(0 \le X \le \frac{1}{2}\right) = \underset{\substack{\text{shaded}\\ \text{region}}}{\int\int} f(x,y)\,dx\,dy = \int_0^{\frac{1}{2}}\int_0^{1-x^2} \frac{5}{4}(x^2+y)\,dy\,dx = \frac{79}{256}.$$

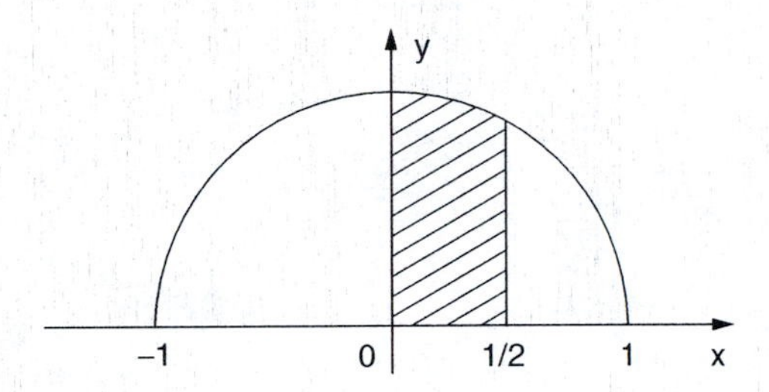

Figure S.3.10: Region of integration for Exercise 5b of Sec. 3.4.

(c) The region over which to integrate is shaded in Fig. S.3.11.

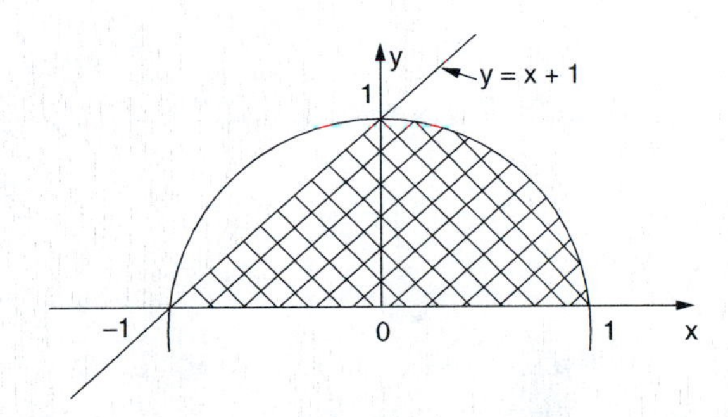

Figure S.3.11: Region of integration for Exercise 5c of Sec. 3.4.

$$\begin{aligned}\Pr(Y \le X+1) &= \underset{\substack{\text{shaded}\\ \text{region}}}{\int\int} f(x,y)\,dx\,dy = 1 - \underset{\substack{\text{unshaded}\\ \text{region}}}{\int\int} f(x,y)\,dx\,dy \\ &= 1 - \int_{-1}^{0}\int_{x+1}^{1-x^2} \frac{5}{4}(x^2+y)\,dy\,dx = \frac{13}{16}.\end{aligned}$$

(d) The probability that (X, Y) will lie on the curve $y = x^2$ is 0 for every continuous joint distribution.

7. (a) $\Pr(X \le 1/4)$ will be equal to the sum of the probabilities of the corners (0, 0) and (0, 1) and the probability that the point is an interior point of the square and lies in the shaded region in Fig. S.3.12. The probability that the point will be an interior point of the square rather than one of the four corners is $1-(0.1+0.2+0.4+0.1) = 0.2$. The probability that it will lie in the shaded region, given that it is an interior point is 1/4. Therefore,

$$\Pr\left(X \le \frac{1}{4}\right) = 0.1 + 0.4 + (0.2)\left(\frac{1}{4}\right) = 0.55.$$

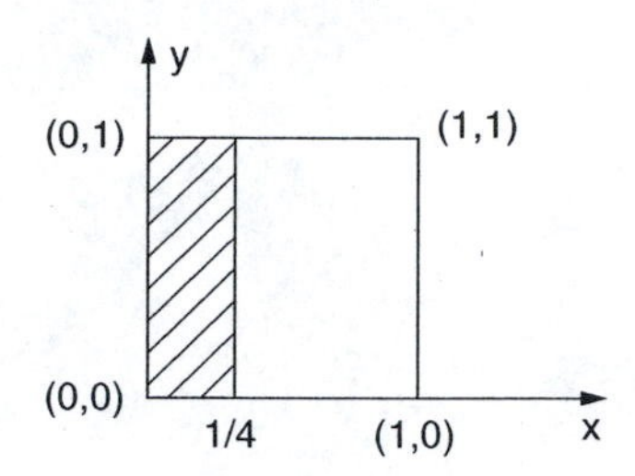

Figure S.3.12: Figure for Exercise 7a of Sec. 3.4.

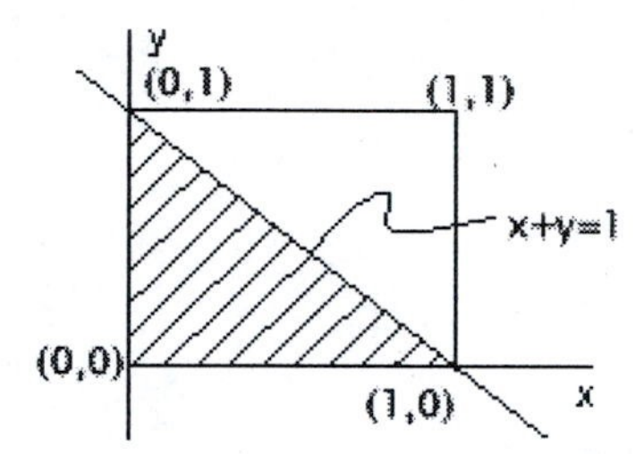

Figure S.3.13: Figure for Exercise 7b of Sec. 3.4.

(b) The region over which to integrate is shaded in Fig. S.3.13.

$$\Pr(X+Y \le 1) = 0.1 + 0.2 + 0.4 + (0.2)\left(\frac{1}{2}\right) = 0.8.$$

9. The joint p.d.f. of water demand X and electricy demand Y is in (3.4.2), and is repeated here:

$$f(x,y) = \begin{cases} 1/29204 & \text{if } 4 \le x \le 200 \text{ and } 1 \le y \le 150, \\ 0 & \text{otherwise.} \end{cases}$$

We need to integrate this function over the set where $x > y$. That region can be written as $\{(x,y) : 4 < x < 200, 1 < y < \min\{x, 150\}\}$. The reason for the complicated upper limit on y is that we require both $y < x$ and $y < 150$.

$$\begin{aligned} \int_4^{200} \int_1^{\min\{x,150\}} \frac{1}{29204} dydx &= \int_4^{200} \frac{\min\{x-1, 149\}}{29204} dx \\ &= \int_4^{150} \frac{x-1}{29204} dx + \int_{150}^{200} \frac{149}{29204} dx \\ &= \left.\frac{(x-1)^2}{2 \times 29204}\right|_{x=4}^{150} + \frac{50 \times 149}{29204} \\ &= \frac{149^2 - 3^2}{58408} + \frac{7450}{29204} = 0.63505. \end{aligned}$$

11. Let $f(x,y)$ stand for the joint p.f. in Table 3.3 in the text for $x = 0, 1$ and $y = 1, 2, 3, 4$.

(a) We are asked for the probability for the set $\{Y \in \{2,3\}\} \cap \{X = 1\}$, which is $f(1,2) + f(1,3) = 0.166 + 0.107 = 0.273$.

(b) This time, we want $\Pr(X = 0) = f(0,1) + f(0,2) + f(0,3) + f(0,4) = 0.513$.

3.5 Marginal Distributions

1. The joint p.d.f. is constant over a rectangle with sides parallel to the coordinate axes. So, for each x, the integral over y will equal the constant times the length of the interval of y values, namely $d - c$. Similarly, for each y, the integral over x will equal the constant times the length of the interval of x values, namely $b - a$. Of course the constant k must equal one over the area of the rectangle. So $k = 1/[(b-a)(d-c)]$. So the marginal p.d.f.'s of X and Y are

$$f_1(x) = \begin{cases} \dfrac{1}{b-a} & \text{for } a \le x \le b, \\ 0 & \text{otherwise,} \end{cases}$$

$$f_2(y) = \begin{cases} \dfrac{1}{d-c} & \text{for } c \le y \le d, \\ 0 & \text{otherwise.} \end{cases}$$

3. (a) For $0 \le x \le 2$, we have

$$f_1(x) = \int_0^1 f(x,y)\,dy = \frac{1}{2}.$$

Also, $f_1(x) = 0$ for x outside the interval $0 \le x \le 2$. Similarly, for $0 \le y \le 1$,

$$f_2(y) = \int_0^2 f(x,y)\,dx = 3y^2.$$

Also, $f_2(y) = 0$ for y outside the interval $0 \le y \le 1$.

(b) X and Y are independent because $f(x,y) = f_1(x)\, f_2(y)$ for $-\infty < x < \infty$ and $-\infty < y < \infty$.

(c) We have

$$\begin{aligned} \Pr\left(X < 1 \text{ and } Y \ge \frac{1}{2}\right) &= \int_0^1 \int_{1/2}^1 f(x,y)\,dx\,dy \\ &= \int_0^1 \int_{1/2}^1 f_1(x) f_2(y)\,dx\,dy \\ &= \int_0^1 f_1(x)\,dx \int_{1/2}^1 f_2(y)\,dy = \Pr(X<1)\Pr\left(Y > \frac{1}{2}\right). \end{aligned}$$

Therefore, by the definition of the independence of two events (Definition 2.2.1), the two given events are independent.

We can also reach this answer, without carrying out the above calculation, by reasoning as follows: Since the random variables X and Y are independent, and since the occurrence or nonoccurence of the event $\{X < 1\}$ depends on the value of X only while the occurrence or nonoccurence of the event $\{Y \ge 1/2\}$ depends on the value of Y only, it follows that these two events must be independent.

5. (a) Since X and Y are independent,

$$f(x,y) = \Pr(X = x \text{ and } Y = y) = \Pr(X = x)\Pr(Y = y) = p_x p_y.$$

(b) $\Pr(X = Y) = \sum_{i=0}^{3} f(i,i) = \sum_{i=0}^{3} p_i^2 = 0.3.$

(c) $\Pr(X > Y) = f(1,0) + f(2,0) + f(3,0) + f(2,1) + f(3,1) + f(3,2) = 0.35.$

7. Since $f(x, y) = 0$ outside a rectangle and $f(x, y)$ can be factored as in Eq. (3.5.7) inside the rectangle (use $h_1(x) = 2x$ and $h_2(y) = \exp(-y)$), it follows that X and Y are independent.

9. (a) Since $f(x, y)$ is constant over the rectangle S and the area of S is 6 units, it follows that $f(x, y) = 1/6$ inside S and $f(x, y) = 0$ outside S. Next, for $0 \le x \le 2$,

$$f_1(x) = \int_{-\infty}^{\infty} f(x, y)\, dy = \int_1^4 \frac{1}{6} dy = \frac{1}{2}.$$

Also, $f_1(x) = 0$ otherwise. Similarly, for $1 \le y \le 4$,

$$f_2(y) = \int_0^2 \frac{1}{6}\, dx = \frac{1}{3}.$$

Also, $f_2(y) = 0$ otherwise. Thus, the marginal distribution of both X and Y are uniform distributions.

(b) Since $f(x, y) = f_1(x)\, f_2(y)$ for all values of x and y, it follows that X and Y are independent.

11. Let X and Y denote the arrival times of the two persons, measured in terms of the number of minutes after 5 P.M. Then X and Y each have the uniform distribution on the interval (0, 60) and they are independent. Therefore, the joint p.d.f. of X and Y is

$$f(x, y) = \begin{cases} \dfrac{1}{3600} & \text{for } 0 < x < 60,\ 0 < y < 60, \\ 0 & \text{otherwise.} \end{cases}$$

We must calculate $\Pr(|X - Y| < 10)$, which is equal to the probability that the point (X, Y) lies in the shaded region in Fig. S.3.14. Since the joint p.d.f. of X and Y is constant over the entire square, this

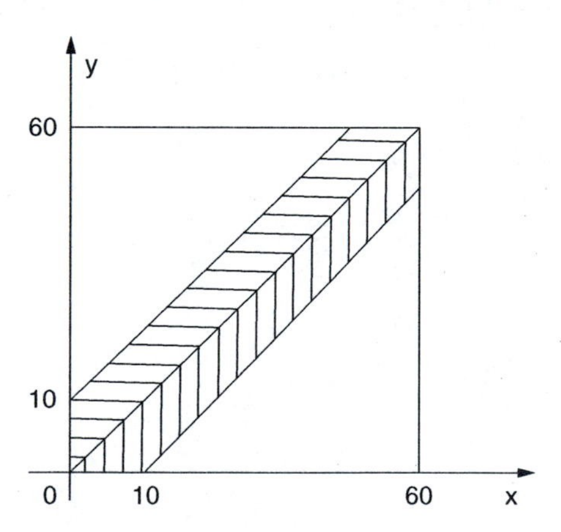

Figure S.3.14: Figure for Exercise 11 of Sec. 3.5.

probability is equal to (area of shaded region)/3600. The area of the shaded region is 1100. Therefore, the required probability is $1100/3600 = 11/36$.

13. Since $f(x, y) = f(y, x)$ for all (x, y), it follows that the marginal p.d.f.'s will be the same. Each of those marginals will equal the integral of $f(x, y)$ over the other variable. For example, to find $f_1(x)$, note that for each x, the values of y such that $f(x, y) > 0$ form the interval $[-\sqrt{1-x^2}, \sqrt{1-x^2}]$. Then, for $-1 \le x \le 1$,

$$f_1(x) \quad = \quad \int f(x, y) dy$$

$$= \int_{-\sqrt{1-x^2}}^{\sqrt{1-x^2}} kx^2y^2dy$$
$$= kx^2 \left. \frac{y^3}{3} \right|_{y=-\sqrt{1-x^2}}^{\sqrt{1-x^2}}$$
$$= 2kx^2(1-x^2)^{3/2}/3.$$

15. (a) Figure S.3.15 shows the region where $f(x, y) > 0$ as the union of two shaded rectangles. Although the region is not a rectangle, it is a *product set*. That is, it has the form $\{(x, y) : x \in A, y \in B\}$ for two sets A and B of real numbers.

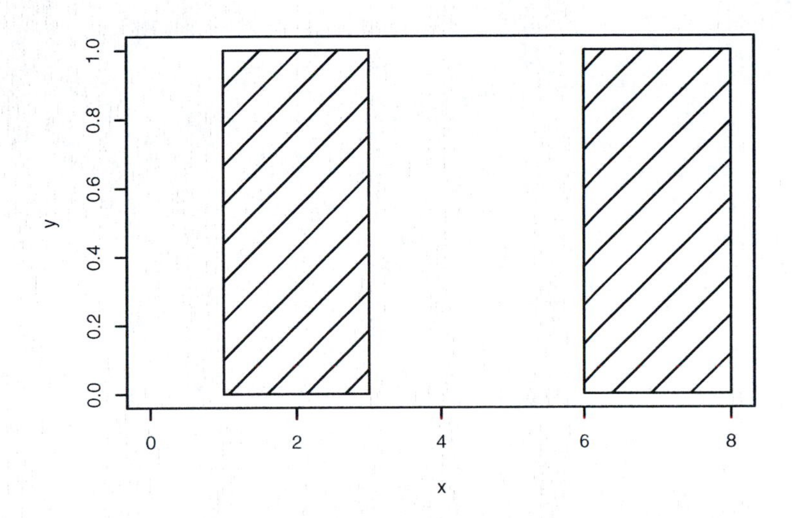

Figure S.3.15: Region of positive p.d.f. for Exercise 15a of Sec. 3.5.

(b) The marginal p.d.f. of X is

$$f_1(x) = \int_0^1 f(x, y)dy = \begin{cases} \frac{1}{3} & \text{if } 1 < x < 3, \\ \frac{1}{6} & \text{if } 6 < x < 8. \end{cases}$$

The marginal p.d.f. of Y is

$$f_2(y) = \int_1^3 \frac{1}{3}dx + \int_6^8 \frac{1}{6}dx = 1,$$

for $0 < y < 1$. The distribution of Y is the uniform distribution on the interval $[0, 1]$.

(c) The product of the two marginal p.d.f.'s is

$$f_1(x)f_2(y) = \begin{cases} \frac{1}{3} & \text{if } 1 < x < 3 \text{ and } 0 < y < 1, \\ \frac{1}{6} & \text{if } 6 < x < 8 \text{ and } 0 < y < 1, \\ 0 & \text{otherwise,} \end{cases}$$

which is the same as $f(x, y)$, hence the two random variables are independent. Although the region where $f(x, y) > 0$ is not a rectangle, it is a product set as we saw in part (a). Although it is sufficient in Theorem 3.5.6 for the region where $f(x, y) > 0$ to be a rectangle, it is necessary that the region be a product set. Technically, it is necessary that there is a version of the p.d.f.

that is strictly positive on a product set. For continuous joint distributions, one can set the p.d.f. to arbitrary values on arbitrary one-dimensional curves without changing it's being a joint p.d.f.

3.6 Conditional Distributions

1. We begin by finding the marginal p.d.f. of Y. The set of x values for which $f(x, y) > 0$ is the interval $[-(1-y^2)^{1/2}, (1-y^2)^{1/2}]$. So, the marginal p.d.f. of Y is, for $-1 \le y \le 1$,

$$f_2(y) = \int_{-(1-y^2)^{1/2}}^{(1-y^2)^{1/2}} kx^2y^2 dx = \frac{ky^2}{3}x^3 \Bigg|_{x=-(1-y^2)^{1/2}}^{(1-y^2)^{1/2}} = \frac{2k}{3}y^2(1-y^2)^{3/2},$$

and 0 otherwise. The conditional p.d.f. of X given $Y = y$ is the ratio of the joint p.d.f. to the marginal p.d.f. just found.

$$g_1(x|y) = \begin{cases} \dfrac{3x^2}{2(1-y^2)^{3/2}} & \text{for } -(1-y^2)^{1/2} \le x \le (1-y^2)^{1/2}, \\ 0 & \text{otherwise.} \end{cases}$$

3. The joint p.d.f. of X and Y is positive for all points inside the circle S shown in the sketch. Since the area of S is 9π and the joint p.d.f. of X and Y is constant over S, this joint p.d.f. must have the form:

$$f(x, y) = \begin{cases} \dfrac{1}{9\pi} & \text{for } (x, y) \in S, \\ 0 & \text{otherwise.} \end{cases}$$

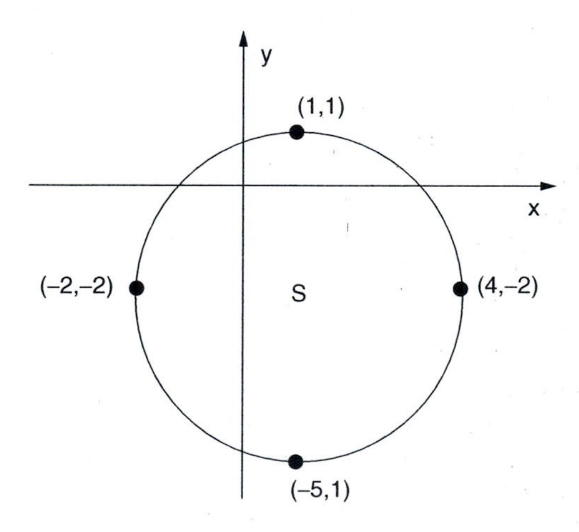

Figure S.3.16: Figure for Exercise 3 of Sec. 3.6.

It can be seen from Fig. S.3.16 that the possible values of X lie between -2 and 4. Therefore, for $-2 < x < 4$,

$$f_1(x) = \int_{-2-[9-(x-1)^2]^{1/2}}^{-2+[9-(x-1)^2]^{1/2}} \frac{1}{9\pi} dy = \frac{2}{9\pi}[9-(x-1)^2]^{1/2}.$$

(a) It follows that for $-2 < x < 4$ and $-2-[9-(x-1)^2]^{1/2} < y < -2+[9-(x-1)^2]^{1/2}$,

$$g_2(y \mid x) = \frac{f(x, y)}{f_1(x)} = \frac{1}{2}[9-(x-1)^2]^{-1/2}.$$

(b) When $X = 2$, it follows from part (a) that

$$g_2(y \mid x = 2) = \begin{cases} \dfrac{1}{2\sqrt{8}} & \text{for } -2 - \sqrt{8} < y < -2 + \sqrt{8} \\ 0 & \text{otherwise.} \end{cases}$$

Therefore,

$$\Pr(Y > 0 \mid X = 2) = \int_0^{-2+\sqrt{8}} g_2(y \mid x = 2)\, dy = \frac{-2+\sqrt{8}}{2\sqrt{8}} = \frac{2-\sqrt{2}}{4}.$$

5. (a) The joint p.d.f. $f(x, y)$ is given by Eq. (3.6.15) and the marginal p.d.f. $f_2(y)$ was also given in Example 3.6.10. Hence, for $0 < y < 1$ and $0 < x < y$, we have

$$g_1(x \mid y) = \frac{f(x, y)}{f_2(y)} = \frac{-1}{(1-x)\,\log(1-y)}.$$

(b) When $Y = 3/4$, it follows from part (a) that

$$g_1\left(x \mid y = \frac{3}{4}\right) = \begin{cases} \dfrac{1}{(1-x)\log 4} & \text{for } 0 < x < \frac{3}{4}, \\ 0 & \text{otherwise.} \end{cases}$$

Therefore,

$$\Pr\left(X > \frac{1}{2} \mid Y = \frac{3}{4}\right) = \int_{1/2}^{3/4} g_1\left(x \mid y = \frac{3}{4}\right) dx = \frac{\log 4 - \log 2}{\log 4} = \frac{1}{2}.$$

7. The joint p.d.f. of X and Y is positive inside the triangle S shown in Fig. S.3.17. It is seen from Fig. S.3.17 that the possible values of X lie between 0 and 2. Hence, for $0 < x < 2$,

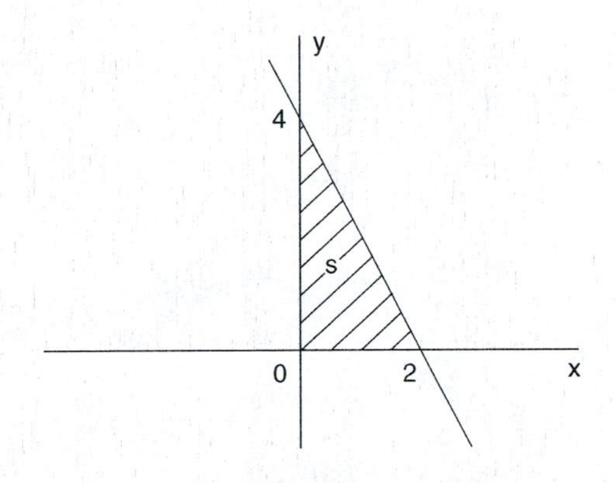

Figure S.3.17: Figure for Exercise 7 of Sec. 3.6.

$$f_1(x) = \int_0^{4-2x} f(x, y)\, dy = \frac{3}{8}(x-2)^2.$$

(a) It follows that for $0 < x < 2$ and $0 < y < 4 - 2x$,

$$g_2(y \mid x) = \frac{f(x, y)}{f_1(x)} = \frac{4 - 2x - y}{2(x-2)^2}.$$

(b) When $X = 1/2$, it follows from part (a) that

$$g_2\left(y \,|\, x = \frac{1}{2}\right) = \begin{cases} \frac{2}{9}(3-y) & \text{for } 0 < y < 3, \\ 0 & \text{otherwise.} \end{cases}$$

Therefore,

$$\Pr\left(Y \geq 2 \,|\, X = \frac{1}{2}\right) = \int_2^3 g_2\left(y \,|\, x = \frac{1}{2}\right) dy = \frac{1}{9}.$$

9. Let Y denote the instrument that is chosen. Then $\Pr(Y = 1) = \Pr(Y = 2) = 1/2$. In this exercise the distribution of X is continuous and the distribution of Y is discrete. Hence, the joint distribution of X and Y is a mixed distribution, as described in Sec. 3.4. In this case, the joint p.f./p.d.f. of X and Y is as follows:

$$f(x,y) = \begin{cases} \frac{1}{2} \cdot 2x = x & \text{for } \ y = 1 \text{ and } \ 0 < x < 1, \\ \frac{1}{2} \cdot 3x^2 = \frac{3}{2}x^2 & \text{for } \ y = 2 \text{ and } \ 0 < x < 1, \\ 0 & \text{otherwise.} \end{cases}$$

(a) It follows that for $0 < x < 1$,

$$f_1(x) = \sum_{y=1}^{2} f(x,y) = x + \frac{3}{2}x^2,$$

and $f_1(x) = 0$ otherwise.

(b) For $y = 1, 2$ and $0 < x < 1$, we have

$$\Pr(Y = y \,|\, X = x) = g_2(y \,|\, x) = \frac{f(x,y)}{f_1(x)}.$$

Hence,

$$\Pr\left(Y = 1 \,|\, X = \frac{1}{4}\right) = \frac{f\left(\frac{1}{4}, 1\right)}{f_1\left(\frac{1}{4}\right)} = \frac{\frac{1}{4}}{\frac{1}{4} + \frac{3}{2} \cdot \frac{1}{16}} = \frac{8}{11}.$$

11. Let F_2 be the c.d.f. of Y. Since f_2 is continuous at both y_0 and y_1, we can write, for $i = 0, 1$,

$$\Pr(Y \in A_i) = F_2(y_i + \epsilon) - F_2(y_i - \epsilon) = 2\epsilon f_2(y_i'),$$

where y_i' is within ϵ of y_i. This last equation follows from the mean value theorem of calculus. So

$$\frac{\Pr(Y \in A_0)}{\Pr(Y \in A_1)} = \frac{f_2(y_0')}{f_2(y_1')}. \tag{S.3.1}$$

Since f_2 is continuous, $\lim_{\epsilon \to 0} f_2(y_i') = f_2(y_i)$, and the limit of (S.3.1) is $0/f_2(y_1) = 0$.

13. There are four different treatments on which we are asked to condition. The marginal p.f. of treatment Y is given in the bottom row of Table 3.6 in the text. The conditional p.f. of response given each treatment is the ratio of the two rows above that to the bottom row:

$$g_1(x|1) = \begin{cases} \frac{0.120}{0.267} = 0.4494 & \text{if } x = 0, \\ \frac{0.147}{0.267} = 0.5506 & \text{if } x = 1. \end{cases}$$

$$g_1(x|2) = \begin{cases} \frac{0.087}{0.253} = 0.3439 & \text{if } x = 0, \\ \frac{0.166}{0.253} = 0.6561 & \text{if } x = 1. \end{cases}$$

$$g_1(x|3) = \begin{cases} \frac{0.146}{0.253} = 0.5771 & \text{if } x = 0, \\ \frac{0.107}{0.253} = 0.4229 & \text{if } x = 1. \end{cases}$$

$$g_1(x|4) = \begin{cases} \frac{0.160}{0.227} = 0.7048 & \text{if } x = 0, \\ \frac{0.067}{0.227} = 0.2952 & \text{if } x = 1. \end{cases}$$

The fourth one looks quite different from the others, especially from the second.

3.7 Multivariate Distributions

1. (a) We have

$$\int_0^1 \int_0^1 \int_0^1 f(x_1, d_2, x_3)\, dx_1\, dx_2\, dx_3 = 3c.$$

Since the value of this integral must be equal to 1, it follows that $c = 1/3$.

(b) For $0 \le x_1 \le 1$ and $0 \le x_3 \le 1$,

$$f_{13}(x_1, x_3) = \int_0^1 f(x_1, x_2, x_3)\, dx_2 = \frac{1}{3}(x_1 + 1 + 3x_3)\,.$$

(c) The conditional p.d.f. of x_3 given that $x_1 = 1/4$ and $x_2 = 3/4$ is, for $0 \le x_3 \le 1$,

$$g_3\left(x_3 \,\middle|\, x_1 = \frac{1}{4}, x_2 = \frac{3}{4}\right) = \frac{f\left(\frac{1}{4}, \frac{3}{4}, x_3\right)}{f_{12}\left(\frac{1}{4}, \frac{3}{4}\right)} = \frac{7}{13} + \frac{12}{13}x_3\,.$$

Therefore,

$$\Pr\left(X_3 < \frac{1}{2} \,\middle|\, X_1 = \frac{1}{4}, X_2 = \frac{3}{4}\right) = \int_0^{\frac{1}{2}} \left(\frac{7}{13} + \frac{12}{13}x_3\right) dx_3 = \frac{5}{13}.$$

3. The p.d.f. should be positive for all $x_i > 0$ not just for all $x_i > 1$ as stated in early printings. This will match the answers in the back of the text.

(a) We have

$$\int_0^\infty \int_0^\infty \int_0^\infty f(x_1, x_2, x_3)\, dx_1\, dx_2\, dx_3 = \frac{1}{6}c.$$

Since the value of this integral must be equal to 1, it follows that c = 6. If one used $x_i > 1$ instead, then the integral would equal $\exp(-6)/6$, so that $c = 6\exp(6)$.

(b) For $x_1 > 0$, $x_3 > 0$,

$$f_{13}(x_1, x_3) \int_0^\infty f(x_1, x_2, x_3)\, dx_2 = 3\exp[-(x_1 + 3x_3)].$$

If one used $x_i > 1$ instead, then for $x_1 > 1$ and $x_3 > 1$, $f_{13}(x_1, x_3) = 3\exp(-x_1 - 3x_3 + 4)$.

(c) It is helpful at this stage to recognize that the random variables X_1, X_2, and X_3 are independent because their joint p.d.f. $f(x_1, x_2, x_3)$ can be factored as in Eq. (3.7.7); i.e., for $x_i > 0$ $(i = 1, 2, 3)$,

$$f(x_1, x_2, x_3) = (\exp(-x_1)) = (2\exp(-x_2))(3\exp(-x_3)).$$

It follows that

$$\Pr(X_1 < 1 \mid X_2 = 2, X_3 = 1) = Pr(X_1 < 1) = \int_0^1 f_1(x_1)\,dx_1 = \int_0^1 \exp(-x_1)dx_1 = 1 - \frac{1}{e}.$$

This answer could also be obtained without explicitly using the independence of X_1, X_2, and X_3 by calculating first the marginal joint p.d.f.

$$f_{23}(x_2, x_3) = \int_0^\infty f(x_1, x_2, x_3)\,dx_1,$$

then calculating the conditional p.d.f.

$$g_1(x_1 \mid x_2 = 2, x_3 = 1) = \frac{f(x_1, 2, 1)}{f_{2,3}(2, 1)},$$

and finally calculating the probability

$$\Pr(X_1 < 1 \mid X_2 = 2, X_3 = 1) = \int_0^1 g_1(x_1 \mid x_2 = 2, x_3 = 1)dx_1.$$

If one used $x_i > 1$ instead, then the probability in this part is 0.

5. (a) The probability that all n independent components will function properly is the product of their individual probabilities and is therefore equal to $\prod_{i=1}^{n} p_i$.

   (b) The probability that all n independent components will not function properly is the product of their individual probabilities of not functioning properly and is therefore equal to $\prod_{i=1}^{n}(1 - p_i)$. The probability that at least one component will function properly is $1 - \prod_{i=1}^{n}(1 - p_i)$.

7. The probability that a particular variable X_i will lie in the interval (a, b) is $p = \int_a^b f(x)\,dx$. Since the variables $X_1, \ldots, X_n$ are independent, the probability that exactly i of these variables will lie in the interval (a, b) is $\binom{n}{i}p^i(1-p)^{n-i}$. Therefore, the required probability is

$$\sum_{i=k}^{n}\binom{n}{i}p^i(1-p)^{n-i}.$$

9. (a) Since $X_i = X$ for $i = 1, 2$, we know that X_i has the same distribution as X. Since X has a continuous distribution, then so does X_i for $i = 1, 2$.

   (b) We know that $\Pr(X_1 = X_2) = 1$. Let $A = \{(x_1, x_2) : x_1 = x_2\}$. Then $\Pr((X_1, X_2) \in A) = 1$. However, for every function f, $\int_A\int f(x_1, x_2)dx_1dx_2 = 0$. So there is no possible joint p.d.f.

11. Since $X_1, \ldots, X_n$ are independent, their joint p.f., p.d.f., or p.f./p.d.f. factors as

$$f(x_1, \ldots, x_n) = f_1(x_1)\cdots f_n(x_n),$$

where each f_i is a p.f. or p.d.f. If we sum or integrate over all x_j such that $j \notin \{i_1, \ldots, i_k\}$ we obtain the joint p.f., p.d.f., or p.f./p.d.f. of $X_{i_1}, \ldots, X_{i_k}$ equal to $f_{i_1}(x_{i_1})\cdots f_{i_k}(x_{i_k})$, which is factored in a way that makes it clear that $X_{i_1}, \ldots, X_{i_k}$ are independent.

13. Let $f(x_1, x_2, x_3, z)$ be the joint p.d.f. of (X_1, X_2, X_3, Z). Let $f_{12}(x_1, x_2)$ be the marginal joint p.d.f. of (X_1, X_2). The the conditional p.d.f. of X_3 given $(X_1, X_2) = (x_1, x_2)$ is

$$\frac{\int f(x_1, x_2, x_3, z)dz}{f_{12}(x_1, x_2)} = \frac{\int g(x_1|z)g(x_2|z)g(x_3|z)f_0(z)dz}{f_{12}(x_1, x_2)} = \int g(x_3|z)\frac{g(x_1|z)g(x_2|z)f_0(z)}{f_{12}(x_1, x_2)}dz.$$

According to Bayes' theorem for random variables, the fraction in this last integral is $g_0(z|x_1, x_2)$. Using the specific formulas in the text, we can calculate the last integral as

$$\begin{aligned}
&\int_0^\infty z\exp(-zx_3)\frac{1}{2}(2+x_1+x_2)^3 z^2 \exp(-z(2+x_1+x_2))dx \\
&= \frac{(2+x_1+x_2)^3}{2}\int_0^\infty z^3\exp(-z(2+x_1+x_2+x_3))dz \\
&= \frac{(2+x_1+x_2)^3}{2}\frac{6}{(2+x_1+x_2+x_3)^4} = \frac{3(2+x_1+x_2)^3}{(2+x_1+x_2+x_3)^4}.
\end{aligned}$$

The joint p.d.f. of (X_1, X_2, X_3) can be computed in a manner similar to the joint p.d.f. of (X_1, X_2) and it is

$$f_{123}(x_1, x_2, x_3) = \frac{12}{(2+x_1+x_2+x_3)^4}.$$

The ratio of $f_{123}(x_1, x_2, x_3)$ to $f_{12}(x_1, x_2)$ is the conditional p.d.f. calculated above.

15. Let A be an arbitrary n-dimensional set. Because $\Pr(W = c) = 1$, we have

$$\Pr((X_1, \ldots, X_n) \in A, W = w) = \begin{cases} \Pr(X_1, \ldots, X_n) \in A) & \text{if } w = c, \\ 0 & \text{otherwise.} \end{cases}$$

It follows that

$$\Pr((X_1, \ldots, X_n) \in A|W = w) = \begin{cases} \Pr(X_1, \ldots, X_n) \in A) & \text{if } w = c, \\ 0 & \text{otherwise.} \end{cases}$$

Hence the conditional joint distribution of $X_1, \ldots, X_n$ given W is the same as the unconditional joint distribution of $X_1, \ldots, X_n$, which is the distribution of independent random variables.

3.8 Functions of a Random Variable

1. The inverse transformation is $x = (1-y)^{1/2}$, whose derivative is $-(1-y)^{-1/2}/2$. The p.d.f. of Y is then

$$g(y) = f([1-y]^{1/2})(1-y)^{-1/2}/2 = \frac{3}{2}(1-y)^{1/2},$$

for $0 < y < 1$.

3. It is seen from Fig. S.3.18 that as x varies over the interval $0 < x < 2$, y varies over the interval

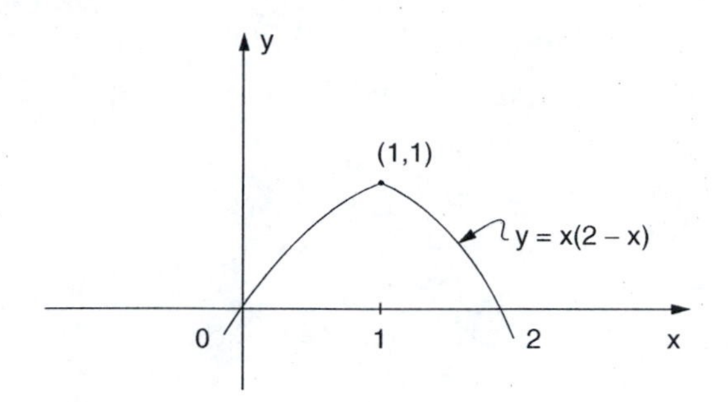

Figure S.3.18: Figure for Exercise 3 of Sec. 3.8.

$0 < y \le 1$. Therefore, for $0 < y \le 1$,

$$\begin{aligned} G(y) &= \Pr(Y \le y) = \Pr[X(2-X) \le y] = \Pr(X^2 - 2X \ge -y) \\ &= \Pr(X^2 - 2X + 1 \ge 1 - y) = \Pr[(X-1)^2 \ge 1 - y] \\ &= \Pr(X - 1 \le -\sqrt{1-y}) + \Pr(X - 1 \ge \sqrt{1-y}) \\ &= \Pr(X \le 1 - \sqrt{1-y}) + \Pr(X \ge 1 + \sqrt{1-y}) \\ &= \int_0^{1-\sqrt{1-y}} \frac{1}{2} x\, dx + \int_{1+\sqrt{1-y}}^{2} \frac{1}{2} x\, dx \\ &= 1 - \sqrt{1-y}. \end{aligned}$$

It follows that, for $0 < y < 1$,

$$g(y) = \frac{dG(y)}{dy} = \frac{1}{2(1-y)^{1/2}}.$$

5. If $y = ax + b$, the inverse function is $x = (y-b)/a$ and $dx/dy = 1/a$. Therefore,

$$g(y) = f\left[\frac{1}{a}(y-b)\right]\left|\frac{dx}{dy}\right| = \frac{1}{|a|} f\left(\frac{y-b}{a}\right).$$

7. (a) If $y = x^2$, then as x varies over the interval $(0,1)$, y also varies over the interval $(0,1)$. Also, $x = y^{1/2}$ and $dx/dy = y^{-1/2}/2$. Hence, for $0 < y < 1$,

$$g(y) = f(y^{1/2})\left|\frac{dx}{dy}\right| = 1 \cdot \frac{1}{2} y^{-1/2} = \frac{1}{2} y^{-1/2}.$$

(b) If $y = -x^3$, then as x varies over the interval $(0,1)$, y varies over the interval $(-1,0)$. Also, $x = -y^{1/3}$ and $dx/dy = -y^{-2/3}/3$. Hence, for $-1 < y < 0$,

$$g(y) = f(-y^{1/3})\left|\frac{dx}{dy}\right| = \frac{1}{3} |y|^{-2/3}.$$

(c) If $y = x^{1/2}$, then as x varies over the interval $(0,1)$, y also varies over the interval $(0,1)$. Also, $x = y^2$ and $dx/dy = 2y$. Hence, for $0 < y < 1$, $g(y) = f(y^2)2y = 2y$.

9. The c.d.f. $G(y)$ corresponding to the p.d.f. $g(y)$ is, for $0 < y < 2$,

$$G(y) = \int_0^y g(t)dt = \int_0^y \frac{3}{8} t^2 dt = \frac{1}{8} y^3.$$

We know that the c.d.f. of the random variable $Y = G^{-1}(X)$ will be G. We must therefore determine the inverse function G^{-1}. If $X = G(Y) = Y^3/8$ then $Y = G^{-1}(X) = 2X^{1/3}$. It follows that $Y = 2X^{1/3}$.

11. We can use the probability integral transformation if we can find the inverse of the c.d.f. The c.d.f. is, for $0 < y < 1$,

$$G(y) = \int_{-\infty}^{y} g(t)dt = \frac{1}{2}\int_{0}^{y}(2t+1)dt = \frac{1}{2}(y^2+y).$$

The inverse of this function can be found by setting $G(y) = p$ and solving for y.

$$\frac{1}{2}(y^2+y) = p; \quad y^2+y-2p = 0; \quad y = \frac{-1+(1+8p)^{1/2}}{2}.$$

So, we should generate four independent uniform pseudo-random variables P_1, P_2, P_3, P_4 and let $Y_i = [-1+(1+8P_i)^{1/2}]/2$ for $i = 1, 2, 3, 4$.

13. The inverse transformation is $z = 1/t$ with derivative $-1/t^2$. the p.d.f. of T is

$$g(t) = f(1/t)/t^2 = 2\exp(-2/t)/t^2,$$

for $t > 0$.

15. Let F be the c.d.f. of X. First, find the c.d.f. of Y, namely, for $y > 0$,

$$\Pr(Y \le y) = \Pr(X^2 \le y) = \Pr(-y^{1/2} \le X \le y^{1/2}) = F(y^{1/2} - F(-y^{1/2}).$$

Now, the p.d.f. of Y is the derivative of the above expression, namely,

$$g(y) = \frac{d}{dy}[F(y^{1/2}) - F(-y^{1/2})] = \frac{f(y^{1/2})}{2y^{1/2}} + \frac{f(-y^{1/2})}{2y^{1/2}}.$$

This equals the expression in the exercise.

17. (a) According to the problem description, $Y = 0$ if $X \le 100$, $Y = X - 100$ if $100 < X \le 5100$, and $Y = 5000$ if $X > 5100$. So, $Y = r(X)$, where

$$r(x) = \begin{cases} 0 & \text{if } x \le 100, \\ x - 100 & \text{if } 100 < x \le 5100, \\ 5000 & \text{if } x > 5100. \end{cases}$$

(b) Let G be the c.d.f. of Y. Then $G(y) = 0$ for $y < 0$, and $G(y) = 1$ for $y \ge 5000$. For $0 \le y < 5000$,

$$\begin{aligned} \Pr(Y \le y) &= \Pr(r(X) \le y) \\ &= \Pr(X \le y + 100) \\ &= \int_{0}^{y+100} \frac{dx}{(1+x)^2} \\ &= 1 - \frac{1}{y+101}. \end{aligned}$$

In summary,

$$G(y) = \begin{cases} 0 & \text{if } y < 0, \\ 1 - \frac{1}{y+101} & \text{if } 0 \le y < 5000, \\ 1 & \text{if } y \ge 5000. \end{cases}$$

(c) There is positive probability that $Y = 5000$, but the rest of the distribution of Y is spread out in a continuous manner between 0 and 5000.

3.9 Functions of Two or More Random Variables

1. The joint p.d.f. of X_1 and X_2 is

$$f(x_1, x_2) = \begin{cases} 1 & \text{for } 0 < x_1 < 1, 0 < x_2 < 1, \\ 0 & \text{otherwise.} \end{cases}$$

By Eq. (3.9.5), the p.d.f. of Y is

$$g(y) = \int_{-\infty}^{\infty} f(y - z, z) dz.$$

The integrand is positive only for $0 < y - z < 1$ and $0 < z < 1$. Therefore, for $0 < y \leq 1$ it is positive only for $0 < z < y$ and we have

$$g(y) = \int_0^y 1 \cdot dz = y.$$

For $1 < y < 2$. the integrand is positive only for $y - 1 < z < 1$ and we have

$$g(y) = \int_{y-1}^1 1 \cdot dz = 2 - y.$$

3. The inverse transformation is:

$$\begin{aligned} x_1 &= y_1, \\ x_2 &= y_2/y_1, \\ x_3 &= y_3/y_2. \end{aligned}$$

Furthermore, the set S where $0 < x_i < 1$ for $i = 1, 2, 3$ corresponds to the set T where $0 < y_3 < y_2 < y_1 < 1$. We also have

$$J = \det \begin{bmatrix} \frac{\partial x_1}{\partial y_1} & \frac{\partial x_1}{\partial y_2} & \frac{\partial x_1}{\partial y_3} \\ \frac{\partial x_2}{\partial y_1} & \frac{\partial x_2}{\partial y_2} & \frac{\partial x_2}{\partial y_3} \\ \frac{\partial x_3}{\partial y_1} & \frac{\partial x_3}{\partial y_2} & \frac{\partial x_3}{\partial y_3} \end{bmatrix} = \det \begin{bmatrix} 1 & 0 & 0 \\ -\frac{y_2}{y_1^2} & \frac{1}{y_1} & 0 \\ 0 & -\frac{y_3}{y_2^2} & \frac{1}{y_2} \end{bmatrix} = \frac{1}{y_1 y_2}.$$

Therefore, for $0 < y_3 < y_2 < y_1 < 1$, the joint p.d.f. of $Y_1, Y_2,$ *and* Y_3 is

$$\begin{aligned} g(y_1, y_2, y_3) &= f\left(y_1, \frac{y_2}{y_1}, \frac{y_3}{y_2}\right) |J| \\ &= 8y_1 \frac{y_2}{y_1} \frac{y_3}{y_2} \frac{1}{y_1 y_2} = \frac{8y_3}{y_1 y_2}. \end{aligned}$$

5. As a convenient device let $Y = X_2$. Then the transformation from X_1 *and* X_2 to Y and Z is a one-to-one transformation between the set S where $0 < x_1 < 1$ *and* $0 < x_2 < 1$ and the set T where $0 < y < 1$ *and* $0 < yz < 1$. The inverse transformation is

$$\begin{aligned} x_1 &= yz, \\ x_2 &= y. \end{aligned}$$

Therefore,

$$J = \det \begin{bmatrix} z & y \\ 1 & 0 \end{bmatrix} = -y.$$

The region where the p.d.f. of (Z, Y) is positive is in Fig. S.3.19. For $0 < y < 1$ *and* $0 < yz < 1$, the

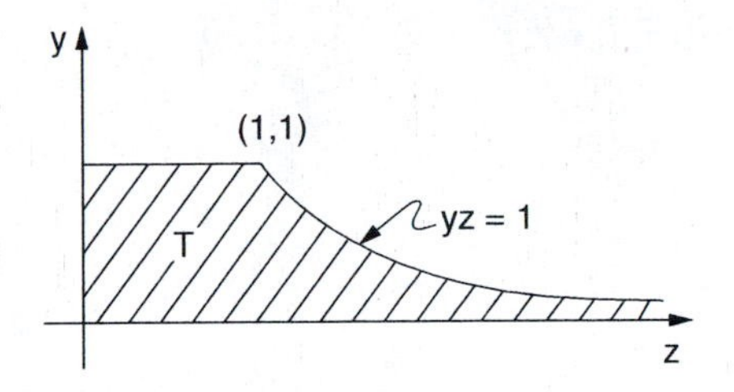

Figure S.3.19: The region where the p.d.f. of (Z, Y) is positive in Exercise 5 of Sec. 3.9.

joint p.d.f. of Y and Z is

$$g(y, z) = f(yz, y)\,|\,J\,| = (yz + y)(y).$$

It follows that for $0 < z \leq 1$, the marginal p.d.f. of Z is

$$g_2(z) = \int_0^1 g(y, z)\, dy = \frac{1}{3}(z + 1).$$

Also, for $z > 1$,

$$g_2(z) = \int_0^{\frac{1}{z}} g(y, z)\, dy = \frac{1}{3z^3}(z + 1).$$

7. Let $Z = -X_2$. Then the p.d.f. of Z is

$$f_2(z) = \begin{cases} \exp(z) & \text{for } z < 0, \\ 0 & \text{for } z \geq 0. \end{cases}$$

Since X_1 and Z are independent, the joint p.d.f. of X_1 and Z is

$$f(x_1, z) = \begin{cases} \exp(-(x - z)) & \text{for } x > 0,\ z < 0, \\ 0 & \text{otherwise.} \end{cases}$$

It now follows from Eq. (3.9.5) that the p.d.f. of $Y = X_1 - X_2 = X_1 + Z$ is

$$g(y) = \int_{-\infty}^{\infty} f(y - z, z)\, dz.$$

The integrand is positive only for $y - z > 0$ and $z < 0$. Therefore, for $y \leq 0$,

$$g(y) = \int_{-\infty}^{y} \exp(-(y - 2z))\, dz = \frac{1}{2}\exp(y).$$

Also, for $y > 0$,

$$g(y) = \int_{-\infty}^{0} \exp(-(y - 2z))\, dz = \frac{1}{2}\exp(-y).$$

9. It was shown in this section that the joint c.d.f. of Y_1 and Y_n is, for $-\infty < y_1 < y_n < \infty$,

$$G(y_1, y_n) = [F(y_n)]^n - [F(y_n) - F(y_1)]^n.$$

Since $F(y) = y$ for the given uniform distribution, we have

$$\Pr(Y_1 \le 0.1, Y_n \le 0.8) = G(0.1, 0.8) = (0.8)^n - (0.7)^n.$$

11. The required probability is equal to

$$\Pr\left(\text{All } n \text{ observations } < \frac{1}{3}\right) + \Pr\left(\text{All } n \text{ observations } > \frac{1}{3}\right) = \left(\frac{1}{3}\right)^n + \left(\frac{2}{3}\right)^n.$$

This exercise could also be solved by using techniques similar to those used in Exercise 10.

13. If X has the uniform distribution on the interval $[0, 1]$, then $aX + b$ $(a > 0)$ has the uniform distribution on the interval $[b, a + b]$. Therefore, $8X - 3$ has the uniform distribution on the interval $[-3, 5]$. It follows that if $X_1, \ldots, X_n$ form a random sample from the uniform distribution on the interval $[0, 1]$, then the n random variables $8X_1 - 3, \ldots, 8X_n - 3$ will have the same joint distribution as a random sample from the uniform distribution on the interval $[-3, 5]$.

   Next, it follows that if the range of the sample $X_1, \ldots, X_n$ is W, then the range of the sample $8X_1 - 3, \ldots, 8X_n - 3$ will be $8W$. Therefore, if W is the range of a random sample from the uniform distribution on the interval $[0, 1]$, then $Z = 8W$ will have the same distribution as the range of a random sample from the uniform distribution on the interval $[-3, 5]$.

   The p.d.f. $h(w)$ of W was given in Example 3.9.8. Therefore, the p.d.f. $f(z)$ of $Z = 8W$ is

$$g(z) = h\left(\frac{z}{8}\right) \cdot \frac{1}{8} = \frac{n(n-1)}{8} \left(\frac{z}{8}\right)^{n-2} \left(1 - \frac{z}{8}\right),$$

   for $-3 < z < 5$.

   This p.d.f. $g(z)$ could also have been derived from first principles as in Example 3.9.8.

15. For any n sets of real numbers $A_1, \ldots, A_n$, we have

$$\begin{aligned}
\Pr(Y_1 \in A_1, \ldots, Y_n \in A_n) &= \Pr[r_1(X_1) \in A_1, \ldots, r_n(X_n) \in A_n] \\
&= \Pr[r_1(X_1) \in A_1] \ldots \Pr[r_n(X_n) \in A_n] \\
&= \Pr(Y_1 \in A_1) \ldots \Pr(Y_n \in A_n).
\end{aligned}$$

   Therefore, $Y_1, \ldots, Y_n$ are independent by Definition 3.5.2.

17. We need to transform (X, Y) to (Z, W), where $Z = XY$ and $W = Y$. The joint p.d.f. of (X, Y) is

$$f(x, y) = \begin{cases} y \exp(-xy) f_2(y) & \text{if } x > 0, \\ 0 & \text{otherwise.} \end{cases}$$

   The inverse transformation is $x = z/w$ and $y = w$. The Jacobian is

$$J = \det \begin{pmatrix} 1/w & -z/w^2 \\ 0 & 1 \end{pmatrix} = \frac{1}{w}.$$

The joint p.d.f. of (Z, W) is

$$g(z, w) = f(z/w, w)/w = w\exp(-z)f_2(w)/w = \exp(-z)f_2(w), \text{ for } z > 0.$$

This is clearly factored in the appropriate way to show that Z and W are independent. Indeed, if we integrate $g(z, w)$ over w, we obtain the marginal p.d.f. of Z, namely $g_1(z) = \exp(-z)$, for $z > 0$. This is the same as the function in (3.9.18).

19. This is a convolution. Let g be the p.d.f. of Y. By (3.9.5) we have, for $y > 0$,

$$\begin{aligned} g(y) &= \int f(y-z)f(z)dx \\ &= \int_0^y e^{z-y}e^{-z}dz \\ &= ye^{-y}. \end{aligned}$$

Clearly, $g(y) = 0$ for $y < 0$, so the p.d.f.øf Y is

$$g(y) = \begin{cases} ye^{-y} & \text{for } y > 0, \\ 0 & \text{otherwise.} \end{cases}$$

21. Transforming to $Z_1 = X_1/X_2$ and $Z_2 = X_1$ has the inverse $X_1 = Z_2$ and $X_2 = Z_2/Z_1$. The set of values where the joint p.d.f. of Z_1 and Z_2 is positive is where $0 < z_2 < 1$ and $0 < z_2/z_1 < 1$. This can be written as $0 < z_2 < \min\{1, z_1\}$. The Jacobian is the determinant of the matrix

$$\begin{pmatrix} 0 & 1 \\ -z_2/z_1^2 & 1/z_1 \end{pmatrix},$$

which is $|z_2/z_1^2|$. The joint p.d.f. of Z_1 and Z_2 is then

$$g(z_1, z_2) = \left|\frac{z_2}{z_1^2}\right| 4z_2\frac{z_2}{z_1} = 4z_2^3z_1^3,$$

for $0 < z_2 < \min\{1, z_1\}$. Integrating z_2 out of this yields, for $z_1 > 0$,

$$\begin{aligned} g_1(z_1) &= \int_0^{\min\{1,z_1\}} 4\frac{z_2^3}{z_1^3}dz_2 \\ &= \frac{\min\{z_1, 1\}^4}{z_1^3} \\ &= \begin{cases} z_1 & \text{if } z_1 < 1, \\ z_1^{-3} & \text{if } z_1 \geq 1. \end{cases} \end{aligned}$$

This is the same thing we got in Example 3.9.11.

3.10 Markov Chains

1. The transition matrix for this Markov chain is

$$\boldsymbol{P} = \begin{bmatrix} \frac{1}{3} & \frac{2}{3} \\ \frac{2}{3} & \frac{1}{3} \end{bmatrix}.$$

   (a) If we multiply the initial probability vector by this matrix we get

$$\boldsymbol{v}\boldsymbol{P} = \left(\frac{1}{2}\frac{1}{3} + \frac{1}{2}\frac{2}{3}, \frac{1}{2}\frac{2}{3} + \frac{1}{2}\frac{1}{3}\right) = \left(\frac{1}{2}, \frac{1}{2}\right).$$

   (b) The two-step transition matrix is $\boldsymbol{P}^2$, namely

$$\begin{bmatrix} \frac{1}{3}\frac{1}{3} + \frac{2}{3}\frac{2}{3} & \frac{1}{3}\frac{2}{3} + \frac{2}{3}\frac{1}{3} \\ \frac{2}{3}\frac{1}{3} + \frac{1}{3}\frac{2}{3} & \frac{2}{3}\frac{2}{3} + \frac{1}{3}\frac{1}{3} \end{bmatrix} = \begin{bmatrix} \frac{5}{9} & \frac{4}{9} \\ \frac{4}{9} & \frac{5}{9} \end{bmatrix}.$$

3. Saturday is three days after Wednesday, so we first compute

$$\boldsymbol{P}^3 = \begin{bmatrix} 0.667 & 0.333 \\ 0.666 & 0.334 \end{bmatrix}.$$

   Therefore, the answers are (a) 0.667 and (b) 0.666.

5. Let $\boldsymbol{v} = (0.2, 0.8)$.

   (a) The answer will be the second component of the vector $\boldsymbol{v}\boldsymbol{P}$. We easily compute $\boldsymbol{v}\boldsymbol{P} = (0.62, 0.38)$, so the probability is 0.38.

   (b) The answer will be the second component of $\boldsymbol{v}\boldsymbol{P}^2$. We can compute $\boldsymbol{v}\boldsymbol{P}^2$ by multiplying $\boldsymbol{v}\boldsymbol{P}$ by $\boldsymbol{P}$ to get $(0.662, 0.338)$, so the probability is 0.338.

   (c) The answer will be the second component of of $\boldsymbol{v}\boldsymbol{P}^3$. Since $\boldsymbol{v}\boldsymbol{P}^3 = (0.6662, 0.3338)$, the answer is 0.3338.

7. Using the matrix in Exercise 6, it is found that

$$\boldsymbol{P}^3 = \begin{bmatrix} 0.368 & 0.632 \\ 0.395 & 0.605 \end{bmatrix}.$$

   Therefore, the answers are (a) 0.632 and (b) 0.605.

9. (a) It is found that

$$\boldsymbol{P}^2 = \begin{bmatrix} \frac{3}{16} & \frac{7}{16} & \frac{2}{16} & \frac{4}{16} \\ 0 & 1 & 0 & 0 \\ \frac{3}{8} & \frac{1}{8} & \frac{2}{8} & \frac{2}{8} \\ \frac{4}{16} & \frac{6}{16} & \frac{3}{16} & \frac{3}{16} \end{bmatrix}.$$

   The answer is given by the element in the third row and second column.

(b) The answer is the element in the first row and third column of $\boldsymbol{P}^3$, namely 0.125.

11. The transition matrix for the states A and B is

$$\begin{bmatrix} \frac{1}{3} & \frac{2}{3} \\ \frac{2}{3} & \frac{1}{3} \end{bmatrix}.$$

It is found that

$$\boldsymbol{P}^4 = \begin{bmatrix} \frac{41}{81} & \frac{40}{81} \\ \frac{40}{81} & \frac{41}{81} \end{bmatrix}.$$

Therefore, the answers are (a) $\frac{40}{81}$ and (b) $\frac{41}{81}$.

13. The states are triples of possible outcomes: (HHH), (HHT), (HTH), etc. There are a total of eight such triples. The conditional probabilities of the possible values of the outcomes on trials $(n-1, n, n+1)$ given all trials up to time n depend only on the trials $(n-2, n-1, n)$ and not on n itself, hence we have a Markov chain with stationary transition probabilities. Every row of the transition matrix has the following form except the two corresponding to (HHH) and (TTT). Let a, b, c stand for three arbitrary elements of $\{H, T\}$, not all equal. The row for (abc) has 0 in every column except for the two columns (abH) and (abT), which have 1/2 in each. In the (HHH) row, every column has 0 except the (HHT) column, which has 1. In the (TTT) row, every column has 0 except the (TTH) column which has 1.

15. We are asked to verify the numbers in the second and fifth rows of the matrix in Example 3.10.6. For the second row, the parents have genotypes AA and Aa, so that the only possible offspring are AA and Aa. Each of these occurs with probability 1/2 because they are determined by which allele comes from the Aa parent. Since the two offspring in the second generation are independent, we will get $\{AA, AA\}$ with probability $(1/2)^2 = 1/4$ and we will get $\{Aa, Aa\}$ with probability 1/4 also. The remaining probability, 1/2, is the probability of $\{AA, Aa\}$. For the fifth row, the parent have genotypes Aa and aa. The only possible offspring are Aa and aa. Indeed, the situation is identical to the second row with a and A switched. The resulting probabilities are also the same after this same switch.

17. (a) We are asked to find the conditional distribution of X_n given $X_{n-1} = \{Aa, aa\}$ and $X_{n+1} = \{AA, aa\}$. For each possible state x_n, we can find

$$\Pr(X_n = x_n | X_{n-1} = \{Aa, aa\}, X_{n+1} = \{AA, aa\}) = \frac{\Pr(X_n = x_n, X_{n+1} = \{AA, aa\} | X_{n-1} = \{Aa, aa\})}{\Pr(X_{n+1} = \{AA, aa\} | X_{n-1} = \{Aa, aa\})}. \quad \text{(S.3.2)}$$

The denominator is 0.0313 from the 2-step transition matrix in Example 3.10.9. The numerator is the product of two terms from the 1-step transition matrix: one from $\{Aa, aa\}$ to x_n and the other from x_n to $\{AA, aa\}$. These products are as follows:

x_n

$\{AA, AA\}$	$\{AA, Aa\}$	$\{AA, aa\}$	$\{Aa, Aa\}$	$\{Aa, aa\}$	$\{aa, aa\}$
0	0	0	0.25×0.125	0	0

Plugging these into (S.3.2) gives

$$\Pr(X_n = \{Aa, Aa\} | X_{n+1} = \{Aa, aa\}, X_{n+1} = \{AA, aa\}) = 1,$$

and all other states have probability 0.

(b) This time, we want

$$\begin{aligned}
&\Pr(X_n = x_n | X_{n-1} = \{Aa, aa\}, X_{n+1} = \{aa, aa\}) \\
&= \frac{\Pr(X_n = x_n, X_{n+1} = \{aa, aa\} | X_{n-1} = \{Aa, aa\})}{\Pr(X_{n+1} = \{aa, aa\} | X_{n-1} = \{Aa, aa\})}.
\end{aligned}$$

The denominator is 0.3906. The numerator products and their ratios to the denominator are:

x_n	$\{AA, AA\}$	$\{AA, Aa\}$	$\{AA, aa\}$	$\{Aa, Aa\}$	$\{Aa, aa\}$	$\{aa, aa\}$
Numerator	0	0	0	0.25×0.0625	0.5×0.25	0.25×1
Ratio	0	0	0	0.0400	0.3200	0.6400

This time, we get

$$\Pr(X_n = x_n | X_{n-1} = \{Aa, aa\}, X_{n+1} = \{Aa, Aa\}) = \begin{cases} 0.04 & \text{if } x_n = \{Aa, Aa\}, \\ 0.32 & \text{if } x_n = \{Aa, aa\}, \\ 0.64 & \text{if } x_n = \{aa, aa\}, \end{cases}$$

and all others are 0.

19. The matrix $\boldsymbol{G}$ and its inverse are

$$\begin{aligned}
\boldsymbol{G} &= \begin{pmatrix} -0.3 & 1 \\ 0.6 & 1 \end{pmatrix}, \\
\boldsymbol{G}^{-1} &= -\frac{10}{9}\begin{pmatrix} 1 & -1 \\ -0.6 & -0.3 \end{pmatrix}.
\end{aligned}$$

The bottom row of $\boldsymbol{G}^{-1}$ is $(2/3, 1/3)$, the unique stationary distribution.

3.11 Supplementary Exercises

1. We can calculate the c.d.f. of Z directly.

$$\begin{aligned}
F(z) &= \Pr(Z \le z) = \Pr(Z = X)\Pr(X \le z) + \Pr(Z = Y)\Pr(Y \le z) \\
&= \frac{1}{2}\Pr(X \le z) + \frac{1}{2}\Pr(Y \le z)
\end{aligned}$$

The graph is in Fig. S.3.20.

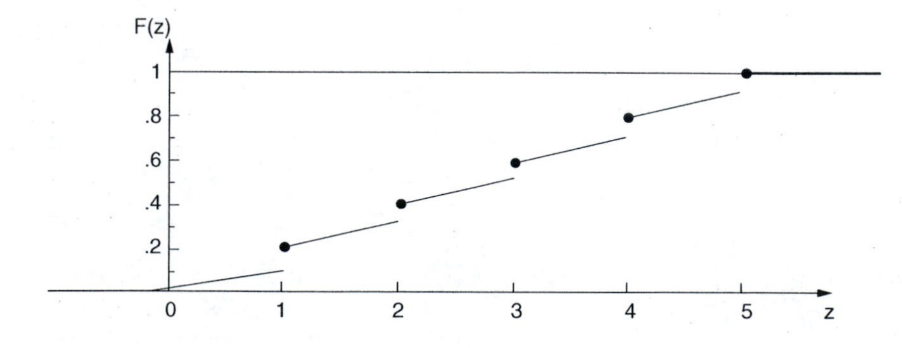

Figure S.3.20: Graph of c.d.f. for Exercise 1 of Sec. 3.11.

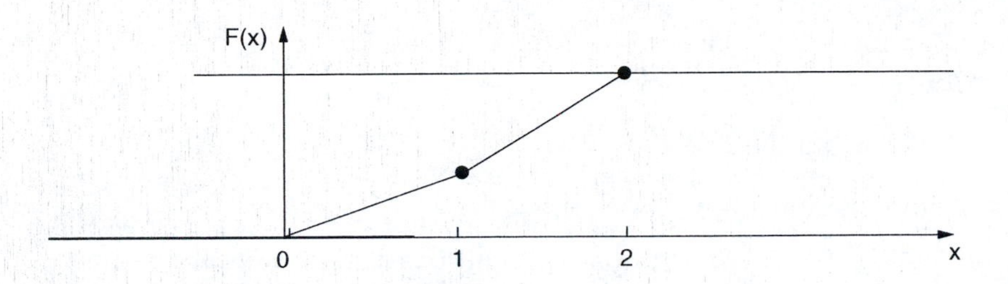

Figure S.3.21: Graph of c.d.f. for Exercise 3 of Sec. 3.11.

3. Since $F(x)$ is continuous and differentiable everywhere except at the points $x = 0$, 1, and 2,

$$f(x) = \frac{dF(x)}{dx} \begin{cases} \frac{2}{5} & \text{for } 0 < x < 1, \\ \frac{3}{5} & \text{for } 1 < x < 2, \\ 0 & \text{otherwise.} \end{cases}$$

5. X_1 and X_2 have the uniform distribution over the square, which has area 1. The area of the quarter circle in Fig. S.3.22, which is the required probability, is $\pi/4$.

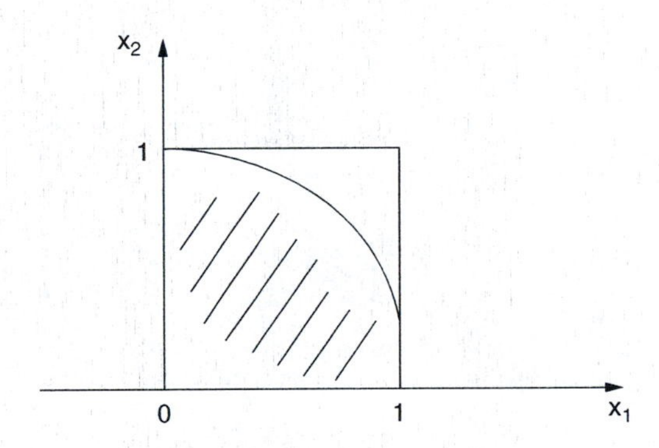

Figure S.3.22: Region for Exercise 5 of Sec. 3.11.

7.

$$\begin{aligned} \Pr(X + X_2 \text{ even}) &= \Pr(X_1 \text{ even}) \Pr(X_2 \text{ even}) + \Pr(X_1 \text{ odd}) \Pr(X_2 \text{ odd}) \\ &= \left(\frac{1}{2^p}\right)\left(\frac{1}{2^p}\right) + \left(1 - \frac{1}{2^p}\right)\left(1 - \frac{1}{2^p}\right) \\ &= 1 - \frac{1}{2^{p-1}} + \frac{1}{2^{2p-1}}. \end{aligned}$$

9. Let A denote the event that the tack will land with its point up on all three tosses. Then $\Pr(A \mid X = x) = x^3$. Hence,

$$\Pr(A) = \int_0^1 x^3 f(x)\, dx = \frac{1}{10}.$$

11. $F(x) = 1 - \exp(-2x)$ for $x > 0$. Therefore, by the probability integral transformation, $F(X)$ will have the uniform distribution on the interval $[0, 1]$. Therefore,

$$Y = 5F(X) = 5(1 - \exp(-2X))$$

will have the uniform distribution on the interval $[0, 5]$.

It might be noted that if Z has the uniform distribution on the interval $[0, 1]$, then $1 - Z$ has the same uniform distribution. Therefore,

$$Y = 5[1 - F(X)] = 5\exp(-2X)$$

will also have the uniform distribution on the interval $[0, 5]$.

13. Only in (c) and (d) is the joint p.d.f. of X and Y positive over a rectangle with sides parallel to the axes, so only in (c) and (d) is there the possibility of X and Y being independent. Since the uniform density is constant, it can be regarded as being factored in the form of Eq. (3.5.7). Hence, X and Y are independent in (c) and (d).

15. This problem is similar to Exercise 11 of Sec. 3.5, but now we have Fig. S.3.23. The area of the shaded

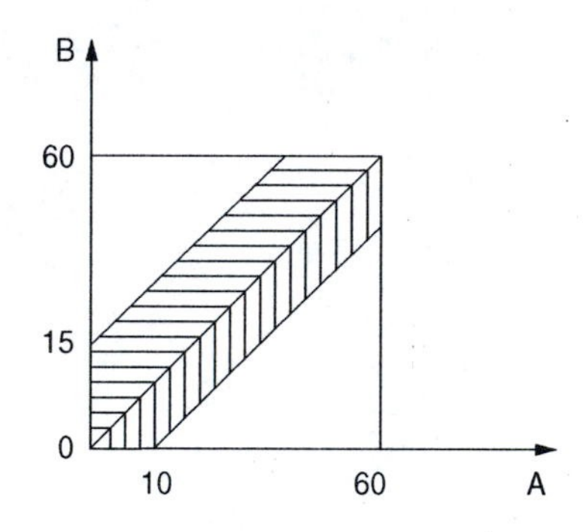

Figure S.3.23: Figure for Exercise 15 of Sec. 3.11.

region is now $550 + 787.5 = 1337.5$. Hence, the required probability is $\dfrac{1337.5}{3600} = .3715$

17. $f(x, y) = f(x)\,g(y \mid x) = \dfrac{9y^2}{x}$ for $0 < y < x < 1$.

Hence,

$$f_2(y) = \int_y^1 f(x, y)\,dx = -9y^2 \log(y) \quad \text{for} \quad 0 < y < 1$$

and

$$g_1(x \mid y) = \frac{f(x, y)}{f_2(y)} = -\frac{1}{x \log(y)} \qquad \text{for} \quad 0 < y < x < 1.$$

19.

$$\begin{aligned} f_1(x) &= \int_x^1 \int_y^1 6\,dz\,dy = 3 - 6x + 3x^2 = 3(1 - x)^2 \quad \text{for } 0 < x < 1, \\ f_2(y) &= \int_0^y \int_y^1 6\,dz\,dx = 6y(1 - y) \quad \text{for } 0 < y < 1. \\ f_3(z) &= \int_0^z \int_0^y 6\,dx\,dy = 3z^2 \quad \text{for} \quad 0 < z < 1. \end{aligned}$$

21. (a) $$f(x,y) = \begin{cases} \exp(-(x+y)) & \text{for } x>0,\ y>0, \\ 0 & \text{otherwise.} \end{cases}$$

Also, $x = uv$ and $y = (1-u)v$, so

$$J = \begin{vmatrix} v & u \\ -v & 1-u \end{vmatrix} = v > 0.$$

Therefore,

$$g(u,v) = f(uv, [1-u]v)\,|\,J\,| = \begin{cases} v\exp(-v) & \text{for } 0<u<1,\ v>0, \\ 0 & \text{otherwise.} \end{cases}$$

(b) Because g can be appropriately factored (the factor involving u is constant) and it is positive over an appropriate rectangle, it follows that U and V are independent.

23. Here, $f(x) = \dfrac{dF(x)}{dx} = \exp(-x)$ for $x > 0$. It follows from the results in Sec. 3.9 that

$$g(y_1, y_n) = n(n-1)(\exp(-y_1) - \exp(-y_n))^{n-2}\exp(-(y_1+y_n))$$

for $0 < y_1 < y_n$. Also, the marginal p.d.f. of Y_n is

$$g_n(y_n) = n(1-\exp(-y_n))^{n-1}\exp(-y_n) \quad \text{for } y_n > 0.$$

Hence,

$$h(y_1\,|\,y_n) = \frac{(n-1)(\exp(-y_1) - \exp(-y_n))^{n-2}\exp(-y_1)}{(1-\exp(-yn))^{n-1}} \quad \text{for } 0< y_1 < y_n.$$

25. (a) Let f_2 be the marginal p.d.f. of Y. We approximate

$$\Pr(y-\epsilon < Y \le y+\epsilon) = \int_{y-\epsilon}^{y+\epsilon} f_2(t)dt \approx 2\epsilon f_2(y).$$

(b) For each s, we approximate

$$\int_{y-\epsilon}^{y+\epsilon} f(s,t)dt \approx 2\epsilon f(s,y).$$

Using this, we can approximate

$$\Pr(X \le x, y-\epsilon < Y \le y+\epsilon) = \int_{-\infty}^{x}\int_{y-\epsilon}^{y+\epsilon} f(s,t)dtds \approx 2\epsilon\int_{-\infty}^{x} f(s,y)ds.$$

(c) Taking the ratio of the approximation in part (b) to the approximation in part (a), we obtain

$$\begin{aligned} \Pr(X \le x | y-\epsilon < Y \le y+\epsilon) &= \frac{\Pr(X \le x, y-\epsilon < Y \le y+\epsilon)}{\Pr(y-\epsilon < Y \le y+\epsilon)} \\ &\approx \frac{\int_{-\infty}^{x} f(s,y)ds}{f_2(y)} \\ &= \int_{-\infty}^{x} g_1(s|y)ds. \end{aligned}$$

27. The transition matrix is as follows:

		Players in game $n+1$		
		(A,B)	(A,C)	(B,C)
Players in game n	(A,B)	0	0.3	0.7
	(A,C)	0.6	0	0.4
	(B,C)	0.8	0.2	0

29. The matrix $\boldsymbol{G}$ and its inverse are

$$\boldsymbol{G} = \begin{pmatrix} -1.0 & 0.3 & 1.0 \\ 0.6 & -1.0 & 1.0 \\ 0.8 & 0.2 & 1.0 \end{pmatrix},$$

$$\boldsymbol{G}^{-1} = \begin{pmatrix} -0.5505 & -0.4587 & 0.5963 \\ 0.0917 & -0.8257 & 0.7339 \\ 0.4220 & 0.2018 & 0.3761 \end{pmatrix}.$$

The bottom row of $\boldsymbol{G}^{-1}$ is the unique stationary distribution, $(0.4220, 0.2018, 0.3761)$.

11. The p.d.f.'s of Y_1 and Y_n were found in Sec. 3.9. For the given uniform distribution, the p.d.f. of Y_1 is

$$g_1(y) = \begin{cases} n(1-y)^{n-1} & \text{for } 0 < y < 1, \\ 0 & \text{otherwise.} \end{cases}$$

Therefore,

$$E(Y_1) = \int_0^1 yn(1-y)^{n-1}dy = \frac{1}{n+1}.$$

The p.d.f. of Y_n is

$$g_n(y) = \begin{cases} ny^{n-1} & \text{for } 0 < y < 1, \\ 0 & \text{otherwise.} \end{cases}$$

Therefore,

$$E(Y_n) = \int_0^1 y \cdot ny^{n-1}dy = \frac{n}{n+1}.$$

13. Let $p = \Pr(X = 300)$. Then $E(X) = 300p + 100(1-p) = 200p + 100$. For risk-neutrality, we need $E(X) = 110 * (1.058) = 116.38$. Setting $200p + 100 = 116.38$ yields $p = 0.0819$. The option has a value of 150 if $X = 300$ and it has a value of 0 if $X = 100$, so the mean of the option value is $150p = 12.285$. The present value of this amount is $12.285/1.058 = 11.61$, the risk-neutral price of the option.

15. The value of the option is 0 if $X = 260$ and it is 40 if $X = 180$, so the expected value of the option is $40(1-p) = 40 \times 0.65 = 26$. The present value of this amount is $26/1.04 = 25$.

4.2 Properties of Expectations

1. The random variable Y is equal to $10(R - 1.5)$ in dollars. The mean of Y is $10[E(R) - 1.5]$. From Exercise 1 in Sec. 4.1, we know that $E(R) = (-3 + 7)/2 = 2$, so $E(Y) = 5$.

3.

$$\begin{aligned} E[(X_1 - 2X_2 + X_3)^2] &= E(X_1^2 + 4X_2^2 + X_3^2 - 4X_1X_2 + 2X_1X_3 - 4X_2X_3) \\ &= E(X_1^2) + 4E(X_2^2) + E(X_3^2) - 4E(X_1X_2) \\ &\quad + 2E(X_1X_3) - 4E(X_2X_3). \end{aligned}$$

Since X_1, X_2, and X_3 are independent,

$$E(X_iX_j) = E(X_i)E(X_j) \qquad \text{for } i \neq j.$$

Therefore, the above expectation can be written in the form:

$$E(X_1^2) + 4E(X_2^2) + E(X_3^2) - 4E(X_1)E(X_2) + 2E(X_1)E(X_3) - 4E(X_2)E(X_3).$$

Also, since each X_i has the uniform distribution on the interval $[0, 1]$, then $E(X_i) = \frac{1}{2}$ and

$$E(X_i^2) = \int_0^1 x^2dx = \frac{1}{3}.$$

Hence, the desired expectation has the value 1/2.

Chapter 4

Expectation

4.1 The Expectation of a Random Variable

1. The mean of X is

$$E(X) = \int xf(x)dx = \int_a^b \frac{x}{b-a}dx = \frac{b^2-a^2}{2(b-a)} = \frac{a+b}{2}.$$

3. The total number of students is 50. Therefore,

$$E(X) = 18\left(\frac{20}{50}\right) + 19\left(\frac{22}{50}\right) + 20\left(\frac{4}{50}\right) + 21\left(\frac{3}{50}\right) + 25\left(\frac{1}{50}\right) = 18.92.$$

5. There are 30 letters and they are each equally probable:

   2 letters appear in the only two-letter word;

   15 letters appear in three-letter words;

   4 letters appear in the only four-letter word;

   9 letters appear in the only nine-letter word.

   Therefore, the possible values of Y and their probabilities are as follows:

y	$g(y)$
2	2/30
3	15/30
4	4/30
9	9/30

$$E(Y) = 2\left(\frac{2}{30}\right) + 3\left(\frac{15}{30}\right) + 4\left(\frac{4}{30}\right) + 9\left(\frac{9}{30}\right) = 4.867.$$

7. $E\left(\frac{1}{X}\right) = \int_0^1 \frac{1}{x}\,\mathrm{dx} = -\lim_{\mathrm{x}\to 0}\log(\mathrm{x}) = \infty$. Since the integral is not finite, $E\left(\frac{1}{X}\right)$ does not exist.

9. If X denotes the point at which the stick is broken, then X has the uniform distribution on the interval $[0,1]$. If Y denotes the length of the longer piece, then $Y = \max\{X, 1-X\}$. Therefore,

$$E(Y) = \int_0^1 \max(x, 1-x)dx = \int_0^{1/2}(1-x)dx + \int_{1/2}^1 x dx = \frac{3}{4}.$$

5. For $i = 1, \dots, n$, let $Y_i = 1$ if the observation X_i falls within the interval (a, b), and let $Y_i = 0$ otherwise. Then $E(Y_i) = \Pr(Y_i = 1) = \int_a^b f(x)dx$. The total number of observations that fall within the interval (a, b) is $Y_1 + \cdots + Y_n$, and

$$E(Y_1 + \cdots + Y_n) = E(Y_1) + \cdots + E(Y_n) = n \int_a^b f(x)dx.$$

7. For $i = 1, \dots, n$, let $X_i = 2$ if the gambler's fortune is doubled on the ith play of the game and let $X_i = 1/2$ if his fortune is cut in half on the ith play. Then

$$E(X_i) = 2\left(\frac{1}{2}\right) + \left(\frac{1}{2}\right)\left(\frac{1}{2}\right) = \frac{5}{4}.$$

After the first play of the game, the gambler's fortune will be cX_1, after the second play it will be $(cX_1)X_2$, and by continuing in this way it is seen that after n plays the gambler's fortune will be $cX_1X_2 \dots X_n$. Since $X_1, \dots, X_n$ are independent,

$$E(cX_1 \dots X_n) = cE(X_1) \dots E(X_n) = c\left(\frac{5}{4}\right)^n.$$

9. We know that $E(X) = np$. Since $Y = n - X$, $E(X - Y) = E(2X - n) = 2E(X) - n = n(2p - 1)$.

11. We shall use the notation presented in the hint for this exercise. It follows from part (a) of Exercise 10 that $E(X_i) = 2$ for $i = 1, \dots, k$. Therefore,

$$E(X) = E(X_1) + \cdots + E(X_k) = 2k.$$

13. Use Taylor's theorem with remainder to write

$$g(X) = g(\mu) + (X - \mu)g'(\mu) + \frac{(X - \mu)^2}{2}g''(Y), \tag{S.4.1}$$

where $\mu = E(X)$ and Y is between X and μ. Take the mean of both sides of (S.4.1). We get

$$E[g(X)] = g(\mu) + 0 + E\left(\frac{(X - \mu)^2}{2}g''(Y)\right).$$

The random variable whose mean is on the far right is nonnegative, hence the mean is nonnegative and $E[g(X)] \geq g(\mu)$.

4.3 Variance

1. We found in Exercise 1 of Sec. 4.1 that $E(X) = (0 + 1)/2 = 1/2$. We can find

$$E(X^2) = \int_0^1 x^2 dx = \frac{1}{3}.$$

So $\text{Var}(X) = 1/3 - (1/2)^2 = 1/12$.

3. The p.d.f. of this distribution is

$$f(x) = \begin{cases} \dfrac{1}{b-a} & \text{for } a < x < b, \\ 0 & \text{otherwise.} \end{cases}$$

Therefore, $E(X) = \dfrac{b+a}{2}$ and

$$E(X^2) = \int_a^b x^2 \frac{1}{b-a} dx = \frac{b^3 - a^3}{3(b-a)} = \frac{1}{3}(b^2 + ab + a^2).$$

It follows that $\text{Var}(X) = E(X^2) - [E(X)]^2 = \frac{1}{12}(b-a)^2$.

5. $E[(X-c)^2] = E(X^2) - 2cE(X) + c^2 = \text{Var}(X) + [E(X)]^2 - 2c\mu + c^2 = \sigma^2 + \mu^2 - 2c\mu + c^2 = \sigma^2 + (\mu - c)^2$.

7. (a) Since X and Y are independent, $\text{Var}(X - Y) = \text{Var}(X) + \text{Var}(Y) = 3 + 3 = 6$.
   (b) $\text{Var}(2X - 3Y + 1) = 2^2 \text{Var}(X) + 3^2 \text{Var}(Y) = 4(3) + 9(3) = 39$.

9. The mean of X is $(n+1)/2$, and the mean of X^2 is $\sum_{k=1}^{n} k^2/n = (n+1)(2n+1)/6$. So,

$$\text{Var}(X) = \frac{(n+1)(2n+1)}{6} - \frac{(n+1)^2}{4} = \frac{n^2-1}{12}.$$

11. The quantile function of X can be found from Example 3.3.8 with $a = 0$ and $b = 1$. It is $F^{-1}(p) = p$. So, the IQR is $0.75 - 0.25 = 0.5$.

13. From Table 3.1, we find the 0.25 and 0.75 quantiles of the distribution of X to be 1 and 2 respectively. This makes the IQR equal to $2 - 1 = 1$.

4.4 Moments

1. Since the uniform p.d.f. is symmetric with respect to its mean $\mu = (a+b)/2$, it follows that $E[(X-\mu)^5] = 0$.

3. $E[(X-\mu)^3] = E[(X-1)^3] = E(X^3 - 3X^2 + 3X - 1) = 5 - 3(2) + 3(1) - 1 = 1$.

5. Let $Y = (X-\mu)^2$. Then by Exercise 4,

$$E(Y^2) = E[(X-\mu)^4] \geq [E(Y)]^2 = [\text{Var}(X)]^2 = \sigma^4.$$

7. $\psi'(t) = \dfrac{1}{4}(3\exp(t) - \exp(-t))$ and $\psi''(t) = \dfrac{1}{4}(3\exp(t) + \exp(-t))$. Therefore, $\mu = \psi'(0) = 1/2$ and $\sigma^2 = \psi''(0) - \mu^2 = 1 - \left(\dfrac{1}{2}\right)^2 = \dfrac{3}{4}$.

9. $\psi_2'(t) = c\psi_1'(t)\exp(c[\psi_1(t) - 1])$ and $\psi''{}_2(t) = \{[c\psi_1'(t)]^2 + c\psi_1''(t)\}\exp(c[\psi_1(t) - 1])$. We know that

$$\psi_1(0) = 1, \psi_1'(0) = \mu, \text{ and } \psi_1''(0) = \sigma^2 + \mu^2.$$

Therefore, $E(Y) = \psi_2'(0) = c\mu$ and

$$\text{Var}(Y) = \psi_2''(0) - [E(Y)]^2 = \{(c\mu)^2 + c(\sigma^2 + \mu^2)\} - (c\mu)^2 = c(\sigma^2 + \mu^2).$$

11. If X can take only a finite number of values $x_1, \dots, x_k$ with probabilities $p_1, \dots, p_k$, respectively, then the m.g.f. of X will be

$$\psi(t) = p_1 \exp(tx_1) + p_2 \exp(tx_2) + \cdots + p_k \exp(tx_k).$$

By matching this expression for $\psi(t)$ with the expression given in the exercise, it can be seen that X can take only the three values 1, 4, and 8, and that $f(1) = 1/5$, $f(4) = 2/5$, and $f(8) = 2/5$.

13. The m.g.f. of a Cauchy random variable would be

$$\psi(t) = \int_{-\infty}^{\infty} \frac{\exp(tx)}{\pi(1+x^2)} dx. \tag{S.4.2}$$

If $t > 0$, $\lim_{x\to\infty} \exp(tx)/(1+x^2) = \infty$, so the integral in Eq. (S.4.2) is infinite. Similarly, if $t < 0$, $\lim_{x\to-\infty} \exp(tx)/(1+x^2) = \infty$, so the integral is still infinite. Only for $t = 0$ is the integral finite, and that value is $\psi(0) = 1$ as it is for every random variable.

15. Let X have a discrete distribution with p.f. $f(x)$. Assume that $E(|X|^a) < \infty$ for some $a > 0$. Let $0 < b < a$. Then

$$\begin{aligned} E(|X|^b) &= \sum_x |x|^b f(x) = \sum_{|x|\le 1} |x|^b f(x) + \sum_{|x|>1} |x|^b f(x) \\ &\le 1 + \sum_{|x|>1} |x|^a f(x) \le 1 + E(|X|^a) < \infty, \end{aligned}$$

where the first inequality follows from the fact that $0 \le |x|^b \le 1$ for all $|x| \le 1$ and $|x|^b < |x|^a$ for all $|x| > 1$. The next-to-last inequality follows from the fact that the final sum is only part of the sum that makes up $E(|X|^a)$.

17. We already computed the mean $\mu = 1$ and variance $\sigma^2 = 1$ in Example 4.4.3. Using the m.g.f., the third moment is computed from the third derivative:

$$\psi'''(t) = \frac{6}{(1-t)^4}.$$

The third moment is 6. The third central moment is

$$E([X-1]^3) = E(X^3) - 3E(X^2) + 3E(X) - 1 = 6 - 6 + 3 - 1 = 2.$$

The skewness is then $2/1 = 2$.

4.5 The Mean and the Median

1. The 1/2 quantile defined in Definition 3.3.2 applies to a continuous random variable whose c.d.f. is one-to-one. The 1/2 quantile is then $x_0 = F^{-1}(1/2)$. That is, $F(x_0) = 1/2$. In order for a number m to be a median as define in this section, it must be that $\Pr(X \le m) \ge 1/2$ and $\Pr(X \ge m) \ge 1/2$. If X has a continuous distribution, then $\Pr(X \le m) = F(m)$ and $\Pr(X \ge m) = 1 - F(m)$. Since $F(x_0) = 1/2$, $m = x_0$ is a median.

3. A median m must satisfy the equation

$$\int_0^m \exp(-x)dx = \frac{1}{2}.$$

Therefore, $1 - \exp(-m) = 1/2$. It follows that $m = \log\ 2$ is the unique median of this distribution.

5. The p.d.f. of X will be $h(x) = [f(x) + g(x)]/2$ for $-\infty < x < \infty$. Therefore,

$$E(X) = \frac{1}{2}\int_{-\infty}^{\infty} x[f(x) + g(x)]dx = \frac{1}{2}(\mu_f + \mu_g).$$

Since $\int_{-\infty}^{1} h(X)dx = \int_0^1 \frac{1}{2}f(x)dx = \frac{1}{2}$ and $\int_2^{\infty} h(x)dx = \int_2^4 \frac{1}{2}g(x)dx = \frac{1}{2}$, it follows that every value of m in the interval $1 \le m \le 2$ will be a median.

7. (a) The required value is $E(X)$, and

$$E(X) = \int_0^1 x\left(x + \frac{1}{2}\right)dx = \frac{7}{12}.$$

(b) The required value is the median m, where

$$\int_0^m \left(x + \frac{1}{2}\right)dx = \frac{1}{2}.$$

Therefore, $m = (\sqrt{5} - 1)/2$.

9. (a) The required point is the mean $E(X)$, and

$$E(X) = (0.2)(-3) + (0.1)(-1) + (0.1)(0) + (0.4)(1) + (0.2)(2) = 0.1.$$

(b) The required point is the median m. Since $\Pr(X \le 1) = 0.8$ and $\Pr(X \ge 1) = 0.6$, the point 1 is the unique median.

11. The M.S.E. of any prediction is a minimum when the prediction is equal to the mean of the variable being predicted, and the minimum value of the M.S.E. is then the variance of the variable. It was shown in the derivation of Eq. (4.3.3) that the variance of the binomial distribution with parameters n and p is $np(1 - p)$. Therefore, the minimum M.S.E. that can be attained when predicting X is $\text{Var}(X) = 7(1/4)(3/4) = 21/16$ and the minimum M.S.E. that can be attained when predicting Y is $\text{Var}(Y) = 5(1/2)(1/2) = 5/4 = 20/16$. Thus, Y can be predicted with the smaller M.S.E.

13. To say that the distribution of X is symmetric around m, means that X and $2m - X$ have the same distribution. That is, $\Pr(X \le x) = \Pr(2m - X \le x)$ for all x. This can be rewritten as $\Pr(X \le x) = \Pr(X \ge 2m - x)$. With $x = m$, we see that $\Pr(X \le m) = \Pr(X \ge m)$. If $\Pr(X \le m) < 1/2$, then $\Pr(X \le m) + \Pr(X > m) < 1$, which is impossible. Hence $\Pr(X \le m) \ge 1/2$ and $\Pr(X \ge m) \ge 1/2$, and m is a median.

15. (a) Since a is assumed to be a median, $F(a) = \Pr(X \le a) \ge 1/2$. Since $b > a$ is assumed to be a median $\Pr(X \ge b) \ge 1/2$. If $\Pr(X \le a) > 1/2$, then $\Pr(X \le a) + \Pr(X \ge b) > 1$. But $\{X \le a\}$ and $\{X \ge b\}$ are disjoint events, so the sum of their probabilities can't be greater than 1. This means that $F(a) > 1/2$ is impossible, so $F(a) = 1/2$.

(b) The c.d.f. F is nondecreasing, so $A = \{x : F(x) = 1/2\}$ is an interval. Since F is continuous from the right, the lower endpoint c of the interval A must also be in A. For every x, $\Pr(X \leq x) + \Pr(X \geq x) \geq 1$. For every $x \in A$, $\Pr(X \leq x) = 1/2$, hence it must be that $\Pr(X \geq x) \geq 1/2$ and x is a median. Let d be the upper endpoint of the interval A. We need to show that d is also a median. Since F is not necessarily continuous from the left, $F(d) > 1/2$ is possible. If $F(d) = 1/2$, then $d \in A$ and d is a median by the argument just given. If $F(d) > 1/2$, then $\Pr(X = d) = F(d) - 1/2$. This makes

$$\Pr(X \geq d) = \Pr(X > d) + \Pr(X = d) = 1 - F(d) + F(d) - 1/2 = 1/2.$$

Hence d is also a median

(c) If X has a discrete distribution, then clearly F must be discontinuous at d otherwise $F(x) = 1/2$ even for some $x > d$ and d would not be the right endpoint of A.

17. As in the previous problem, $1 = \Pr(X < m) + \Pr(X = m) + \Pr(X > m)$. Since $\Pr(X < m) < 1/2$ and $\Pr(X > m) < 1/2$, we have $\Pr(X \geq m) = 1 - \Pr(X < m) > 1/2$ and $\Pr(X \leq m) = 1 - \Pr(X > m) > 1/2$. Hence m is a median. Let $k > m$. Then $\Pr(X \geq k) \leq \Pr(X > m) < 1/2$, and k is not a median. Similarly, if $k < m$, then $\Pr(X \leq k) \leq \Pr(X < m) < 1/2$, and k is not a median. So, m is the unique median.

4.6 Covariance and Correlation

1. The location of the circle makes no difference since it only affects the means of X and Y. So, we shall assume that the circle is centered at (0,0). As in Example 4.6.5, $\text{Cov}(X, Y) = 0$. It follows that $\rho(X, Y) = 0$ also.

3. Since the p.d.f. of X is symmetric with respect to 0, it follows that $E(X) = 0$ and that $E(X^k) = 0$ for every odd positive integer k. Therefore, $E(XY) = E(X^7) = 0$. Since $E(XY) = 0$ and $E(X)E(Y) = 0$, it follows that $\text{Cov}(X, Y) = 0$ and $\rho(X, Y) = 0$.

5. We have $E(aX + b) = a\mu_X + b$ and $E(cY + d) = c\mu_Y + d$. Therefore,

$$\begin{aligned}
\text{Cov}(aX + b, cY + d) &= E[(aX + b - a\mu_X - b)(cY + d - c\mu_Y - d)] \\
&= E[ac(X - \mu_X)(Y - \mu_Y)] = ac\,\text{Cov}(X, Y).
\end{aligned}$$

7. We have $E(aX + bY + c) = a\mu_X + b\mu_Y + c$. Therefore,

$$\begin{aligned}
\text{Cov}(aX + bY + c, Z) &= E[(aX + bY + c - a\mu_X - b\mu_Y - c)(Z - \mu_Z)] \\
&= E\{[a(X - \mu_X) + b(Y - \mu_Y)](Z - \mu_Z)\} \\
&= aE[(X - \mu_X)(Z - \mu_Z)] + bE[(Y - \mu_Y)(Z - \mu_Z)] \\
&= a\,\text{Cov}(X, Z) + b\,\text{Cov}(Y, Z).
\end{aligned}$$

9. Let $U = X + Y$ and $V = X - Y$. Then

$$E(UV) = E[(X + Y)(X - Y)] = E(X^2 - Y^2) = E(X^2) - E(Y^2).$$

Also,

$$E(U)E(V) = E(X + Y)E(X - Y) = (\mu_X + \mu_Y)(\mu_X - \mu_Y) = \mu_X^2 - \mu_Y^2.$$

Therefore,

$$\begin{aligned}\text{Cov}(U,V) &= E(UV) - E(U)E(V) = [E(X^2) - \mu_X^2] - [E(Y^2) - \mu_Y^2] \\ &= \text{Var}(X) - \text{Var}(Y) = 0.\end{aligned}$$

It follows that $\rho(U,V) = 0$.

11. For the given values,

$$\begin{aligned}\text{Var}(X) &= E(X^2) - [E(X)]^2 = 10 - 9 = 1, \\ \text{Var}(Y) &= E(Y^2) - [E(Y)]^2 = 29 - 4 = 25, \\ \text{Cov}(X,Y) &= E(XY) - E(X)E(Y) = 0 - 6 = -6.\end{aligned}$$

Therefore,

$$\rho(X,Y) = \frac{-6}{(1)(5)} = -\frac{6}{5}, \text{ which is impossible.}$$

13. $\text{Cov}(X,Y) = \rho(X,Y)\sigma_X\sigma_Y = -\frac{1}{6}(3)(2) = -1.$

(a) $\text{Var}(X+Y) = \text{Var}(X) + \text{Var}(Y) + 2\,\text{Cov}(X,Y) = 11.$

(b) $\text{Var}(X - 3Y + 4) = \text{Var}(X) + 9\,\text{Var}(Y) - (2)(3)\,\text{Cov}(X,Y) = 51.$

15. Since each variance is equal to 1 and each covariance is equal to 1/4,

$$\begin{aligned}\text{Var}(X_1 + \cdots + X_n) &= \sum_i \text{Var}(X_i) + 2\sum\sum_{i<j} \text{Cov}(X_i, X_j) \\ &= n(1) + 2.\frac{n(n-1)}{2}\left(\frac{1}{4}\right) = n + \frac{n(n-1)}{4}.\end{aligned}$$

17. Let $\mu_X = E(X)$ and $\mu_Y = E(Y)$. Apply Theorem 4.6.2 with $U = X - \mu_X$ and $V = Y - \mu_Y$. Then (4.6.4) becomes

$$\text{Cov}(X,Y)^2 \le \text{Var}(X)\,\text{Var}(Y). \tag{S.4.3}$$

Now $|\rho(X,Y)| = 1$ is equivalent to equality in (S.4.3). According to Theorem 4.6.2, we get equality in (4.6.4) and (S.4.3) if and only if there exist constants a and b such that $aU + bV = 0$, that is $a(X - \mu_X) + b(Y - \mu_Y) = 0$, with probability 1. So $|\rho(X,Y)| = 1$ implies $aX + bY = a\mu_X = b\mu_Y$ with probability 1.

4.7 Conditional Expectation

1. The M.S.E. after observing $X = 18$ is $\text{Var}(P|18) = 19 \times (41 - 18)/[42^2 \times 43] = 0.00576$. This is about seven percent of the marginal M.S.E.

3. Since $E(X \mid Y) = c$, then $E(X) = E[E(X \mid Y)] = c$ and
$E(XY) = E[E(XY \mid Y)] = E(YE(X \mid Y)] = E(cY) = cE(Y)$.
Therefore,

$$\text{Cov}(X, Y) = E(XY) - E(X)E(Y) = cE(Y) - cE(Y) = 0.$$

5. For any given value x_{n-1} of X_{n-1}, $E(X_n \mid x_{n-1})$ will be the midpoint of the interval $(x_{n-1}, 1)$. Therefore,
$E(X_n \mid X_{n-1}) = \dfrac{1 + X_{n-1}}{2}$.
It follows that

$$E(X_n) = E[E(X_n \mid X_{n-1})] = \frac{1}{2} + \frac{1}{2}E(X_{n-1}).$$

Similarly, $E(X_{n-1}) = \dfrac{1}{2} + \dfrac{1}{2}E(X_{n-2})$, etc. Since $E(X_1) = \dfrac{1}{2}$, we obtain

$$E(X_n) = \frac{1}{2} + \frac{1}{4} + \frac{1}{8} + \cdots + \frac{1}{2^n} = 1 - \frac{1}{2^n}.$$

7. The marginal p.d.f. of X is

$$f_1(x) = \int_0^1 (x + y)dy = x + \frac{1}{2} \qquad \text{for} \quad 0 \le x \le 1.$$

Therefore, for $0 \le x \le 1$, the conditional p.d.f. of Y given that $X = x$ is

$$g(y \mid x) = \frac{f(x, y)}{f_1(x)} = \frac{2(x + y)}{2x + 1} \qquad \text{for } 0 \le y \le 1.$$

Hence,

$$\begin{aligned} E(Y \mid x) &= \int_0^1 \frac{2(xy + y^2)}{2x + 1} dy = \frac{3x + 2}{3(2x + 1)}, \\ E(Y^2 \mid x) &= \int_0^1 \frac{2(xy^2 + y^3)}{2x + 1} dy = \frac{4x + 3}{6(2x + 1)}, \end{aligned}$$

and

$$\text{Var}(Y \mid x) = \frac{4x + 3}{6(2x + 1)} - \left[\frac{3x + 2}{3(2x + 1)}\right]^2 = \frac{1}{36}\left[3 - \frac{1}{(2x + 1)^2}\right].$$

9. The overall M.S.E. is

$$E[\text{Var}(Y \mid X)] = \int_0^1 \frac{1}{36}\left[3 - \frac{1}{(2x + 1)^2}\right] f_1(x)dx.$$

It was found in the solution of Exercise 7 that

$$f_1(x) = x + \frac{1}{2} \qquad \text{for } 0 \le x \le 1.$$

Therefore, it can be found that $E[\text{Var}(Y \mid X)] = \dfrac{1}{12} - \dfrac{\log 3}{144}$.

11. Let $E(Y) = \mu_Y$. Then

$$\begin{aligned}\text{Var}(Y) &= E[(Y-\mu_Y)^2] = E\{[(Y-E(Y\mid X)) + (E(Y\mid X)-\mu_Y)]^2\}\\ &= E\{[Y-E(Y\mid X)]^2\} + 2E\{[Y-E(Y\mid X)][E(Y\mid X)-\mu_Y]\}\\ &\quad +E\{[E(Y\mid X)-\mu_Y]^2\}.\end{aligned}$$

We shall now consider further each of the three expectations in the final sum. First,

$$E\{[Y-E(Y\mid X)]^2\} = E(E\{[Y-E(Y\mid X]^2\mid X\}) = E[\text{Var}(Y\mid X)].$$

Next,

$$\begin{aligned}E\{[Y-E(Y\mid X)][E(Y\mid X)-\mu_Y]\} &= E(E\{[Y-E(Y\mid X)][E(Y\mid X)-\mu_Y]\mid X\})\\ &= E([E(Y\mid X)-\mu_Y]E\{Y-E(Y\mid X)\mid X\})\\ &= E([E(Y\mid X)-\mu_Y]\cdot 0)\\ &= 0.\end{aligned}$$

Finally, since the mean of $E(Y\mid X)$ is $E[E(Y\mid X)] = \mu_Y$, we have

$$E\{[E(Y\mid X)-\mu_Y]^2\} = \text{Var}[E(Y\mid X)].$$

It now follows that

$$\text{Var}(Y) = E[\text{Var}(Y\mid X)] + \text{Var}[E(Y\mid X)].$$

13. (a) The prediction is the mean of Y:

$$E(Y) = \int_0^1\int_0^1 y\cdot\frac{2}{5}(2x+3y)dx\,dy = \frac{3}{5}.$$

(b) The prediction is the median m of X. The marginal p.d.f. of X is

$$f_1(x) = \int_0^1 \frac{2}{5}(2x+3y)dy = \frac{1}{5}(4x+3) \qquad \text{for } 0\le x\le 1.$$

We must have

$$\int_0^m \frac{1}{5}(4x+3)dx = \frac{1}{2}.$$

Therefore, $4m^2+6m-5=0$ and $m = \dfrac{\sqrt{29}-3}{4}$.

15. (a) For $0\le x\le 1$ and $0\le y\le 1$, the conditional p.d.f. of Y given that $X=x$ is

$$g(y\mid x) = \frac{f(x,y)}{f_1(x)} = \frac{2(2x+3y)}{4x+3}.$$

When $X=0.8$, the prediction of Y is

$$E(Y\mid X=0.8) = \int_0^1 yg(y\mid x=0.8)dy = \int_0^1\frac{y(1.6+3y)}{3.1}dy = \frac{18}{31}.$$

(b) The marginal p.d.f. of Y is

$$f_2(y) = \int_0^1 \frac{2}{5}(2x+3y)dx = \frac{2}{5}(1+3y) \qquad \text{for } 0 \le y \le 1.$$

Therefore, for $0 \le x \le 1$ and $0 \le y \le 1$, the conditional p.d.f. of X given that $Y = y$ is

$$h(x \mid y) = \frac{f(x,y)}{f_2(y)} = \frac{2x+3y}{1+3y}.$$

When $Y = 1/3$, the prediction of X is the median m of the conditional p.d.f. $h(x \mid y = 1/3)$. We must have

$$\int_0^m \frac{2x+1}{2}\,dx = \frac{1}{2}.$$

Hence, $m^2 + m = 1$ and $m = (\sqrt{5}-1)/2$.

17. Let $Z = r(X,Y)$, and let (X,Y) have joint p.f. $f(x,y)$. Also, let $W = r(x_0, Y)$, for some possible value x_0 of X. We need to show that the conditional p.f. of Z given $X = x_0$ is the same as the conditional p.f. of W given $X = x_0$ for all x_0.

Let $f_1(x)$ be the marginal p.f. of X. For each possible value (z,x) of (Z,X), define $B_{(z,x)} = \{y : r(x,y) = z\}$. Then, $(Z,X) = (z,x)$ if and only if $X = x$ and $Y \in B_{(z,x)}$. The joint p.f. of (Z,X) is then

$$g(z,x) = \sum_{y \in B_{(z,x)}} f(x,y).$$

The conditional p.f. of Z given $X = x_0$ is $g_1(z|x_0) = g(z,x_0)/f_1(x_0)$ for all z and all x_0.

Next, notice that $(W,X) = (w,x)$ if and only if $X = x$ and $w \in B_{(w,x_0)}$. The joint p.f. of (W,X) is then

$$h(w,x) = \sum_{y \in B_{(w,x_0)}} f(x,y).$$

The conditional p.f. of W given $X = x$ is $h_1(w|x) = h(w,x)/f_1(x)$. Now, for $x = x_0$, we get $h_1(w|x_0) = h(w,x_0)/f_1(x_0)$. But $h(w,x_0) = g(w,x_0)$ for all w and all x_0. Hence $h_1(w|x_0) = g_1(w|x_0)$ for all w and all x_0. This is the desired conclusion.

4.8 Utility

1. The utility of not buying the ticket is $U(0) = 0$. If the decision maker buys the ticket, the utility is $U(499)$ if the ticket is a winner and and $U(-1)$ if the ticket is a loser. That is the utility is 499^α with probability 0.001 and it is -1 with probability 0.999. The expected utility is then $0.001 \times 499^\alpha - 0.999$. The decision maker prefers buying the ticket if this expected utility is greater than 0. Setting the expected utility greater than 0 means $499^\alpha > 999$. Taking logarithms of both sides yields $\alpha > 1.11$.

3.

$$\begin{aligned} E[U(X)] &= \frac{1}{2}\sqrt{5} + \frac{1}{2}\sqrt{25} = 3.618, \\ E[U(Y)] &= \frac{1}{2}\sqrt{10} + \frac{1}{2}\sqrt{20} = 3.817, \\ E[U(Z)] &= \sqrt{15} = 3.873. \end{aligned}$$

Hence, Z is preferred.

5. Since the person is indifferent between the gamble and the sure thing,

$$U(50) = \frac{1}{3}U(0) + \frac{2}{3}U(100) = \frac{1}{3}\cdot 0 + \frac{2}{3}\cdot 1 = \frac{2}{3}.$$

7. For any given values of a,

$$E[U(X)] = p\log\ a + (1-p)\log(1-a).$$

The maximum of this expected utility can be found by elementary differentiation. We have

$$\frac{\partial E[U(X)]}{\partial a} = \frac{p}{a} - \frac{1-p}{1-a}.$$

When this derivative is set equal to 0, we find that $a = p$. Since

$$\frac{\partial^2 E[U(X)]}{\partial a^2} = -\frac{p}{a^2} - \frac{1-p}{(1-a)^2} < 0,$$

It follows that E[U(X)] is a maximum when $a = p$.

9. For any given value of a,

$$E[U(X)] = pa + (1-p)(1-a).$$

This is a linear function of a. If $p < 1/2$, it has the form shown in sketch (i) of Fig. S.4.1.

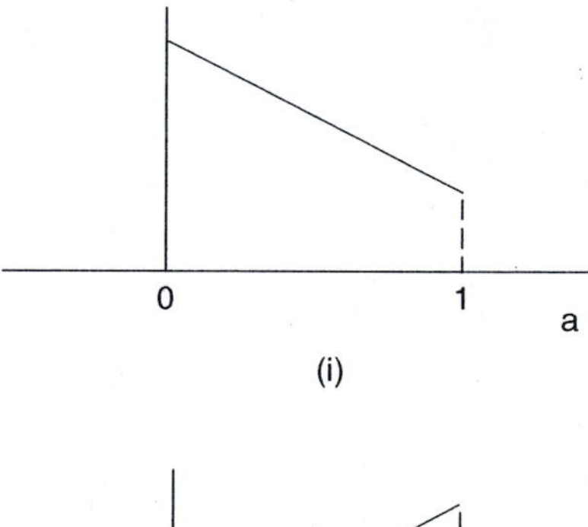

(i)

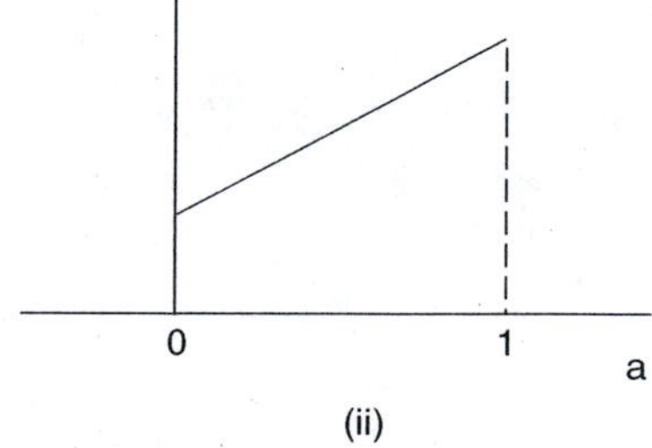

(ii)

Figure S.4.1: Figure for Exercise 9 of Sec. 4.8.

Therefore, $E[U(X)]$ is a maximum when $a = 0$. If $p > 1/2$, it has the form shown in sketch (ii) of Fig. S.4.1. Therefore, $E[U(X)]$ is a maximum when $a = 1$. If $p = 1/2$, then $E[U(X)] = 1/2$ for all values of a.

11. For any given value of b,

$$E[U(X)] = p\log(A+b) + (1-p)\log(A-b).$$

Therefore,

$$\frac{\partial E[U(X)]}{\partial b} = \frac{p}{A+b} - \frac{1-p}{A-b}.$$

When this derivative is set equal to 0, we find that

$$b = (2p-1)A.$$

Since $\frac{\partial^2 E[U(X)]}{\partial b^2} < 0$, this value of b does yield a maximum value of $E[U(X)]$. If $p \geq 1/2$, this value of b lies between 0 and A as required. However, if $p < 1/2$, this value of b is negative and not permissible. In this case, it can be shown that the maximum value of $E[U(X)]$ for $0 \leq b \leq A$ occurs when $b = 0$; that is, when the person does not bet at all.

13. For any given value of b,

$$E[U(X)] = p(A+b) + (1-p)(A-b).$$

This is a linear function of b. If $p > 1/2$, it has the form shown in sketch (i) of Fig. S.4.2 and $b = A$ is best. If $p < 1/2$, it has the form shown in sketch (ii) of Fig. S.4.2 and $b = 0$ is best. If $p = 1/2$,

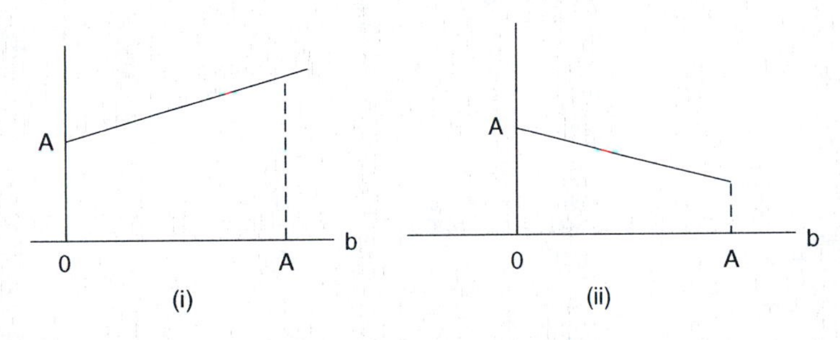

Figure S.4.2: Figure for Exercise 13 of Sec. 4.8.

$E[U(X)] = A$ for all values of b.

15. The expected utility for the lottery ticket is

$$E[U(X)] = \int_0^4 x^\alpha \frac{1}{4} dx = \frac{4^\alpha}{\alpha+1}.$$

The utility of accepting x_0 dollars instead of the lottery ticket is $U(x_0) = x_0^\alpha$. Therefore, the person will prefer to sell the lottery ticket for x_0 dollars if

$$x_0^\alpha > \frac{4^\alpha}{\alpha+1} \quad \text{or if} \quad x_0 > \frac{4}{(\alpha+1)^{1/\alpha}}.$$

It can be shown that the right-hand side of this last inequality is an increasing function of α.

17. The gain is 10^6 if $P > 1/2$ and -10^6 if $P \leq 1/2$. The utility of continuing to promote is then $10^{5.4}$ if $P > 1/2$ and -10^6 if $P \leq 1/2$. To find the expected utility, we need $\Pr(P \leq 1/2)$. Using the stated p.d.f. for P, we get $\Pr(P \leq 1/2) = \int_0^{1/2} 56p^6(1-p)dp = 0.03516$. The expected utility is then $10^{5.4} \times (1-0.03516) - 10^6 \times 0.03516 = 207197$. This is greater than 0, so we would continue to promote the treatment.

4.9 Supplementary Exercises

1. If $u \geq 0$,

$$\int_u^\infty xf(x)dx \geq u \int_u^\infty f(x)dx = u[1 - F(u)].$$

Since

$$\lim_{u\to\infty} \int_{-\infty}^u xf(x)dx = E(X) = \int_{-\infty}^\infty xf(x)dx < \infty,$$

it follows that

$$\lim_{u\to\infty} \left[E(X) - \int_{-\infty}^u xf(x)dx\right] = \lim_{u\to\infty} \int_u^\infty xf(x)dx = 0.$$

3. Let $x_1, x_2, \ldots$ denote the possible values of X. Since $F(X)$ is a step function, the integral given in Exercise 1 becomes the following sum:

$$\begin{aligned} &(x_1 - 0) + [1 - f(x_1)](x_2 - x_1) + [1 - f(x_1) - f(x_2)](x_3 - x_2) + \cdots \\ &= x_1 f(x_1) + x_2\, f(x_2) + x_3 f(x_3) + \cdots \\ &= E(X). \end{aligned}$$

5. We need $E(Y) = a\mu + b = 0$ and

$$\text{Var}(Y) = a^2\sigma^2 = 1.$$

Therefore, $a = \pm\,\frac{1}{\sigma}$ and $b = -a\mu$.

7. The dealer's expected gain is

$$E(Y - X) = \frac{1}{36}\int_0^6 \int_0^y (y - x)x\, dx\, dy = \frac{3}{2}.$$

9. Suppose first that $r(X)$ is nondecreasing. Then

$$\Pr[Y \geq r(m)] = \Pr[r(X) \geq r(m)] \geq \Pr(X \geq m) \geq \frac{1}{2},$$

and

$$\Pr[Y \leq r(m)] = \Pr[r(X) \leq r(m)] \geq \Pr(X \leq m) \geq \frac{1}{2}.$$

Hence, $r(m)$ is a median of the distribution of Y. If $r(X)$ is nonincreasing, then

$$\Pr[Y \geq r(m)] \geq \Pr(X \leq m) \geq \frac{1}{2}$$

and

$$\Pr[Y \leq r(m)] \geq \Pr(X \geq m) \geq \frac{1}{2}.$$

11. Suppose that you order s liters. If the demand is $x < s$, you will make a profit of gx cents on the x liters sold and suffer a loss of $c(s-x)$ cents on the $s-x$ liters that you do not sell. Therefore, your net profit will be $gx - c(s-x) = (g+c)x - cs$. If the demand is $x \geq s$, then you will sell all s liters and make a profit of gs cents. Hence, your expected net gain is

$$\begin{aligned} E &= \int_0^s [(g+c)x - cs]f(x)dx + gs\int_s^\infty f(x)dx \\ &= \int_0^s (g+c)x\ f(x)dx - csF(s) + gs[1-F(s)]. \end{aligned}$$

To find the value of s that maximizes E, we find, after some calculations, that

$$\frac{dE}{ds} = g - (g+c)\ \ F(s).$$

Thus, $\frac{dE}{ds} = 0$ and E is maximized when s is chosen so that $F(s) = g/(g+c)$.

13. $E(Z) = 5(3) - 1 + 15 = 29$ in all three parts of this exercise. Also,

$$\mathrm{Var}(Z) = 25\,\mathrm{Var}(X) + \mathrm{Var}(Y) - 10\,\mathrm{Cov}(X,Y) = 109 - 10\,\mathrm{Cov}(X,Y).$$

Hence, $\mathrm{Var}(Z) = 109$ in parts (a) and (b). In part (c),

$$\mathrm{Cov}(X,Y) = \rho\sigma_X\sigma_Y = (.25)(2)(3) = 1.5$$

so $\qquad \mathrm{Var}(Z) = 94.$

15. Let $v^2 = \mathrm{Var}(X_1 + \cdots + X_n) = \sum_i \mathrm{Var}(X_i) + 2\sum_{i<j}\mathrm{Cov}(X_i, X_j)$. In this problem $\mathrm{Var}(X_i) = \sigma^2$ for all i and $\mathrm{Cov}(X_i, X_j) = \rho\sigma^2$ for all $i \neq j$. Therefore,

$$v^2 = n\sigma^2 + n(n-1)\rho\sigma^2.$$

Since $v^2 \geq 0$, it follows that $\rho \geq -1/(n-1)$.

17. For $i = 1, \ldots, n$, let $X_i = 1$ if the ith letter is placed in the correct envelope and let $X_i = 0$ otherwise. Then $E(X_i) = 1/n$ and, for $i \neq j$,

$$E(X_iX_j) = \Pr(X_iX_j = 1) = \Pr(X_i = 1 \text{ and } X_j = 1) = \frac{1}{n(n-1)}.$$

Also, $E(X_i^2) = E(X_i) = 1/n$. Hence,

$$\mathrm{Var}(X_i) = \frac{1}{n} - \frac{1}{n^2} = \frac{n-1}{n^2}$$

and $\mathrm{Cov}(X_i, X_j) = \dfrac{1}{n(n-1)} - \dfrac{1}{n^2} = \dfrac{1}{n^2(n-1)}$. The total number of correct matches is $X = \sum_{i=1}^n X_i$. Therefore,

$$\mathrm{Var}(X) = \sum_{i=1}^n \mathrm{Var}(X_i) + 2\sum_{i<j}\mathrm{Cov}(X_i, X_j) = n\cdot\frac{n-1}{n^2} + n(n-1)\cdot\frac{1}{n^2(n-1)} = 1.$$

19. $$c'(t) = \frac{\psi'(t)}{\psi(t)} \quad \text{and} \quad c''(t) = \frac{\psi(t)\psi''(t) - [\psi'(t)]^2}{[\psi(t)]^2}$$

Since $\psi(0) = 1$, $\psi'(0) = \mu$, and $\psi''(0) = E(X^2) = \sigma^2 + \mu^2$, it follows that $c'(0) = \mu$ and $c''(0) = \sigma^2$.

21. Since the coefficient of X in $E(Y \mid X)$ is negative, it follows from Exercise 20 that $\rho < 0$. Furthermore, it follows from Exercise 20 that the product of the coefficients of X and Y in $E(Y \mid X)$ and $E(X \mid Y)$ must be ρ^2. Hence, $\rho^2 = 1/4$ and, since $\rho < 0$, $\rho = -1/2$.

23.

$$\begin{aligned} \text{Cov}(X, X + bY) &= \text{Var}(X) + b\,\text{Cov}(X, Y) \\ &= 1 + b\rho. \\ \text{Var}(X) &= 1, \text{Var}(X + bY) = 1 + b^2 + 2b\rho. \end{aligned}$$

Hence,

$$\rho(X, X + bY) = \frac{1 + b\rho}{(1 + b^2 + 2b\rho)^{1/2}}.$$

If we set this quantity equal to ρ, square both sides, and solve for b, we obtain $b = -1/(2\rho)$.

25. (a) The marginal p.d.f. of X is

$$f_1(x) = \int_0^x 8xy\, dy = 4x^3 \qquad \text{for } 0 < x < 1.$$

Therefore, the conditional p.d.f. of Y given that $X = .2$ is

$$g_1(y \mid X = .2) = \frac{f(.2, y)}{f_1(.2)} = 50y \quad \text{for} \quad 0 < y < .2\,.$$

The mean of this distribution is

$$E(Y \mid X = .2) = \frac{2}{15} = .1333.$$

(b) The median of $g_1(y \mid X = .2)$ is $m = \left(\frac{1}{50}\right)^{1/2} = .1414$.

27. Let N be the number of balls in the box. Since the proportion of red balls is p, there are Np red balls in the box. (Clearly, p must be an integer multiple of $1/N$.) There are $N(1 - p)$ blue balls in the box. Let $K = Np$ so that there are $N - K$ blue balls and K red balls. If $n > K$, then $\Pr(Y = n) = 0$ since there are not enough red balls. Since $\Pr(X = n) > 0$ for all n, the result is true if $n > K$. For $n \leq K$, let $X_i = 1$ if the ith ball is red for $i = 1, \ldots, n$. For sampling without replacement,

$$\Pr(Y = n) = \Pr(X_1 = 1) \prod_{i=2}^{n} \Pr(X_i = 1 | X_1 = 1, \ldots, X_{i-1} = 1) = \frac{K}{N}\frac{K-1}{N-1} \cdots \frac{K-n+1}{N-n+1}. \quad \text{(S.4.4)}$$

For sampling with replacement, the X_i's are independent, so

$$\Pr(X = n) = \prod_{i=1}^{n} \Pr(X_i = 1) = \left(\frac{K}{N}\right)^n. \quad \text{(S.4.5)}$$

For $j = 1, \ldots, n-1$, $KN - jN < KN - jK$, so $(K - j)/(N - j) < K/N$. Hence the product in (S.4.4) is smaller than the product in (S.4.5). This argument makes sense only if N is finite. If N is infinite, then sampling with and without replacement are equivalent.

29. The expected utility from allocating the amounts a and $m - a$ is

$$\begin{aligned} E &= p\log(g_1 a) + (1-p)\log[g_2(m-a)] \\ &= p\log\ a + (1-p)\log(m-a) \\ &+ p\log\ g_1 + (1-p)\log\ g_2. \end{aligned}$$

The maximum over all values of a can now be found by elementary differentiation, as in Exercise 7 of Sec.4.8, and we obtain $a = pm$.

Chapter 5

Special Distributions

5.2 The Bernoulli and Binomial Distributions

1. Since $E(X^k)$ has the same value for every positive integer k, we might try to find a random variable X such that X, X^2, X^3, X^4,... all have the same distribution. If X can take only the values 0 and 1, then $X^k = X$ for every positive integer k since $0^k = 0$ and $1^k = 1$. If $\Pr(X = 1) = p = 1 - \Pr(X = 0)$, then in order for $E(X^k) = 1/3$, as required, we must have $p = 1/3$. Therefore, a random variable X such that $\Pr(X = 1) = 1/3$ and $\Pr(X = 0) = 2/3$ satisfies the required conditions.

3. Let X be the number of heads obtained. Then strictly more heads than tails are obtained if $X \in \{6, 7, 8, 9, 10\}$. The probability of this event is the sum of the numbers in the binomial table corresponding to $p = 0.5$ and $n = 10$ for $k = 6, \ldots, 10$. By the symmetry of this binomial distribution, we can also compute the sum as $(1 - \Pr(X = 5))/2 = (1 - 0.2461)/2 = 0.37695$.

5. The tables do not include the value $p = 0.6$, so we must use the trick described in Exercise 7 of Sec. 3.1. The number of *tails* X will have the binomial distribution with parameters $n = 9$ and $p = 0.4$. Therefore,

$$\begin{aligned}
\Pr(\text{Even number of heads}) &= \Pr(\text{Odd number of tails}) \\
&= \Pr(X = 1) + \Pr(X = 3) + \Pr(X = 5) + \Pr(X = 7) + \Pr(X = 9) \\
&= .0605 + .2508 + .1672 + .0212 + .0003 \\
&= .5000.
\end{aligned}$$

7. If we assume that N_A, N_B, and N_C are independent, then

$$\begin{aligned}
\text{Var}(N_A + N_B + N_C) &= \text{Var}(N_A) + \text{Var}(N_B) + \text{Var}(N_C) \\
&= 3 \cdot \frac{1}{8} \cdot \frac{7}{8} + 5 \cdot \frac{1}{4} \cdot \frac{3}{4} + 2 \cdot \frac{1}{2} \cdot \frac{1}{2} = \frac{113}{64}.
\end{aligned}$$

9. $$\Pr\left(X_1 = 1 \,\middle|\, \sum_{i=1}^{n} X_i = k\right) = \frac{\Pr\left(X_1 = 1 \text{ and } \sum_{i=1}^{n} X_i = k\right)}{\Pr\left(\sum_{i=1}^{n} X_i = k\right)} = \frac{\Pr\left(X_1 = 1 \text{ and } \sum_{i=2}^{n} X_i = k - 1\right)}{\Pr\left(\sum_{i=1}^{n} X_i = k\right)}.$$

Since the random variables $X_1, \ldots, X_n$ are independent, it follows that X_1 and $\sum_{i=2}^n X_i$ are independent. Therefore, the final expression can be rewritten as

$$\frac{\Pr(X_1 = 1)\Pr\left(\sum_{i=2}^n X_i = k-1\right)}{\Pr\left(\sum_{i=1}^n X_i = k\right)}.$$

The sum $\sum_{i=2}^n X_i$ has the binomial distribution with parameters $n-1$ and p, and the sum $\sum_{i=1}^n X_i$ has the binomial distribution with parameters n and p. Therefore,

$$\Pr\left(\sum_{i=2}^n X_i = k-1\right) = \binom{n-1}{k-1}p^{k-1}(1-p)^{(n-1)-(k-1)} = \binom{n-1}{k-1}p^{k-1}(1-p)^{n-k},$$

and

$$\Pr\left(\sum_{i=1}^n X_i = k\right) = \binom{n}{k}p^k(1-p)^{n-k}.$$

Also, $\Pr(X_1 = 1) = p$. It now follows that

$$\Pr\left(X_1 = 1 \middle| \sum_{i=1}^n X_i = k\right) = \frac{\binom{n-1}{k-1}p^k(1-p)^{n-k}}{\binom{n}{k}p^k(1-p)^{n-k}} = \frac{k}{n}.$$

11. Since the value of the term being summed here will be 0 for $x = 0$ and for $x = 1$, we may change the lower limit of the summation from $x = 2$ to $x = 0$, without affecting the value of the sum. The summation can then be rewritten as

$$\sum_{x=0}^n x^2\binom{n}{x}p^x(1-p)^{n-x} - \sum_{x=0}^n x\binom{n}{x}p^x(1-p)^{n-x}.$$

If X has the binomial distribution with parameters n and p, then the first summation is simply $E(X^2)$ and the second summation is simply $E(X)$. Finally,

$$E(X^2) - E(X) = \text{Var}(X) + [E(X)]^2 - E(X) = np(1-p) + (np)^2 - np = n(n-1)p^2.$$

13. Let X be the number of successes in the group with probability 0.5 of success. Let Y be the number of successes in the group with probability 0.6 of success. We want $\Pr(X \geq Y)$. Both X and Y have discrete (binomial) distributions with possible values $0, \ldots, 5$. There are 36 possible (X, Y) pairs and we need the sum of the probabilities of the 21 of them for which $X \geq Y$. To save time, we shall calculate the probabilities of the 15 other ones and subtract the total from 1. Since X and Y are independent, we can write $\Pr(X = x, Y = y) = \Pr(X = x)\Pr(Y = y)$, and find each of the factors in the binomial table in the back of the book. For example, for $x = 1$ and $y = 2$, we get $0.1562 \times 0.2304 = 0.03599$. Adding up all 15 of these and subtracting from 1 we get 0.4957.

15. We need the maximum number of tests if and only if every first-stage and second-stage subgroup has at least one positive result. In that case, we would need $10 + 100 + 1000 = 1110$ total tests. The probability that we have to run this many tests is the probability that every $Y_{2,i,k} = 1$, which in turn is the probability that every $Z_{2,i,k} > 0$. The $Z_{2,i,k}$'s are independent binomial random variables with parameters 10 and 0.002, and there are 100 of them altogether. The probability that each is positive is 0.0198, as computed in Example 5.2.7. The probability that they are all positive is $(0.0198)^{100} = 4.64 \times 10^{-171}$.

5.3 The Hypergeometric Distributions

1. Using Eq. (5.3.1) with the parameters $A = 10$, $B = 24$, and $n = 11$, we obtain the desired probability

$$\Pr(X = 10) = \frac{\binom{10}{10}\binom{24}{1}}{\binom{34}{11}} = 8.389 \times 10^{-8}.$$

3. As in Exercise 2, let X denote the number of red balls in the sample. Then, by Eqs. (5.3.3) and (5.3.4),

$$E(X) = \frac{nA}{A+B} = \frac{7}{3} \quad \text{and} \quad \text{Var}(X) = \frac{nAB}{(A+B)^2} \cdot \frac{A+B-n}{A+B-1} = \frac{8}{9}.$$

Since $\overline{X} = X/n$,

$$E(\overline{X}) = \frac{1}{n}E(X) = \frac{1}{3} \quad \text{and} \quad \text{Var}(\overline{X}) = \frac{1}{n^2}\text{Var}(X) = \frac{8}{441}.$$

5. By Eq. (5.3.4),

$$\text{Var}(X) = \frac{A(T-A)}{T^2(T-1)} n(T-n).$$

If T is an even integer, then the quadratic function $n(T-n)$ is a maximum when $n = T/2$. If T is an odd integer, then the maximum value of $n(T-n)$, for $n = 0, 1, 2, \ldots, T$, occurs at the two integers $(T-1)/2$ and $(T+1)/2$.

7. (a) The probability of obtaining exactly x defective items is

$$\frac{\binom{0.3T}{x}\binom{0.7T}{10-x}}{\binom{T}{10}}.$$

Therefore, the probability of obtaining not more than one defective item is the sum of these probabilities for $x = 0$ and $x = 1$.

Since

$$\binom{0.3T}{0} = 1 \quad \text{and} \quad \binom{0.3T}{1} = 0.3T,$$

this sum is equal to

$$\frac{\binom{0.7T}{10} + 0.3T\binom{0.7T}{9}}{\binom{T}{10}}.$$

(b) The probability of obtaining exactly x defectives according to the binomial distribution, is

$$\binom{10}{x}(0.3)^x(0.7)^{10-x}.$$

The desired probability is the sum of these probabilities for $x = 0$ and $x = 1$, which is

$$(0.7)^{10} + 10(0.3)(0.7)^9.$$

For a large value of T, the answers in (a) and (b) will be very close to each other, although this fact is not obvious from the different forms of the two answers.

9. By Eq. (5.3.14),

$$\binom{3/2}{4} = \frac{(3/2)(1/2)(-1/2)(-3/2)}{4!} = \frac{3}{128}.$$

11. Write $(1+a_n)^{c_n}e^{-a_nc_n} = \exp[c_n\log(1+a_n) - a_nc_n]$. The result is proven if we can show that

$$\lim_{n\to\infty}[c_n\log(1+a_n) - a_nc_n] = 0. \tag{S.5.1}$$

Use Taylor's theorem with remainder to write

$$\log(1+a_n) = a_n - \frac{a_n^2}{2(1+y_n)^2},$$

where y_n is between 0 and a_n. It follows that

$$c_n\log(1+a_n) - a_nc_n = c_na_n - \frac{c_na_n^2}{2(1+y_n)^2} - a_nc_n = -\frac{c_na_n^2}{2(1+y_n)^2}.$$

We have assumed that $c_na_n^2$ goes to 0. Since y_n is between 0 and a_n, and a_n goes to 0, we have $1/[2(1+y_n)^2]$ goes to 0. This establishes (S.5.1).

5.4 The Poisson Distributions

1. The number of oocysts X in $t = 100$ liters of water has the Poisson distribution with mean $0.2 \times 0.1 \times 100 = 2$. Using the Poisson distribution table in the back of the book, we find

$$\Pr(X \geq 2) = 1 - \Pr(X \leq 1) = 1 - 0.1353 - 0.2707 = 0.594.$$

3. Since the number of defects on each bolt has the Poisson distribution with mean 0.4, and the observations for the five bolts are independent, the sum for the numbers of defects on five bolts will have the Poisson distribution with mean $5(0.4) = 2$. It is found from the table of the Poisson distribution that

$$\Pr(X \geq 6) = .0120 + .0034 + .0009 + .0002 + .0000 = .0165.$$

There is some rounding error in this, and 0.0166 is closer.

5. Let Y denote the number of misprints on a given page. Then the probability p that a given page will contain more than k misprints is

$$p = \Pr(Y > k) = \sum_{i=k+1}^{\infty} f(i \mid \lambda) = \sum_{i=k+1}^{\infty} \frac{\exp(-\lambda)\lambda^i}{i!}.$$

Therefore,

$$1 - p = \sum_{i=0}^{k} f(i \mid \lambda) = \sum_{i=0}^{k} \frac{\exp(-\lambda)\lambda^i}{i!}.$$

Now let X denote the number of pages, among the n pages in the book, on which there are more than k misprints. Then for $x = 0, 1, \ldots, n$,

$$\Pr(X = x) = \binom{n}{x} p^x (1-p)^{n-x}$$

and

$$\Pr(X \geq m) = \sum_{x=m}^{n} \binom{n}{x} p^x (1-p)^{n-x}.$$

7. We shall assume that customers are served in accordance with a Poisson process. Then the number of customers served in a two-hour period will have the Poisson distribution with mean $\mu = 2(15) = 30$. Therefore, the probability that more than 20 customers will be served is

$$\Pr(X > 20) = \sum_{x=21}^{\infty} \frac{\exp(-30)(30)^x}{x!}.$$

9. Let N denote the total number of items produced by the machine and let X denote the number of defective items produced by the machine. Then, for $x = 0, 1, \ldots,$

$$\Pr(X = x) = \sum_{n=0}^{\infty} \Pr(X = x \mid N = n) \Pr(N = n).$$

Clearly, it must be true that $X \leq N$. Therefore, the terms in this summation for $n < x$ will be 0, and we may write

$$\Pr(X = x) = \sum_{n=x} \Pr(X = x \mid N = n) \Pr(N = n).$$

Clearly, $\Pr(X = 0 \mid N = 0) = 1$. Also, given that $N = n > 0$, the conditional distribution of X will be a binomial distribution with parameters n and p. Therefore,

$$\Pr(X = x \mid N = n) = \frac{n!}{x!(n-x)!} p^x (1-p)^{n-x}.$$

Also, since N has the Poisson distribution with mean λ,

$$\Pr(N = n) = \frac{\exp(-\lambda)\lambda^n}{n!}.$$

Hence,

$$\Pr(X = x) = \sum_{n=x}^{\infty} \frac{n!}{x!(n-x)!} p^x (1-p)^{n-x} \frac{\exp(-\lambda)\lambda^n}{n!} = \frac{1}{x!} p^x \exp(-\lambda) \sum_{n=x}^{\infty} \frac{1}{(n-x)!} (1-p)^{n-x} \lambda^n.$$

If we let $t = n - x$, then

$$\begin{aligned}
\Pr(X = x) &= \frac{1}{x!} p^x \exp(-\lambda) \sum_{t=0}^{\infty} \frac{1}{t!} (1-p)^t \lambda^{t+x} \\
&= \frac{1}{x!} (\lambda p)^x \exp(-\lambda) \sum_{t=0}^{\infty} \frac{[\lambda(1-p)]^t}{t!} \\
&= \frac{1}{x!} (\lambda p)^x \exp(-\lambda) \exp(\lambda(1-p)) = \frac{\exp(-\lambda p)(\lambda p)^x}{x!}.
\end{aligned}$$

It can be seen that this final term is the value of the p.f. of the Poisson distribution with mean λp.

11. If $f(x \mid \lambda)$ denotes the p.f. of the Poisson distribution with mean λ, then

$$\frac{f(x+1 \mid \lambda)}{f(x \mid \lambda)} = \frac{\lambda}{x+1}.$$

Therefore, $f(x \mid \lambda) < f(x+1 \mid \lambda)$ if and only if $x + 1 < \lambda$. It follows that if λ is not an integer, then the mode of this distribution will be the largest integer x that is less than λ or, equivalently, the smallest integer x such that $x + 1 > \lambda$. If λ is an integer, then both the values $\lambda - 1$ and λ will be modes.

13. It can be assumed that the exact number of sets of triplets in this hospital is a binomial distribution with parameters $n = 700$ and $p = 0.001$. Therefore, this distribution can be approximated by a Poisson distribution with mean $700(0.001) = 0.7$. It is found from the table of the Poisson distribution that

$$\Pr(X = 1) = 0.3476.$$

15. The joint p.f./p.d.f. of X and λ is the Poisson p.f. with parameter λ times $f(\lambda)$ which equals

$$\exp(-\lambda)\frac{\lambda^x}{x!} 2\exp(-2\lambda) = 2\exp(-3\lambda)\frac{\lambda^x}{x!}. \tag{S.5.2}$$

We need to compute the marginal p.f. of X at $x = 1$ and divide that into (S.5.2) to get the conditional p.d.f. of λ given $X = 1$. The marginal p.f. of X at $x = 1$ is the integral of (S.5.2) over λ when $x = 1$ is plugged in.

$$f_1(1) = \int_0^{\infty} 2\lambda \exp(-3\lambda) d\lambda = \frac{2}{9}.$$

This makes the conditional p.d.f. of λ equal to $9\lambda \exp(-3\lambda)$ for $\lambda > 0$.

17. Because $n_T A_T/(A_T + B_T)$ converges to λ, $n_T/(A_T + B_T)$ goes to 0. Hence, B_T eventually gets larger than n_T. Once B_T is larger than $n_T + x$ and A_T is larger than x, we have

$$\Pr(X_T = x) = \frac{\binom{A_T}{x}\binom{B_T}{n_T - x}}{\binom{A_T + B_T}{n_T}} = \frac{A_T! B_T! n_T! (A_T + B_T - n_T)!}{x!(A_T - x)!(n_T - x)!(B_T - n_T + x)!(A_T + B_T)!}.$$

Apply Stirling's formula to each of the factorials in the above expression except $x!$. A little manipulation gives that

$$\lim_{T\to\infty} \frac{A_T^{A_T+1/2} B_T^{B_T+1/2} n_T^{n_T+1/2} (A_T+B_T-n_T)^{A_T+B_T-n_T+1/2} e^x}{\Pr(X_T=x)x!(A_T-x)^{A_T-x+1/2}(n_T-x)^{n_T-x+1/2}(B_T-n+x)^{B_T-n+x+1/2}(A_T+B_T)^{A_T+B_T+1/2}} = 1. \tag{S.5.3}$$

Each of the following limits follows from Theorem 5.3.3:

$$\begin{aligned}
\lim_{T\to\infty}\left(\frac{A_T}{A_T-x}\right)^{A_T-x+1/2} &= e^x,\\
\lim_{T\to\infty}\left(\frac{B_T}{B_T-n_T+x}\right)^{B_T-n_T+x+1/2} e^{-n_T} &= e^{-x},\\
\lim_{T\to\infty}\left(\frac{A_T+B_T-n_T}{A_T+B_T}\right)^{A_T+B_T-n+1/2} e^{n_T} &= 1,\\
\lim_{T\to\infty}\left(\frac{n_T}{n_T-x}\right)^{n_T-x+1/2} &= e^{-x},\\
\lim_{T\to\infty}\left(\frac{B_T}{A_T+B_T}\right)^{n_T-x} &= e^{-\lambda},
\end{aligned}$$

Inserting these limits in (S.5.3) yields

$$\lim_{T\to\infty}\frac{A_T^x e^{-\lambda} n_T^x}{\Pr(X_T=x)x!(A_T+B_T)^x} = 1. \tag{S.5.4}$$

Since $n_T A_T/(A_T+B_T)$ converges to λ, we have

$$\lim_{T\to\infty}\frac{A_T^x n_T^x}{(A_T+B_T)^x} = \lambda^x. \tag{S.5.5}$$

Together (S.5.4) and (S.5.5) imply that

$$\lim_{T\to\infty}\frac{\lambda^x e^{-\lambda}/x!}{\Pr(X_T=x)} = 1.$$

The numerator of this last expression is $\Pr(Y=x)$, which completes the proof.

5.5 The Negative Binomial Distributions

1. (a) Two particular days in a row have independent draws, and each draw has probability 0.01 of producing triples. So, the probability that two particular days in a row will both have triples is 10^{-4}.

   (b) Since a particular day and the next day are independent, the conditional probability of triples on the next day is 0.01 conditional on whatever happens on the first day.

3. (a) Let X denote the number of tails that are obtained before five heads are obtained, and let Y denote the total number of tosses that are required. Then $Y = X + 5$. Therefore, $E(Y) = E(X) + 5$. It follows from Exercise 2(a) that $E(Y) = 150$.

(b) Suppose $Y = X + 5$, then $\text{Var}(Y) = \text{Var}(X)$. Therefore, it follows from Exercise 2(b) that $\text{Var}(Y) = 4350$.

5. By Eq. (5.5.6), the m.g.f. of X_i is

$$\psi_i(t) = \left(\frac{p}{1-(1-p)\exp(t)}\right)^{r_i} \quad \text{for} \quad t < \log\left(\frac{1}{1-p}\right).$$

Therefore, the m.g.f. of $X_1 + \cdots + X_k$ is

$$\psi(t) = \prod_{i=1}^{k} \psi_i(t) = \left(\frac{p}{1-(1-p)\exp(t)}\right)^{r_1+\cdots+r_k} \quad \text{for} \quad t < \log\left(\frac{1}{1-p}\right).$$

Since $\psi(t)$ is the m.g.f. of the negative binomial distribution with parameters $r_1 + \cdots + r_k$ and p, that must be the distribution of $X_1 + \cdots + X_k$.

7. $\Pr(X \geq k) = \sum_{x=j}^{\infty} p(1-p)^x = p(1-p)^k \sum_{x=j}^{\infty} (1-p)^{x-k}$. If we let $i = x - k$, then

$$\Pr(X \geq k) = p(1-p)^k \sum_{x=j}^{\infty} (1-p)^i = p(1-p)^k \frac{1}{1-[1-p]} = (1-p)^k.$$

9. Since the components are connected in series, the system will function properly only as long as every component functions properly. Let X_i denote the number of periods that component i functions properly, for $i = 1, \ldots, n$, and let X denote the number of periods that system functions properly. Then for any nonnegative integer x,

$$\Pr(X \geq x) = \Pr(X_1 \geq x, \ldots, X_n \geq x) = \Pr(X_1 \geq x) \ldots \Pr(X_n \geq x),$$

because the n components are independent. By Exercise 7,

$$\Pr(X_i \geq x) = (1-p_i)^x = (1-p_i)^x.$$

Therefore, $\Pr(X \geq x) = \prod_{i=1}^{n}(1-p_i)^x$. It follows that

$$\begin{aligned}\Pr(X = x) &= \Pr(X \geq x) - \Pr(X \geq x+1) = \prod_{i=1}^{n}(1-p_i)^x - \prod_{i=1}^{n}(1-p_i)^{x+1} \\ &= \left(1 - \prod_{i=1}^{n}(1-p_i)\right)\left(\prod_{i=1}^{n}(1-p_i)\right)^x.\end{aligned}$$

It can be seen that this is the p.f. of the geometric distribution with $p = 1 - \prod_{i=1}^{n}(1-p_i)$.

11. According to Exercise 10 in Sec. 5.3,

$$\binom{-r}{x} = (-1)^x \binom{r+x-1}{x}.$$

This makes

$$\binom{-r}{x} p^r(-[1-p])^x = \binom{r+x-1}{x} p^r(1-p)^x,$$

which is the proper form of the negative binomial p.f. for $x = 0, 1, 2, \ldots$.

13. (a) The memoryless property says that, for all $k, t \geq 0$,

$$\frac{\Pr(X = k + t)}{1 - F(t - 1)} = \Pr(X = k).$$

(The above version switches the use of k and t from Theorem 5.5.5.) If we sum both sides of this over $k = h, h + 1, \ldots$, we get

$$\frac{1 - F(t + h - 1)}{1 - F(t - 1)} = 1 - F(h - 1).$$

(b) $\ell(t + h) = \log[1 - F(t + h - 1)]$. From part (a), we have

$$1 - F(t + h - 1) = [1 - F(t - 1)][1 - F(h - 1)],$$

Hence

$$\ell(t + h) = \log([1 - F(t - 1)] + \log[1 - F(h - 1)] = \ell(t) + \ell(h).$$

(c) We prove this by induction. Clearly $\ell(1) = 1 \times \ell(1)$, so the result holds for $t = 1$. Assume that the result holds for all $t \leq t_0$. Then $\ell(t_0 + 1) = \ell(t_0) + \ell(1)$ by part (b). By the induction hypothesis, $\ell(t_0) = t_0\ell(1)$, hence $\ell(t_0 + 1) = (t_0 + 1)\ell(1)$, and the result holds for $t = t_0 + 1$.

(d) Since $\ell(1) = \log[1 - F(0)]$, we have $\ell(1) < 0$. Let $p = 1 - \exp[\ell(1)]$, which between 0 and 1. For every integer $x \geq 1$, we have, from part (c) and the definition of ℓ, that

$$F(x - 1) = 1 - \exp[\ell(x)] = 1 - \exp[x\ell(1)] = 1 - (1 - p)^x.$$

Setting $t = x - 1$ for $x \geq 1$, we get

$$F(t) = 1 - (1 - p)^{t+1}, \text{ for } t = 0, 1, \ldots. \qquad \text{(S.5.6)}$$

It is easy to verify that (S.5.6) is the c.d.f. of the geometric distribution with parameter p.

5.6 The Normal Distributions

1. By the symmetry of the standard normal distribution around 0, the 0.5 quantile must be 0. The 0.75 quantile is found by locating 0.75 in the $\Phi(x)$ column of the standard normal table and interpolating in the x column. We find $\Phi(0.67) = 0.7486$ and $\Phi(0.68) = 0.7517$. Interpolating gives the 0.75 quantile as 0.6745. By symmetry, the 0.25 quantile is -0.6745. Similarly we find the 0.9 quantile by interpolation using $\Phi(1.28) = 0.8997$ and $\Phi(1.29) = 0.9015$. The 0.9 quantile is then 1.282 and the 0.1 quantile is -1.282.

3. If X denotes the temperature in degrees Fahrenheit and Y denotes the temperature in degrees Celsius, then $Y = 5(X - 32)/9$. Since Y is a linear function of X, then Y will also have a normal distribution. Also,

$$E(Y) = \frac{5}{9}(68 - 32) = 20 \quad \text{and} \quad \text{Var}(Y) = \left(\frac{5}{9}\right)^2 (16) = \frac{400}{81}.$$

5. Let A_i be the event that chip i lasts at most 290 hours. We want the probability of $\cup_{i=1}^3 A_i^c$, whose probability is

$$1 - \Pr\left(\cap_{i=1}^3 A_i\right) = 1 - \prod_{i=1}^{3} \Pr(A_i).$$

Since the lifetime of each chip has the normal distribution with mean 300 and standard deviation 10, each A_i has probability

$$\Phi([290 - 300]/10) = \Phi(-1) = 1 - 0.8413 = 0.1587.$$

So the probability we want is $1 - 0.1587^3 = 0.9960$.

7. If X is a measurement having the specified normal distribution, and if $Z = (X - 120)/2$, then Z will have the standard normal distribution. Therefore, the probability that a particular measurement will lie in the given interval is

$$p = \Pr(116 < X < 118) = \Pr(-2 < Z < -1) = \Pr(1 < Z < 2) = \Phi(2) - \Phi(1) = 0.1360.$$

The probability that all three measurements will lie in the interval is p^3.

9. The total length of the rod is $X = A + B + C - 4$. Since X is a linear combination of A, B, and C, it will also have the normal distribution with

$$E(X) = 20 + 14 + 26 - 4 = 56$$

and $\text{Var}(X) = 0.04 + 0.01 + 0.04 = 0.09$. If we let $Z = (X - 56)/0.3$, then Z will have the standard normal distribution. Hence,

$$\Pr(55.7 < X < 56.3) = \Pr(-1 \le Z \le 1) = 2\Phi(1) - 1 = 0.6827.$$

11. If we let $Z = \sqrt{n}(\overline{X}_n - \mu)/2$, then Z will have the standard normal distribution. Therefore,

$$\Pr(|\overline{X}_n - \mu| < 0.1) = \Pr(|Z| < 0.05\sqrt{n}) = 2\Phi(0.05\sqrt{n}) - 1.$$

This value will be at least 0.9 if $2\Phi(0.05\sqrt{n}) - 1 \ge 0.9$ or $\Phi(0.05\sqrt{n}) \ge 0.95$. It is found from a table of the values of Φ that we must therefore have $0.05\sqrt{n} \ge 1.645$. The smallest integer n which satisfies this inequality is $n = 1083$.

13. Let X denote the diameter of the bolt and let Y denote the diameter of the nut. The $Y - X$ will have the normal distribution for which

$$E(Y - X) = 2.02 - 2 = 0.02$$

and

$$\text{Var}(Y - X) = 0.0016 + 0.0009 = 0.0025.$$

If we let $Z = (Y - X - 0.02)/0.05$, then Z will have the standard normal distribution. Therefore,

$$\Pr(0 < Y - X \le 0.05) = \Pr(-0.4 < Z \le 0.6) = \Phi(0.6) - [1 - \Phi(0.4)] = 0.3812.$$

15. Let $f_1(x)$ denote the p.d.f. of X if the person has glaucoma and let $f_2(x)$ denote the p.d.f. of X if the person does not have glaucoma. Furthermore, let A_1 denote the event that the person has glaucoma and let $A_2 = A_1^C$ denote the event that the person does not have glaucoma. Then

$$\begin{aligned} \Pr(A_1) &= 0.1, \quad \Pr(A_2) = 0.9, \\ f_1(x) &= \frac{1}{(2\pi)^{1/2}} \exp\left\{-\frac{1}{2}(x-25)^2\right\} \quad \text{for} \quad -\infty < x < \infty, \\ f_2(x) &= \frac{1}{(2\pi)^{1/2}} \exp\left\{-\frac{1}{2}(x-20)^2\right\} \quad \text{for} \quad -\infty < x < \infty. \end{aligned}$$

(a) $$\Pr(A_1 \mid X = x) = \frac{\Pr(A_1)f_1(x)}{\Pr(A_1)f_1(x) + \Pr(A_2)f_2(x)}$$

(b) The value found in part (a) will be greater than 1/2 if and only if

$$\Pr(A_1)f_1(x) > \Pr(A_2)f_2(x).$$

All of the following inequalities are equivalent to this one:
(i) $\exp\{-(x-25)^2/2\} > 9\exp\{-(x-20)^2/2\}$
(ii) $-(x-25)^2/2 > \log 9 - (x-20)^2/2$
(iii) $(x-20)^2 - (x-25)^2 > 2\log 9$
(iv) $10x - 225 > 2\log 9$
(v) $x > 22.5 + \log(9)/5$.

17. If $Y = \log X$, then the p.d.f. of Y is

$$g(y) = \frac{1}{(2\pi)^{1/2}\sigma} \exp\left\{-\frac{1}{2\sigma^2}(y-\mu)^2\right\} \quad \text{for} \quad -\infty < y < \infty.$$

Since $\dfrac{dy}{dx} = \dfrac{1}{x}$, it now follows that the p.d.f. of X, for $x > 0$, is $f(x) = g(\log x)/x$.

19. The conditional p.d.f. of X given μ is

$$g_1(x|\mu) = \frac{1}{(2\pi)^{1/2}} \exp(-(x-\mu)^2/2),$$

while the marginal p.d.f. of μ is $f_2(\mu) = 0.1$ for $5 \le \mu \le 15$. We need the marginal p.d.f. of X, which we get by integrating μ out of the joint p.d.f.

$$g_1(x|\mu)f_2(\mu) = \frac{0.1}{(2\pi)^{1/2}} \exp(-(x-\mu)^2/2), \text{ for } 5 \le \mu \le 15.$$

The integral is

$$f_1(x) = \int_5^{15} \frac{0.1}{(2\pi)^{1/2}} \exp(-(x-\mu)^2/2)d\mu = 0.1[\Phi(15-x) - \Phi(5-x)].$$

With $x = 8$, the value is $0.1[\Phi(7) - \Phi(-3)] = 0.0999$. This makes the conditional p.d.f. of μ given $X = 8$

$$g_2(\mu|8) = \frac{1.0013}{(2\pi)^{1/2}} \exp(-(8-\mu)^2/2), \text{ for } 5 \le \mu \le 15.$$

21. Note that $\log(XY) = \log(X) + \log(Y)$. Since X and Y are independent with normal distributions, we have that $\log(XY)$ has the normal distribution with the sum of the means (4.6) and sum of the variances (10.5). This means that XY has the lognormal distribution with parameters 4.6 and 10.5.

23. Since $\log(3X^{1/2}) = \log(3) + \log(X)/2$, we know that $\log(3X^{1/2})$ has the normal distribution with mean $\log(3) + 4.1/2 = 3.149$ and variance $8/4 = 2$. This means that $3X^{1/2}$ has the lognormal distribution with parameters 3.149 and 2.

25. Divide the time interval of u years into n intervals of length u/n each. At the end of n such intervals, the principal gets multiplied by $(1 + ru/n)^n$. The limit of this as $n \to \infty$ is $\exp(ru)$.

5.7 The Gamma Distributions

1. Let $f(x)$ denote the p.d.f. of X and let $Y = cX$. Then $X = Y/c$. Since $dx = dy/c$, then for $x > 0$,

$$g(y) = \frac{1}{c}f\left(\frac{y}{c}\right) = \frac{1}{c}\frac{\beta^\alpha}{\Gamma(\alpha)}\left(\frac{y}{c}\right)^{\alpha-1}\exp(-\beta(y/c)) = \frac{(\beta/c)^\alpha}{\Gamma(\alpha)}y^{\alpha-i}\exp(-(\beta/c)y).$$

3. The three p.d.f.'s are in Fig. S.5.1.

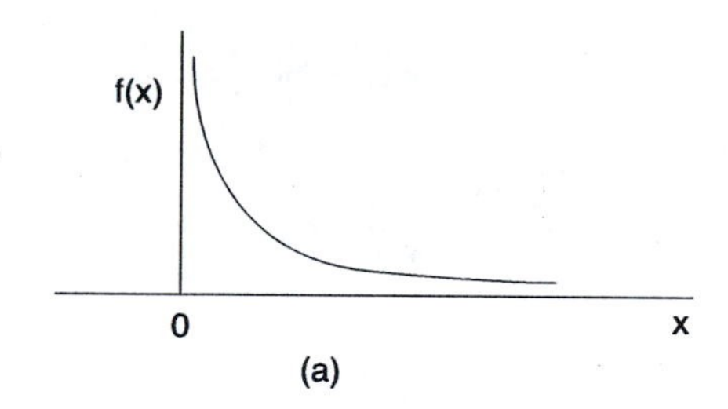

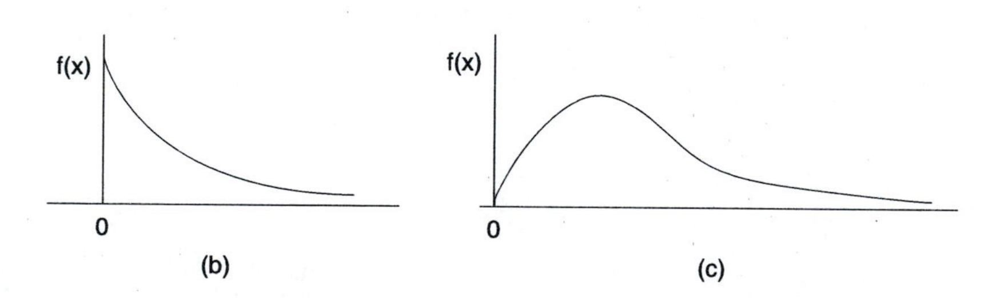

Figure S.5.1: Figure for Exercise 3 of Sec. 5.7.

5. All three p.d.f.'s are in Fig. S.5.2.

7. Let $A_i = \{X_i > t\}$ for $i = 1, 2, 3$. The event that at least one X_i is greater than t is $\bigcup_{i=1}^3 A_i$. We could use the formula in Theorem 1.10.1, or we could use that $\Pr(\bigcup_{i=1}^3 A_i) = 1 - \Pr(\bigcap_{i=1}^3 A_i^c)$. The latter is easier because the X_i are mutually independent and identically distributed.

$$\Pr\left(\bigcap_{i=1}^3 A_i^c\right) = \Pr(A_1^c)^3 = [1 - \exp(-\beta t)]^3.$$

So, the probability we want is $1 - [1 - \exp(-\beta t)]^3$.

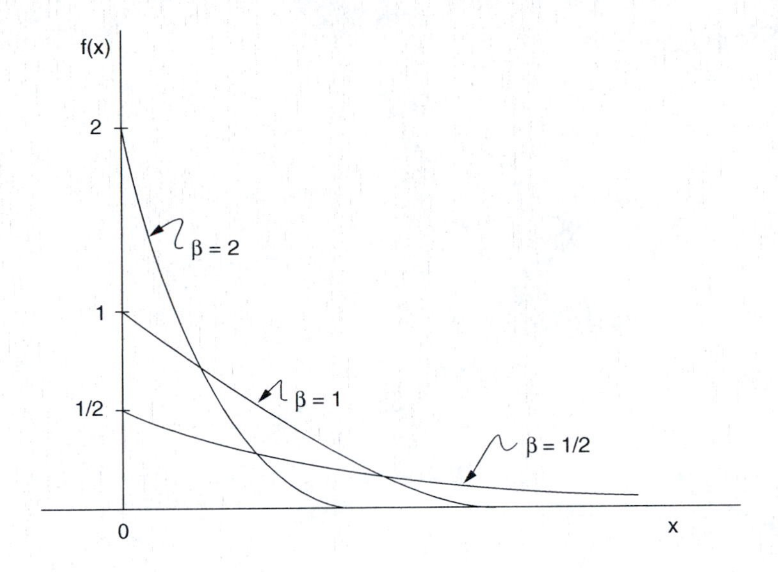

Figure S.5.2: Figure for Exercise 5 of Sec. 5.7.

9. Let Y denote the length of life of the system. Then by Exercise 8, Y has the exponential distribution with parameter $0.001 + 0.003 + 0.006 = 0.01$. Therefore,

$$\Pr(Y > 100) = \exp(-100\ (0.01)) = \frac{1}{e}.$$

11. The length of time Y_1 until one component fails has the exponential distribution with parameter $n\beta$. Therefore, $E(Y_1) = 1/(n\beta)$. The additional length of time Y_2 until a second component fails has the exponential distribution with parameter $(n-1)\beta$. Therefore, $E(Y_2) = 1/[(n-1)\beta]$. Similarly, $E(Y_3) = 1/[(n-2)\beta]$. The total time until three components fail is $Y_1+Y_2+Y_3$ and $E(Y_1+Y_2+Y_3) = \left(\frac{1}{n} + \frac{1}{n-1} + \frac{1}{n-2}\right)\frac{1}{\beta}$.

13. The time Y_1 until one of the students completes the examination has the exponential distribution with parameter $5\beta = 5/80 = 1/16$. Therefore,

$$\Pr(Y_1 < 40) = 1 - \exp(-40/16) = 1 - \exp(-5/2) = 0.9179.$$

15. No matter when the first student completes the examination, the second student to complete the examination will do so at least 10 minutes later than the first student if $Y_2 > 10$. Similarly, the third student to complete the examination will do so at least 10 minutes later than the second student if $Y_3 > 10$. Furthermore, the variables Y_1, Y_2, Y_3, Y_4, and Y_5 are independent. Therefore, the probability that no two students will complete the examination within 10 minutes of each other is

$$\begin{aligned}
\Pr(Y_2 > 10, \dots, Y_5 > 10) &= \Pr(Y_2 > 10)\dots\Pr(Y_5 > 10) \\
&= \exp(-(10)4\beta)\exp(-(10)3\beta)\exp(-(10)2\beta)\exp(-10\beta) \\
&= \exp(-40/80)\exp(-30/80)\exp(-20/80)\exp(-10/80) \\
&= \exp(-5/4) = 0.2865.
\end{aligned}$$

17.

$$\begin{aligned}E[(X-\mu)^{2n}] &= \int_{-\infty}^{\infty}(x-\mu)^{2n}\frac{1}{(2\pi)^{1/2}\sigma}\exp\left\{-\frac{(x-\mu)^2}{2\sigma^2}\right\}dx\\ &= \frac{2}{(2\pi)^{1/2}\sigma}\int_{\mu}^{\infty}(x-\mu)^{2n}\exp\left\{-\frac{(x-\mu)^2}{2\sigma^2}\right\}dx.\end{aligned}$$

Let $y=(x-\mu)^2$. Then $dx = dy/(2y^{1/2})$ and the above integral can be rewritten as

$$\frac{2}{(2\pi)^{1/2}\sigma}\int_0^{\infty} y^n\exp\left\{-\frac{y}{2\sigma^2}\right\}\frac{1}{2y^{1/2}}dy = \frac{1}{(2\pi)^{1/2}\sigma}\int_0^{\infty} y^{n-1/2}\exp\left\{-\frac{y}{2\sigma^2}\right\}dy.$$

The integrand in this integral is the p.d.f. of a gamma distribution with parameters $\alpha = n+1/2$ and $\beta = 1/(2\sigma^2)$, except for the constant factor

$$\frac{\beta^\alpha}{\Gamma(\alpha)} = \frac{1}{(2\sigma^2)^{n+1/2}\Gamma(n+1/2)}.$$

Since the integral of the p.d.f. of the gamma distribution must be equal to 1, it follows that

$$\int_0^{\infty} y^{n-1/2}\exp\left\{-\frac{y}{2\sigma^2}\right\}dy = (2\sigma^2)^{n+1/2}\Gamma(n+1/2).$$

From Eqs. (5.7.6) and (5.7.9), $\Gamma\left(n+\frac{1}{2}\right) = \left(n-\frac{1}{2}\right)\left(n-\frac{3}{2}\right)\cdots\left(\frac{1}{2}\right)\pi^{1/2}$. Therefore,

$$\begin{aligned}E[(X-\mu)^{2n}] &= \frac{1}{(2\pi)^{1/2}\sigma}(2\sigma^2)^{n+1/2}\left(n-\frac{1}{2}\right)\left(n-\frac{3}{2}\right)\cdots\left(\frac{1}{2}\right)\pi^{1/2}\\ &= 2^n\left(n-\frac{1}{2}\right)\left(n-\frac{3}{2}\right)\cdots\left(\frac{1}{2}\right)\sigma^{2n}\\ &= (2n-1)(2n-3)\dots(1)\sigma^{2n}.\end{aligned}$$

19. Let $Y = X^b$. Then $X = Y^{1/b}$ and $dx = \frac{1}{b}y^{(1-b)/b}dy$. Therefore, for $y>0$,

$$g(y) = f(y^{1/b}|a,b)\frac{1}{b}y^{(1-b)/b} = \frac{1}{a^b}\exp(-y/a^b).$$

21. (a) The mean of $1/X$ is

$$\int_0^{\infty}\frac{1}{x}\frac{\beta^\alpha}{\Gamma(\alpha)}x^{\alpha-1}\exp(-\beta x)dx = \frac{\beta^\alpha}{\Gamma(\alpha)}\int_0^{\infty}x^{\alpha-2}\exp(-\beta x)dx = \frac{\beta^\alpha}{\Gamma(\alpha)}\frac{\Gamma(\alpha-1)}{\beta^{\alpha-1}} = \frac{\beta}{\alpha-1}.$$

(b) The mean of $1/X^2$ is

$$\begin{aligned}\int_0^{\infty}\frac{1}{x^2}\frac{\beta^\alpha}{\Gamma(\alpha)}x^{\alpha-1}\exp(-\beta x)dx &= \frac{\beta^\alpha}{\Gamma(\alpha)}\int_0^{\infty}x^{\alpha-3}\exp(-\beta x)dx = \frac{\beta^\alpha}{\Gamma(\alpha)}\frac{\Gamma(\alpha-2)}{\beta^{\alpha-2}}\\ &= \frac{\beta^2}{(\alpha-1)(\alpha-2)}.\end{aligned}$$

This makes the variance of $1/X$ equal to

$$\frac{\beta^2}{(\alpha-1)(\alpha-2)} - \left(\frac{\beta}{\alpha-1}\right) = \frac{\beta^2}{(\alpha-1)^2(\alpha-2)}.$$

23. The memoryless property means that $\Pr(X > t + h | X > t) = \Pr(X > h)$.

(a) In terms of the c.d.f. the memoryless property means

$$\frac{1 - F(t+h)}{1 - F(t)} = 1 - F(h).$$

(b) From (a) we obtain $[1 - F(h)][1 - F(t)] = [1 - F(t+h)]$. Taking logarithms of both sides yields $\ell(h) + \ell(t) = \ell(t+h)$.

(c) Apply the result in part (b) with h and t both replaced by t/m. We obtain $\ell(2t/m) = 2\ell(t/m)$. Repeat with t replaced by $2t/m$ and $h = t/m$. The result is $\ell(3t/m) = 3\ell(t/m)$. After $k - 1$ such applications, we obtain

$$\ell(kt/m) = k\ell(t/m). \tag{S.5.7}$$

In particular, when $k = m$, we get $\ell(t) = m\ell(t/m)$ or $\ell(t/m) = \ell(t)/m$. Substituting this into (S.5.7) we obtain $\ell(kt/m) = (k/m)\ell(t)$.

(d) Let $c > 0$ and let $c_1, c_2, \ldots$ be a sequence of rational numbers that converges to c. Since ℓ is a continuous function, $\ell(c_n t) \to \ell(ct)$. But $\ell(c_n t) = c_n \ell(t)$ by part (c) since c_n is rational. It follows that $c_n \ell(t) \to \ell(ct)$. But, we know that $c_n \ell(t) \to c\ell(t)$. So, $c\ell(t) = \ell(ct)$.

(e) Apply part (d) with $c = 1/t$ to obtain $\ell(t)/t = \ell(1)$, a constant.

(f) Let $\beta = \ell(1)$. According to part (e), $\ell(t) = \beta t$ for all $t > 0$. Then $\log[1 - F(x)] = \beta x$ for $x > 0$. Solving for $F(x)$ gives $F(x) = 1 - \exp(-\beta x)$, which is the c.d.f. the exponential distribution with parameter $\beta = \ell(1)$.

5.8 The Beta Distributions

1. The c.d.f. of the beta distribution with parameters $\alpha > 0$ and $\beta = 1$ is

$$F(x) = \begin{cases} 0 & \text{for } x \le 0, \\ x^\alpha & \text{for } 0 < x < 1, \\ 1 & \text{for } x \ge 1. \end{cases}$$

Setting this equal to p and solving for x yields $F^{-1}(p) = p^{1/\alpha}$.

3. The vertical scale is to be chosen in each part of Fig. S.5.3 so that the area under the curve is 1. The figure in (h) is the mirror image of the figure in (g) with respect to $x = 1/2$.

5.

$$\begin{aligned} E[X^r(1-X)^S] &= \int_0^1 x^r(1-x)^S \frac{\Gamma(\alpha+\beta)}{\Gamma(\alpha)\Gamma(\beta)} x^{\alpha-1}(1-x)^{\beta-1}dx \\ &= \frac{\Gamma(\alpha+\beta)}{\Gamma(\alpha)\Gamma(\beta)} \int_0^1 x^{\alpha+r-1}(1-x)^{\beta+s-1}dx \\ &= \frac{\Gamma(\alpha+\beta)}{\Gamma(\alpha)\Gamma(\beta)} \cdot \frac{\Gamma(\alpha+r)\Gamma(\beta+s)}{\Gamma(\alpha+\beta+r+s)} \\ &= \frac{\Gamma(\alpha+r)}{\Gamma(\alpha)} \cdot \frac{\Gamma(\beta+s)}{\Gamma(\beta)} \cdot \frac{\Gamma(\alpha+\beta)}{\Gamma(\alpha+\beta+r+s)} \\ &= \frac{[\alpha(\alpha+1)\cdots(\alpha+r-1)][\beta(\beta+1)\cdots(\beta+s-1)]}{(\alpha+\beta)(\alpha+\beta+1)\cdots(\alpha+\beta+r+s-1)}. \end{aligned}$$

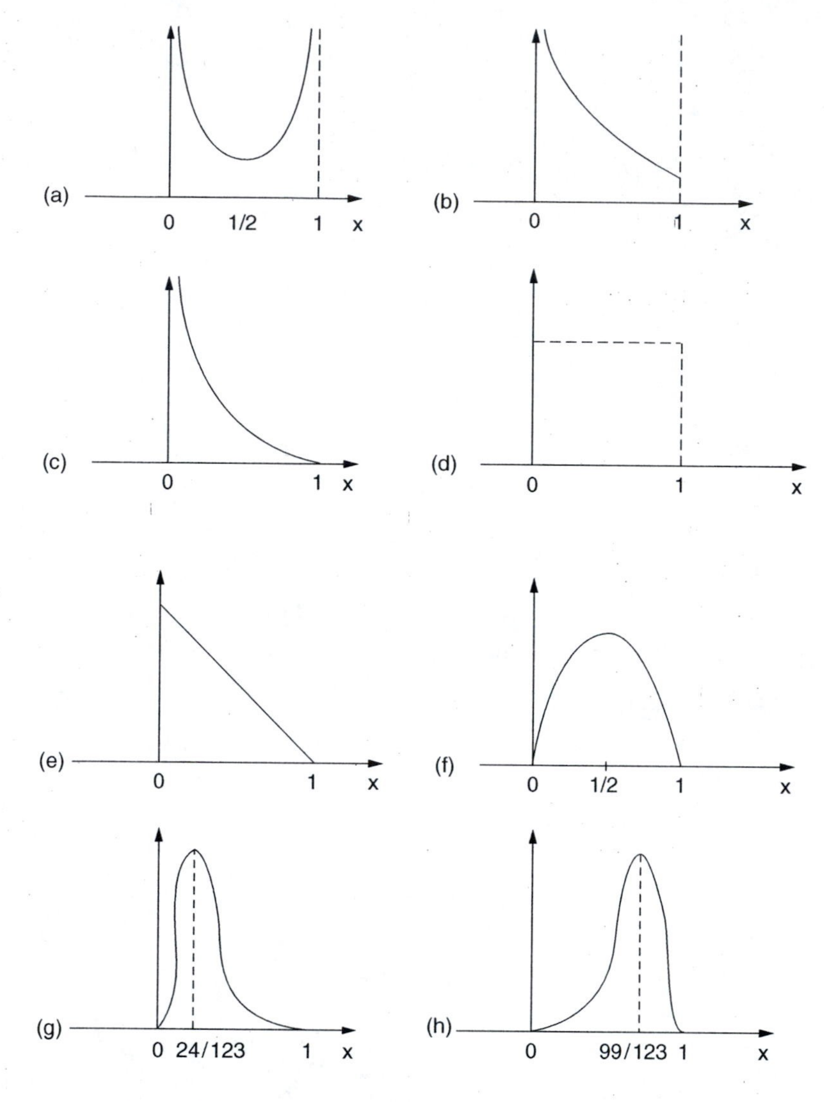

Figure S.5.3: Figure for Exercise 3 of Sec. 5.8.

7. Since X_1 and X_2 each have the gamma distribution with parameters $\alpha = 1$ and β, it follows from Exercise 6 that the distribution of $X_1/(X_1 + X_2)$ will be a beta distribution with parameters $\alpha = 1$ and $\beta = 1$. This beta distribution is the uniform distribution on the interval (0, 1).

9. Prior to observing the sample, the mean of P is $\alpha/(\alpha+\beta) = 0.05$, which means that $\alpha = \beta/19$. If we use the result in the note that follows Example 5.8.3, the distribution of P after finding 10 defectives in a sample of size 10 would be beta with parameters $\alpha+10$ and β, whose mean is $(\alpha+10)/(\alpha+\beta+10) = 0.9$. This means that $\alpha = 9\beta - 10$. So $9\beta - 10 = \beta/19$ and $\beta = 19/17$ so $\alpha = 1/17$. The distribution of P is then a beta distribution with parameters 1/17 and 19/17.

5.9 The Multinomial Distributions

1. Let $Y = X_1 + \cdots + X_\ell$. We shall show that Y has the binomial distribution with parameters n and $p_1 + \cdots + p_\ell$. Let $Z_1, \ldots, Z_n$ be i.i.d. random variables with the p.f.

$$f(z) = \begin{cases} p_i & \text{for } z = i,\ i = 1, \ldots, k, \\ 0 & \text{otherwise.} \end{cases}$$

For each $i = 1, \ldots, k$ and each $j = 1, \ldots, n$, define

$$\begin{aligned} A_{ij} &= \{Z_j = i\}, \\ W_{ij} &= \begin{cases} 1 & \text{if } A_{ij} \text{ occurs,} \\ 0 & \text{if not.} \end{cases} \end{aligned}$$

Finally, define $V_i = \sum_{j=1}^{n} W_{ij}$ for $i = 1, \ldots, k$. It follows from the discussion in the text that $(X_1, \ldots, X_k)$ has the same distribution as $(V_1, \ldots, V_k)$. Hence Y has the same distribution as $U = V_1 + \cdots + V_\ell$. But

$$U = V_1 + \cdots + V_\ell = \sum_{i=1}^{\ell}\sum_{j=1}^{n} W_{ij} = \sum_{j=1}^{n}\sum_{i=1}^{\ell} W_{ij}.$$

Define $U_j = \sum_{i=1}^{\ell} W_{ij}$. It is easy to see that $U_j = 1$ if $\cup_{i=1}^{\ell} A_{ij}$ occurs and $U_j = 0$ if not. Also $\Pr(\cup_{i=1}^{\ell} A_{ij}) = p_1 + \cdots + p_\ell$. Hence, $U_1, \ldots, U_n$ are i.i.d. random variables each having a Bernoulli distribution with parameter $p_1 + \cdots + p_\ell$. Since $U = \sum_{i=1}^{n} U_i$, we know that U has the binomial distribution with parameters n and $p_1 + \cdots + p_\ell$.

3. Let X_1 denote the number of times that the number 1 appears, let X_2 denote the number of times that the number 4 appears, and let X_3 denote the number of times that a number other than 1 or 4 appears. Then the vector (X_1, X_2, X_3) has the multinomial distribution with parameters $n = 5$ and $\boldsymbol{p} = (1/6, 1/6, 4/6)$. Therefore,

$$\begin{aligned} \Pr(X_1 = X_2) &= \Pr(X_1 = 0,\ X_2 = 0,\ X_3 = 5) + \Pr(X_1 = 1,\ X_2 = 1,\ X_3 = 3) \\ &\quad + \Pr(X_1 = 2,\ X_2 = 2,\ X_3 = 1) \\ &= \left(\frac{4}{6}\right)^5 + \frac{5!}{1!1!3!}\left(\frac{1}{6}\right)\left(\frac{1}{6}\right)\left(\frac{4}{6}\right)^3 + \frac{5!}{2!2!1!}\left(\frac{1}{6}\right)^2\left(\frac{1}{6}\right)^2\left(\frac{4}{6}\right) \\ &= \frac{1024}{6^5} + \frac{1280}{6^5} + \frac{120}{6^5} = \frac{2424}{6^5}. \end{aligned}$$

5. The number X of freshman or sophomores selected will have the binomial distribution with parameters $n = 15$ and $p = 0.16 + 0.14 = 0.30$. Therefore, it is found from the table in the back of the book that

$$\Pr(X \geq 8) = .0348 + .0116 + .0030 + .0006 + .0001 = .0501.$$

7. For any nonnegative integers $x_1, \ldots, x_k$ such that $\sum_{i=1}^{k} x_i = n$,

$$\Pr\left(X_1 = x_1, \ldots, X_k = x_k \,\middle|\, \sum_{i=1}^{k} X_i = n\right) = \frac{\Pr(X_1 = x_1, \ldots, X_k = x_k)}{\Pr\left(\sum_{i=1}^{k} X_i = n\right)}.$$

Since $X_1, \ldots, X_k$ are independent,

$$\Pr(X_1 = x_1, \ldots, X_k = x_k) = \Pr(X_1 = x_1) \ldots \Pr(X_k = x_k).$$

Since X_i has the Poisson distribution with mean λ_i,

$$\Pr(X_i = x_i) = \frac{\exp(-\lambda_i)\lambda_i^{x_i}}{x_i!}.$$

Also, by Theorem 5.4.4, the distribution of $\sum_{i=1}^k X_i$ will be a Poisson distribution with mean $\lambda = \sum_{i=1}^k \lambda_i$. Therefore,

$$\Pr\left(\sum_{i=1}^k X_i = n\right) = \frac{\exp(-\lambda)\lambda^n}{n!}.$$

It follows that

$$\Pr\left(X_1 = x_1, \ldots, X_k = x_k \middle| \sum_{i=1}^k X_i = n\right) = \frac{n!}{x_i! \ldots x_k!} \prod_{i=1}^k \left(\frac{\lambda_i}{\lambda}\right)^{x_i}$$

5.10 The Bivariate Normal Distributions

1. The conditional distribution of the height of the wife given that the height of the husband is 72 inches is a normal distribution with mean $66.8 + 0.68 \times 2(72-70)/2 = 68.16$ and variance $(1-0.68^2)2^2 = 2.1504$. The 0.95 quantile of this distribution is

$$68.16 + 2.1504^{1/2}\Phi^{-1}(0.95) = 68.16 + 1.4664 \times 1.645 = 70.57.$$

3. The sum $X_1 + X_2$ will have the normal distribution with mean $85 + 90 = 175$ and variance $(10)^2 + (16)^2 + 2(0.8)(10)(16) = 612$. Therefore, $Z = (X_1 + X_2 - 175)/24.7386$ will have the standard normal distribution. It follows that

$$\Pr(X_1 + X_2 > 200) = \Pr(Z > 1.0106) = 1 - \Phi(1.0106) = 0.1562.$$

5. The predicted value should be the mean of the conditional distribution of X_1 given that $X_2 = 100$. This value is $85 + (0.8)(10)\left(\frac{100-90}{16}\right) = 90$. The M.S.E. for this prediction is the variance of the conditional distribution, which is $(1-0.64)100 = 36$.

7. Since $E(X_1|X_2) = 3.7 - 0.15X_2$, it follows from Eq. (5.10.8) that

(i) $\mu_1 - \rho\frac{\sigma_1}{\sigma_2}\mu_2 = 3.7$,

(ii) $\rho\frac{\sigma_1}{\sigma_2} = -0.15$. Since $E(X_2|X_1) = 0.4 - 0.6X_1$, it follows from Eq. (5.10.6) that

(iii) $\mu_2 - \rho\frac{\sigma_2}{\sigma_1}\mu_1 = 0.4$,

(iv) $\rho\frac{\sigma_2}{\sigma_1} = -0.6$.

Finally, since $\text{Var}(X_2|X_1) = 3.64$, it follows that

(v) $(1-\rho^2)\sigma_2^2 = 3.64$.

By multiplying (ii) and (iv) we find that $\rho^2 = 0.09$. Therefore, $\rho = \pm 0.3$. Since the right side of (ii) is negative, ρ must be negative also. Hence, $\rho = -0.3$. It now follows from (v) that $\sigma_2^2 = 4$. Hence,

$\sigma_2 = 2$ and it is found from (ii) that $\sigma_1 = 1$. By using the values we have obtained, we can rewrite (i) and (iii) as follows:

(i) $\mu_1 + 0.15\mu_2 = 3.7$,

(iii) $0.6\mu_1 + \mu_2 = 0.4$.

By solving these two simultaneous linear equations, we find that $\mu_1 = 4$ and $\mu_2 = -2$.

9. Let a_1 and a_2 be as defined in Exercise 8. If $f(x_1, x_2) = k$, then $a_1^2 - 2\rho a_1 a_2 + a_2^2 = b^2$, where b is a particular positive constant. Suppose first that $\rho = 0$ and $\sigma_1 = \sigma_2 = \sigma$. Then this equation has the form

$$(x_1 - \mu_1)^2 + (x_2 - \mu_2)^2 = b^2\sigma^2.$$

This is the equation of a circle with center at (μ_1, μ_2) and radius $b\sigma$. Suppose next that $\rho = 0$ and $\sigma_1 \neq \sigma_2$. Then the equation has the form

$$\frac{(x_1 - \mu_1)^2}{\sigma_1^2} + \frac{(x_2 - \mu_2)^2}{\sigma_2^2} = b^2.$$

This is the equation of an ellipse for which the center is (μ_1, μ_2) and the major and minor axes are parallel to the x_1 and x_2 axes. Suppose finally that $\rho \neq 0$. It was shown in Exercise 8 that $a_1^2 - 2\rho a_1 a_2 + a_2^2 \geq 0$ for all values of a_1 and a_2. It therefore follows from the methods of analytic geometry and elementary calculus that the set of points which satisfy the equation

$$\frac{(x_1 - \mu_1)^2}{\sigma_1^2} - 2\rho\frac{(x_1 - \mu_1)}{\sigma_1} \cdot \frac{(x_2 - \mu_2)}{\sigma_2} + \frac{(x_2 - \mu_2)^2}{\sigma_2^2} = b^2.$$

will be an ellipse for which the center is (μ_1, μ_2) and for which the major and minor axes are rotated so that they are not parallel to the x_1 and x_2 axes.

11. By Exercise 10, the joint distribution of $X_1 + X_2$ and $X_1 - X_2$ is a bivariate normal distribution. By Exercise 9 of Sec. 4.6, these two variables are uncorrelated. Therefore, they are also independent.

13. The exponent of a bivariate normal p.d.f. can be expressed as $-[ax^2 + by^2 + cxy + ex + gy + h]$, where

$$\begin{aligned}
a &= \frac{1}{2\sigma_1^2(1-\rho^2)}, \\
b &= \frac{1}{2\sigma_2^2(1-\rho^2)}, \\
c &= -\frac{\rho}{\sigma_1\sigma_2(1-\rho^2)}, \\
e &= -\frac{\mu_1}{\sigma_1^2(1-\rho^2)} + \frac{\mu_2\rho}{\sigma_1\sigma_2(1-\rho^2)}, \\
g &= -\frac{\mu_2}{\sigma_2^2(1-\rho^2)} + \frac{\mu_1\rho}{\sigma_1\sigma_2(1-\rho^2)},
\end{aligned}$$

and h is irrelevant because $\exp(-h)$ just provides an additional constant factor that we are ignoring anyway. The only restrictions that the bivariate normal p.d.f. puts on the numbers a, b, c, e, and g are that $a, b > 0$ and whatever is equivalent to $|\rho| < 1$. It is easy to see that, so long as $a, b > 0$, we will have $|\rho| < 1$ if and only if $ab > (c/2)^2$. Hence, every set of numbers that satisfies these inequalities corresponds to a bivariate normal p.d.f. Assuming that these inequalities are satisfied, we can solve the above equations to find the parameters of the bivariate normal distribution.

$$\begin{aligned}
\rho &= -\frac{c/2}{(ab)^{1/2}}, \\
\sigma_1^2 &= \frac{1}{2a - c^2/[2b]}, \\
\sigma_2^2 &= \frac{1}{2b - c^2/[2a]}, \\
\mu_1 &= \frac{cg - 2be}{4ab - c^2}, \\
\mu_2 &= \frac{ce - 2ag}{4ab - c^2}.
\end{aligned}$$

15. (a) Let $Y = \sum_{j \neq i} X_j$. Since $X_1, \ldots, X_n$ are independent, we know that Y is independent of X_i. Since Y is the sum of independent normal random variables it has a normal distribution. The mean and variance of Y are easily seen to be $(n-1)\mu$ and $(n-1)\sigma^2$ respectively. Since Y and X_i are independent, all pairs of linear combinations of them have a bivariate normal distribution. Now write

$$\begin{aligned}
X_i &= 1X_i + 0Y, \\
\overline{X}_n &= \frac{1}{n}X_i + \frac{1}{n}Y.
\end{aligned}$$

Clearly, both X_i and Y have mean μ, and we already know that X_i has variance σ^2 while Y has variance σ^2/n. The correlation can be computed from the covariance of the two linear combinations.

$$\text{Cov}\left(1X_i + 0Y, \frac{1}{n}X_i + \frac{1}{n}Y\right) = \frac{1}{n}\sigma^2.$$

The correlation is then $(\sigma^2/n)/[\sigma^2\sigma^2/n]^{1/2} = 1/n^{1/2}$.

(b) The conditional distribution of X_i given $\overline{X}_n = \overline{x}_n$ is a normal distribution with mean equal to

$$\mu + \frac{1}{n^{1/2}}\frac{\sigma}{\sigma/n^{1/2}}(\overline{x}_n - \mu) = \overline{x}_n.$$

The conditional variance is

$$\sigma^2 - \left(1 - \frac{1}{n}\right).$$

5.11 Supplementary Exercises

1. Let $g_1(x|p)$ be the conditional p.f. of X given $P = p$, which is the binomial p.f. with parameters n and p. Let $f_2(p)$ be the marginal p.d.f. of P, which is beta p.d.f. with parameters 1 and 1, also known as the uniform p.d.f. on the interval $[0,1]$. According to the law of total probability for random variables, the marginal p.f. of X is

$$f_1(x) = \int g_1(x|p)f_2(p)dp = \int_0^1 \binom{n}{x} p^x(1-p)^{n-x}dp = \binom{n}{x}\frac{x!(n-x)!}{(n+1)!} = \frac{1}{n+1},$$

for $x = 0, \ldots, n$. In the above, we used Theorem 5.8.1 and the fact that $\Gamma(k+1) = k!$ for each integer k.

3. Since $\text{Var}(X) = E(X) = \lambda_1$ and $\text{Var}(Y) = E(Y) = \lambda_2$, it follows that $\lambda_1 + \lambda_2 = 5$. Hence, $X + Y$ has the Poisson distribution with mean 5 and

$$\Pr(X + Y < 2) = \Pr(X + Y = 0) + \Pr(X + Y = 1) = \exp(-5) + 5\exp(-5) = .0067 + .0337 = .0404.$$

5. The event $\{\overline{X} < 1/2\}$ can occur only if all four observations are 0, which has probability $(\exp(-\lambda))^4$, or three of the observations are 0 and the other is 1, which has probability $4(\lambda\exp(-\lambda))(\exp(-\lambda))^3$. Hence, the total probability is as given in this exercise.

7. It follows from Exercise 18 of Sec. 4.9 that

$$E[(X - \mu)^3] = E(X^3) - 3\mu\sigma^2 - \mu^3.$$

Because of the symmetry of the normal distribution with respect to μ, the left side of this relation is 0. Hence,

$$E(X^3) = 3\mu\sigma^2 + \mu^3.$$

9. The number of men that arrive during the one-minute period has the Poisson distribution with mean 2. The number of women is independent of the number of men and has the Poisson distribution with mean 1. Therefore, the total number of people X that arrive has the Poisson distribution with mean 3. From the table in the back of the book it is found that

$$\Pr(X \le 4) = .0498 + .1494 + .2240 + .2240 + .1680 = .8152.$$

11. The probability that at least one of the two children will be successful on a given Sunday is $(1/3) + (1/5) - (1/3)(1/5) = 7/15$. Therefore, from the geometric distribution, the expected number of Sundays until a successful launch is achieved is $15/7$.

13. By the Poisson approximation, the distribution of X is approximately Poisson with mean $120(1/36) = 10/3$. The probability that such a Poisson random variable equals 3 is $\exp(-10/3)(10/3)^3/3! = 0.2202$. (The actual binomial probability is 0.2229.)

15. (a) $\Pr(T_1 > t) = \Pr(X = 0)$, where X is the number of occurrences between time 0 and time t. Since X has the Poisson distribution with mean $5t$, it follows that $\Pr(T_1 > t) = \exp(-5t)$. Hence, T_1 has the exponential distribution with parameter $\beta = 5$.

    (b) T_k is the sum of k i.i.d. random variables, each of which has the exponential distribution given in part (a). Therefore, the distribution of T_k is a gamma distribution with parameters $\alpha = k$ and $\beta = 5$.

    (c) Let X_i denote the time following the ith occurrence until the $(i + 1)$st occurrence. Then the random variable $X_1, \ldots, X_{k-1}$ are i.i.d., each of which has the exponential distribution given in part (a). Since t is measured in hours in that distribution, the required probability is

$$\Pr\left(X_i > \frac{1}{3}, i = 1, \ldots, k - 1\right) = (\exp(-5/3))^{k-1}.$$

17. $\Pr(X_1 > kX_2) = \int_0^\infty \Pr(X_1 > kX_2 | X_2 = x)(f_2 x)dx = \int_0^\infty \exp(-\beta_1 kx)\beta_2\exp(-\beta_2 x)dx = \dfrac{\beta_2}{k\beta_1 + \beta_2}.$

19. It follows from Eq. (5.3.8) that

$$\text{Var}(\overline{X}) = \frac{1}{n^2}\,\text{Var}(X) = \frac{p(1-p)}{n} \cdot \frac{T-n}{T-1},$$

where T is the population size, p is the proportion of persons in the population who have the characteristic, and $n = 100$. Since $p(1-p) \le 1/4$ for $0 \le p \le 1$ and $(T-n)(T-1) \le 1$ for all values of T, it follows that

$$\text{Var}(\overline{X}_n) \le \frac{1}{400}.$$

Hence, the standard deviation is $\le 1/20 = .05$.

21. Consider the event that there are at least k occurrences between time 0 and time t. The number X of occurrences in this interval has the specified Poisson distribution, so the left side represents the probability of this event. But the event also means that the total waiting time Y until the kth event occurs is $\le t$. It follows from part (b) of Exercise 15 that Y has the specified gamma distribution. Hence, the right side also expresses the probability of this same event.

23. (a) It follows from Theorem 5.9.2 that

$$\rho(X_i, X_j) = \frac{\text{Cov}(X_i, X_j)}{[\text{Var}(X_i)\,\text{Var}(X_j)]^{1/2}} = \frac{-10p_ip_j}{10[p_i(1-p_i)p_j(1-p_j)]^{1/2}} = -\left(\frac{p_i}{1-p_i} \cdot \frac{p_j}{1-p_j}\right)^{1/2}$$

(b) $\rho(X_i, X_j)$ is most negative when p_i and p_j have their largest values; i.e., for $i = 1$ ($p_i = .4$) and $j = 2$ ($p_2 = .3$).

(c) $\rho(X_i, X_j)$ is closest to 0 when p_i and p_j have their smallest values; i.e., for $i = 3$ ($p_3 = .2$) and $j = 4$ ($p_4 = .1$).

25. Since X has a normal distribution and the conditional distribution of Y given X is also normal with a mean that is a linear function of X and constant variance, it follows that X and Y jointly have a bivariate normal distribution. Hence, Y has a normal distribution. From Eq. (5.10.6), $2X - 3 = \mu_2 + \rho\sigma_2 X$. Hence, $\mu_2 = -3$ and $\rho\sigma_2 = 2$. Also, $(1-\rho^2)\sigma_2^2 = 12$. Therefore $\sigma_2^2 = 16$ and $\rho = 1/2$. Thus, Y has the normal distribution with mean -3 and variance 16, and $\rho(X, Y) = 1/2$.

Chapter 6

Large Random Samples

6.1 Introduction

1. The p.d.f. of $Y = X_1 + X_2$ is

$$g(y) = \begin{cases} y & \text{if } 0 < y \le 1, \\ 2 - y & \text{if } 1 < y < 2, \\ 0 & \text{otherwise.} \end{cases}$$

It follows easily from the fact that $\overline{X}_2 = Y/2$ that the p.d.f. of $\overline{X}_2$ is

$$h(x) = \begin{cases} 4x & \text{if } 0 < x \le 1/2, \\ 4 - 4x & \text{if } 1/2 < x < 1, \\ 0 & \text{otherwise.} \end{cases}$$

We easily compute

$$\begin{aligned} \Pr(|X_1 - 0.5| < 0.1) &= 0.6 - 0.4 = 0.2, \\ \Pr(|\overline{X}_2 - 0.5| < 0.1) &= \int_{0.4}^{0.5} 4x dx + \int_{0.5}^{0.6} (4 - 4x) dx \\ &= 2(0.5^2 - 0.4^2) + 4(0.6 - 0.5) - 2(0.6^2 - 0.5^2) = 0.36. \end{aligned}$$

The reason that $\overline{X}_2$ has higher probability of being close to 0.5 is that its p.d.f. is much higher near 0.5 than is the uniform p.d.f. of X_1 (twice as high right at 0.5).

3. To do this by hand we would have to add all of the binomial probabilities corresponding to $W = 80, \ldots, 120$. Most statistical software will do this calculation automatically. The result is 0.9964. It looks like the probability is increasing to 1.

6.2 The Law of Large Numbers

1. Let $\epsilon > 0$. We need to show that

$$\lim_{n\to\infty} \Pr(|X_n - 0| \ge \epsilon) = 0. \tag{S.6.1}$$

Since $X_n \ge 0$, we have $|X_n - 0| \ge \epsilon$ if and only if $X_n \ge \epsilon$. By the Markov inequality $\Pr(X_n \ge \epsilon) \le \mu_n/\epsilon$. Since $\lim_{n\to\infty} \mu_n = 0$, Eq. (S.6.1) holds.

3. By the Chebyshev inequality,

$$\text{Var}(X) \geq 9\Pr(|X - \mu| \geq 3) = 9\Pr(X \leq 7 \text{ or } X \geq 13) = 9(0.2 + 0.3) = \frac{9}{2}.$$

5. By the Chebyshev inequality,

$$\Pr(|X_n - \mu| \leq 2\sigma) \geq 1 - \frac{1}{4n}.$$

Therefore, we must have $1 - \frac{1}{4n} \geq 0.99$. or $n \geq 25$.

7. By the Markov inequality,

$$\Pr(|X - \mu| \geq t) = \Pr(|X - \mu|^4 \geq t^4) \leq \frac{E(|X - \mu|^4)}{t^4} = \frac{\beta_4}{t^4}.$$

9. $E(Z_n) = n^2 \cdot \frac{1}{n} + 0\left(1 - \frac{1}{n}\right) = n$. Hence, $\lim_{n\to\infty} E(Z_n) = \infty$. Also, for any given $\epsilon > 0$,

$$\Pr(|Z_n| < \epsilon) = \Pr(Z_n = 0) = 1 - \frac{1}{n}.$$

Hence, $\lim_{n\to\infty} \Pr(|Z_n| < \epsilon) = 1$, which means that $Z_n \xrightarrow{p} 0$.

11. Suppose that the sequence $Z_1, Z_2, \ldots$ converges to b in the quadratic mean. Since

$$|Z_n - b| \leq |Z_n - E(Z_n)| + |E(Z_n) - b|\ ,$$

then for any value of $\epsilon > 0$,

$$\begin{aligned} \Pr(|Z_n - b| < \epsilon) &\geq \Pr(|Z_n - E(Z_n)| + |E(Z_n) - b| < \epsilon) \\ &= \Pr(|Z_n - E(Z_n)| < \epsilon - |E(Z_n) - b|). \end{aligned}$$

By Exercise 10, we know that $\lim_{n\to\infty} E(Z_n) = b$. Therefore, for sufficiently large values of n, it will be true that $\epsilon - |E(Z_n) - b| > 0$. Hence, by the Chebyshev inequality, the final probability will be at least as large as

$$1 - \frac{\text{Var}(Z_n)}{[\epsilon - |E(Z_n) - b|]^2}\ .$$

Again, by Exercise 10,

$$\lim_{n\to\infty} \text{Var}(Z_n) = 0 \text{ and } \lim_{n\to\infty} [\epsilon - |E(Z_n) - b|]^2 = \epsilon^2.$$

Therefore,

$$\lim_{n\to\infty} \Pr(|Z_n - b| < \epsilon) \geq \lim_{n\to\infty} \left\{1 - \frac{\text{Var}(Z_n)}{[\epsilon - |E(Z_n) - b|]^2}\right\} = 1,$$

which means that $Z_n \xrightarrow{p} b$.

13. (a) For any value of n large enough so that $1/n < \epsilon$, we have

$$\Pr(|Z_n| < \epsilon) = \Pr\left(Z_n = \frac{1}{n}\right) = 1 - \frac{1}{n^2}.$$

Therefore, $\lim_{n\to\infty} \Pr(|Z_n| < \epsilon) = 1$, which means that $Z_n \xrightarrow{p} 0$.

(b) $E(Z_n) = \frac{1}{n}\left(1 - \frac{1}{n^2}\right) + n\left(\frac{1}{n^2}\right) = \frac{2}{n} - \frac{1}{n^3}$. Therefore, $\lim_{n\to\infty} E(Z_n) = 0$. It follows from Exercise 10 that the only possible value for the constant c is $c = 0$, and there will be convergence to this value if and only if $\lim_{n\to\infty} \text{Var}(Z_n) = 0$. But

$$E(Z_n^2) = \frac{1}{n^2}\left(1 - \frac{1}{n^2}\right) + n^2 \cdot \frac{1}{n^2} = 1 + \frac{1}{n^2} - \frac{1}{n^4}.$$

Hence, $\text{Var}(Z_n) = 1 + \frac{1}{n^2} - \frac{1}{n^4} - \left(\frac{2}{n} - \frac{1}{n^3}\right)^2$ and $\lim_{n\to\infty} \text{Var}(Z_n) = 1$.

15. We need to prove that, for every $\epsilon > 0$,

$$\lim_{n\to\infty} \Pr(|g(Z_n) - g(b)| < \epsilon) = 1.$$

Let $\epsilon > 0$. Since g is continuous at b, there exists δ such that $|z - b| < \delta$ implies that $|g(z) - g(b)| < \epsilon$. Also, since $Z_n \xrightarrow{P} b$, we know that

$$\lim_{n\to\infty} \Pr(|Z_n - b| < \delta) = 1.$$

But $\{|Z_n - b| < \delta\} \subset \{|g(Z_n) - g(b)| < \epsilon\}$. So

$$\Pr(|g(Z_n) - g(b)| < \epsilon) \geq \Pr(|Z_n - b| < \delta) \tag{S.6.2}$$

Since the right side of (S.6.2) goes to 1 as $n \to \infty$ so does the left side.

17. (a) The mean of X is np, and the mean of Y is np/k. Since $Z = kY$, the mean of Z is $knp/k = np$.

(b) The variance of X is $np(1-p)$, and the variance of Y is $n(p/k)(1-p/k)$. So, the variance $Z = kY$ is k^2 times the variance of Y, i.e.,

$$\text{Var}(Z) = k^2 n(p/k)(1 - p/k) = knp(1 - p/k).$$

If p is small, then both $1 - p$ and $1 - p/k$ will be close to 1, and $\text{Var}(Z)$ is approximately knp while the variance of X is approximately np.

(c) In Fig. 6.1, each bar has height equal to 0.01 times a binomial random variable with parameters 100 and the probability that X_1 is in the interval under the bar. In Fig. 6.2, each bar has height equal to 0.02 times a binomial random variable with parameters 100 and probability that X_1 is in the interval under the bar. The bars in Fig. 6.2 have approximately one-half of the probability of the bars in Fig. 6.1, but their heights have been multiplied by 2. By part (b), we expect the heights in Fig. 6.2 to have approximately twice the variance of the heights in Fig. 6.1.

19. (a) First, insert s from (6.2.15) into the expression in (6.2.14). We get

$$n\left[\log(p) + \left(\frac{1-p}{p} + u\right)\log\left\{\frac{(1+u)p+1-p}{up+1-p}(1-p)\right\} - \log\left\{1 - \frac{1-p}{\frac{(1+u)p+1-p}{up+1-p}(1-p)}\right\}\right].$$

The last term can be rewritten as

$$-\log\left\{1-\frac{up+1-p}{(1+u)p+1-p}\right\}=-\log(p)+\log\{(1+u)p+1-p\}.$$

The result is then

$$n\left[\left(\frac{1-p}{p}+u\right)\log\left\{\frac{(1+u)p+1-p}{up+1-p}(1-p)\right\}+\log\{(1+u)p+1-p\}\right].$$

This is easily recognized as n times the logarithm of (6.2.16).

(b) For all u, q is given by (6.2.16). For $u=0$, $q=(1-p)^{(1-p)/p}$. Since $0<1-p<1$ and $(1-p)/p>0$, we have $0<q<1$ when $u=0$. For general u, let $x=p(1+u)+1-p$ and rewrite

$$\log(q)=\log(p+x)+\frac{p+x}{p}\log\frac{(1-p)(p+x)}{x}.$$

Since x is a linear increasing function of u, if we show that $\log(q)$ is decreasing in x, then q is decreasing in u. The derivative of $\log(q)$ with respect to x is

$$-\frac{p}{x(p+x)}+\frac{1}{p}\log\frac{(1-p)(p+x)}{x}.$$

The first term is negative, and the second term is negative at $u=0$ $(x=1)$. To be sure that the sum is always negative, examine the second term more closely. The derivative of the second term is

$$\frac{1}{p}\left(\frac{1}{p+x}-\frac{1}{x}\right)=\frac{-1}{x(p+x)}<0.$$

Hence, the derivative is always negative, and q is less than 1 for all u.

21. (a) The m.g.f. of the exponential distribution with parameter 1 is $1/(1-s)$ for $s<1$, hence the m.g.f. of Y_n is $1/(1-s)^n$ for $t<1$. The Chernoff bound is the minimum (over $s>0$) of $e^{-nus}/(1-s)^n$. The logarithm of this is $-n[us+\log(1-s)]$, which is minimized at $s=(u-1)/u$, which is positive if and only if $u>1$. The Chernoff bound is $[u\exp(1-u)]^n$.

    (b) If $u<1$, then the expression in Theorem 6.2.7 is minimized over $s>0$ near $s=0$, which provides a useless bound of 1 for $\Pr(Y_n>nu)$.

23. Each Z_n has the Bernoulli distribution with parameter $1/k_n$, hence $E[(Z_n-0)^2]=1/k_n$, which goes to 0.

6.3 The Central Limit Theorem

1. The length of rope produced in one hour X has a mean of $60\times 4=240$ feet and a standard deviation of $60^{1/2}\times 5=38.73$ inches, which is 3.23 feet. The probability that $X\geq 250$ is approximately the probability that a normal random variable with mean 240 and standard deviation 3.23 is at least 250, namely $1-\Phi([250-240]/3.23)=1-\Phi(3.1)=0.001$.

3. Since the variance of a Poisson distribution is equal to the mean, the number of defects on any bolt has mean 5 and variance 5. Therefore, the distribution of the average number $\overline{X}_n$ on the 125 bolts will be approximately the normal distribution with mean 5 and variance $5/125=1/25$. If we let $Z=(\overline{X}_n-5)/(1/5)$, then the distribution of Z will be approximately a standard normal distribution. Hence,

$$\Pr(\overline{X}_n<5.5)=\Pr(Z<2.5)\simeq\Phi(2.5)=0.9938.$$

5. The distribution of the proportion $\overline{X}_n$ of defective items in the sample will be approximately the normal distribution with mean 0.1 and variance $(0.1)(0.9)/n = 0.09/n$. Therefore, the distribution of $Z = \sqrt{n}(\overline{X}_n - 0.1)/0.3$ will be approximately the standard normal distribution. It follows that

$$\Pr(\overline{X}_n < 0.13) = \Pr(Z < 0.1\sqrt{n}) \simeq \Phi(0.1\sqrt{n}).$$

For this value to be at least 0.99, we must have $0.1\sqrt{n} \geq 2.327$ or, equivalently, $n \geq 541.5$. Hence, the smallest possible value of n is 542.

7. The mean of a random digit X is

$$E(X) = \frac{1}{10}(0 + 1 + \cdots + 9) = 4.5.$$

Also,

$$E(X^2) = \frac{1}{10}(0^2 + 1^2 + \cdots + 9^2) = \frac{1}{10} \cdot \frac{(9)(10)(19)}{6} = 28.5.$$

Therefore, $\text{Var}(X) = 28.5 - (4.5)^2 = 8.25$. The distribution of the average $\overline{X}_n$ of 16 random digits will therefore be approximately the normal distribution with mean 4.5 and variance $8.25/16 = 0.5156$. Hence, the distribution of

$$Z = \frac{\overline{X}_n - 4.5}{\sqrt{0.5156}} = \frac{\overline{X}_n - 4.5}{0.7181}$$

will be approximately a standard normal distribution. It follows that

$$\begin{aligned}\Pr(4 \leq \overline{X}_n \leq 6) &= \Pr(-0.6963 \leq Z \leq 2.0888)\\ &= \Phi(2.0888) - [1 - \Phi(0.6963)]\\ &= 0.9816 - 0.2431 = 0.7385.\end{aligned}$$

9. (a) By Eq. (6.2.4),

$$\Pr\left(|\overline{X}_n - \mu| \geq \frac{\sigma}{4}\right) \leq \frac{\sigma^2}{n} \cdot \frac{16}{\sigma^2} = \frac{16}{25}.$$

Therefore, $\Pr\left(|\overline{X}_n - \mu| \leq \frac{\sigma}{4}\right) \geq 1 - \frac{16}{25} = 0.36.$

(b) The distribution of

$$Z = \frac{\overline{X}_n - \mu}{\sigma/\sqrt{n}} = \frac{5}{\sigma}(\overline{X}_n - \mu)$$

will be approximately a standard normal distribution. Therefore,

$$\Pr\left(|\overline{X}_n - \mu| \leq \frac{\sigma}{4}\right) = \Pr\left(|Z| \leq \frac{5}{4}\right) \simeq 2\Phi(1.25) - 1 = 0.7887.$$

11. For a student chosen at random, the number of parents X who will attend the graduation ceremony has mean $\mu = 0/3 + 1/3 + 2/3 = 1$ and variance $\sigma^2 = E[(X - \mu)^2] = (0 - 1)^2/3 + (1 - 1)^2/3 + (2 - 1)^2/3 = 2/3$. Therefore, the distribution of the total number of parents W who attend the ceremony will be approximately the normal distribution with mean $(600)(1) = 600$ and variance $600(2/3) = 400$. Therefore, the distribution of $Z = (W - 600)(20)$ will be approximately a standard normal distribution. It follows that

$$\Pr(W \leq 650) = \Pr(Z \leq 2.5) = \Phi(2.5) = 0.9938.$$

13. We are asking for the asymptotic distribution of $g(\overline{X}_n)$, where $g(x) = x^3$. The distribution of $\overline{X}_n$ is normal with mean θ and variance σ^2/n. According to the delta method, the asymptotic distribution of $g(\overline{X}_n)$ should be the normal distribution with mean $g(\theta) = \theta^3$ and variance $(\sigma^2/n)[g'(\theta)]^2 = 9\theta^4\sigma^2/n$.

15. (a) Clearly, $Y_n \le y$ if and only if $X_i \le y$ for $i = 1, \ldots, n$. Hence,

$$\Pr(Y_n \le y) = \Pr(X_1 \le y)^n = \begin{cases} (y/\theta)^n \text{if } 0 < y < \theta, & \\ 0 & \text{if } y \le 0, \\ 1 & \text{if } y \ge \theta. \end{cases}$$

(b) The c.d.f. of Z_n is, for $z < 0$,

$$\Pr(Z_n \le z) = \Pr(Y_n \le \theta + z/n) = (1 + z/[n\theta])^n. \tag{S.6.3}$$

Since $Z_n \le 0$, the c.d.f. is 1 for $z \ge 0$. According to Theorem 5.3.3, the expression in (S.6.3) converges to $\exp(z/\theta)$.

(c) Let $\alpha(y) = y^2$. Then $\alpha'(y) = 2y$. We have $n(Y_n - \theta)$ converging in distribution to the c.d.f. in part (b). The delta method says that, for $\theta > 0$, $n(Y_n^2 - \theta^2)/[2\theta]$ converges in distribution to the same c.d.f.

6.4 The Correction for Continuity

1. The mean of X_i is 1 and the mean of X_i^2 is 1.5. So, the variance of X_i is 0.5. The central limit theorem says that $Y = X_1 + \cdots + X_{30}$ has approximately the normal distribution with mean 30 and variance 15. We want the probability that $Y \le 33$. Using the correction for continuity, we would assume that Y has the normal distribution with mean 30 and variance 15 and compute the probability that $Y \le 33.5$. This is $\Phi([33.5 - 30]/15^{1/2}) = \Phi(0.904) = 0.8169$.

3. In the notation of Example 2,

$$\Pr(H > 495) = \Pr(H \ge 495.5) = \Pr\left(Z \ge \frac{495.5 - 450}{15}\right) \approx 1 - \Phi(3.033) \approx .0012.$$

5. Let X denote the total number of defects in the sample. Then X has a Poisson distribution with mean $5(125) = 625$, so σ_X is $(625)^{1/2} = 25$. Hence,

$$\Pr(\overline{X}_n < 5.5) = \Pr[X < 125(5.5)] = \Pr(X < 687.5).$$

Since this final probability is just the value that would be used with the correction for continuity, the probability to be found here is the same as that originally found in Exercise 3 of Sec. 6.3.

7. Let S denote the sum of the 16 digits. Then

$$E(S) = 16(4.5) = 72 \quad \text{and} \quad \sigma_X = [16(8.25)]^{1/2} = 11.49.$$

Hence,

$$\begin{aligned} \Pr(4 \le \overline{X}_n \le 6) &= \Pr(64 \le S \le 96) = \Pr(63.5 \le S \le 96.5) \\ &= \Pr\left(\frac{63.5 - 72}{11.49} \le Z \le \frac{96.5 - 72}{11.49}\right) \\ &\approx \Phi(2.132) - \Phi(-.740) \approx .9835 - .2296 = .7539. \end{aligned}$$

6.5 Supplementary Exercises

1. By the central limit theorem, the distribution of X is approximately normal with mean $(120)(1/6) = 20$ and standard deviation $[120(1/6)(5/6)]^{1/2} = 4.082$. Let $Z = (X - 20)/4.082$. Then from the table of the standard normal distribution we find that $\Pr(|Z| \leq 1.96) = .95$. Hence, $k = (1.96)(4.082) = 8.00$.

3. By the previous exercise, X has approximately a normal distribution with mean 10 and standard deviation $(10)^{1/2} = 3.162$. Thus, without the correction for continuity,

$$\Pr(8 \leq X \leq 12) = \Pr\left(\frac{8-10}{3.162} \leq Z \leq \frac{12-10}{3.162}\right) \approx \Phi(.6325) - \Phi(-.6325) = .473.$$

With the correction for continuity, we find

$$\Pr(7.5 \leq X \leq 12.5) = \Pr\left(-\frac{2.5}{3.162} \leq Z \leq \frac{2.5}{3.162}\right) \approx \Phi(.7906) - \Phi(-.7906) = .571.$$

The exact probability is found from the Poisson table to be

$$(.1126) + (.1251) + (.1251) + (.1137) + (.0948) = .571.$$

Thus, the approximation with the correction for continuity is almost perfect.

5. The central limit theorem says that $\overline{X}_n$ has approximately the normal distribution with mean p and variance $p(1-p)/n$. A variance stabilizing transformation will be

$$\alpha(x) = \int_a^x [p(1-p)]^{-1/2} dp.$$

To perform this integral, transform to $z = p^{1/2}$, that is, $p = z^2$. Then

$$\alpha(x) = \int_{a^{1/2}}^{x^{1/2}} \frac{dz}{(1-z^2)^{1/2}}.$$

Next, transform so that $z = \sin(w)$ or $w = \arcsin(z)$. Then $dz = \cos(w)dw$ and

$$\alpha(x) = \int_{\arcsin a^{1/2}}^{\arcsin x^{1/2}} dw = \arcsin x^{1/2},$$

where we have chosen $a = 0$. The variance stabilizing transformation is $\alpha(x) = \arcsin(x^{1/2})$.

7. Let F_n be the c.d.f. of X_n. The most direct proof is to show that $\lim_{n\to\infty} F_n(x) = F(x)$ for every point at which F is continuous. Since F is the c.d.f. of an integer-valued distribution, the continuity points are all non-integer values of x together with those integer values of x to which F assigns probability 0. It is clear, that it suffices to prove that $\lim_{n\to\infty} F_n(x) = F(x)$ for every non-integer x, because continuity of F from the right and the fact that F is nondecreasing will take care of the integers with zero probability. For each non-integer x, let m_x be the largest integer such that $m < x$. Then

$$F_n(x) = \sum_{k=1}^{m} \Pr(X_n = k) \to \sum_{k=1}^{m} f(k) = F(m) = F(x),$$

where the convergence follows because the sums are finite.

9. Let $X_1, \ldots, X_{16}$ be the times required to serve the 16 customers. The parameter of the exponenital distribution is $1/3$. According to Theorem 5.7.8, the mean and variance of each X_i are 3 and 9 respectively. Let $\sum_{k=1}^{16} X_k = Y$ be the total time. The central limit theorem approximation to the distribution of Y is the normal distribution with mean $16 \times 3 = 48$ and variance $16 \times 9 = 144$. The approximate probablity that $Y > 60$ is

$$1 - \Phi\left(\frac{60 - 48}{(144)^{1/2}}\right) = 1 - \Phi(1) = 0.1587.$$

The actual distribution of Y is the gamma distribution with parameters 16 and $1/3$. Using the gamma c.d.f., the probability is 0.1565.

11. (a) The gamma distribution with parameters n and 3 is the distribution of the sum of n i.i.d. exponential random variables with parameter 3. If n is large, the central limit theorem should apply to approximate the distribution of the sum of n exponentials.

    (b) The mean and variance of each exponential random variable are $1/3$ and $1/9$ respectively. The distribution of the sum of n of these has approximately the normal distribution with mean $n/3$ and variance $n/9$.

Chapter 7

Estimation

7.1 Statistical Inference

1. The random variables of interest are the observables $X_1, X_2, \ldots$ and the hypothetically observable (parameter) P. The X_i's are i.i.d. Bernoulli with parameter p given $P = p$.

3. The random variables of interest are the observables $Z_1, Z_2, \ldots$, the times at which successive particles hit the target, and β, the hypothetically observable (parameter) rate of the Poisson process. The hit times occur according to a Poisson process with rate β conditional on β. Other random variables of interest are the observable inter-arrival times $Y_1 = Z_1$, and $Y_k = Z_k - Z_{k-1}$ for $k \geq 2$.

5. The statement that the interval $(\overline{X}_n - 0.98, \overline{X}_n + 0.98)$ has probability 0.95 of containing μ is an inference.

7. The random variables of interest are Y, the hypothetically observable number of oocysts in t liters, the hypothetically observable indicators $X_1, X_2, \ldots$ of whether each oocyst is counted, X the observable count of oocysts, the probability (parameter) p of each oocyst being counted, and the (parameter) λ the rate of oocysts per liter. We model Y as a Poisson random variable with mean $t\lambda$ given λ. We model $X_1, \ldots, X_y$ as i.i.d. Bernoulli random variables with parameter p given p and given $Y = y$. We define $X = X_1 + \ldots + X_y$.

7.2 Prior and Posterior Distributions

1. We still have $y = 16178$, the sum of the five observed values. The posterior distribution of β is now the gamma distribution with parameters 6 and 21178. So,

$$\begin{aligned} f(x_6|\boldsymbol{x}) &= \int_0^\infty 7.518 \times 10^{23} \beta^5 \exp(-21178\beta)\beta \exp(-x_6\beta) d\beta \\ &= 7.518 \times 10^{23} \int_0^\infty \beta^6 \exp(-\beta[21178 + x_6]) d\beta \\ &= 7.518 \times 10^{23} \frac{\Gamma(7)}{(21178 + x_6)^7} = \frac{5.413 \times 10^{26}}{(21178 + x_6)^7}, \end{aligned}$$

for $x_6 > 0$. We can now compute $\Pr(X_6 > 3000|\boldsymbol{x})$ as

$$\Pr(X_6 > 3000|\boldsymbol{x}) = \int_{3000}^\infty \frac{5.413 \times 10^{26}}{(21178 + x_6)^7} dx_6 = \frac{5.413 \times 10^{26}}{6 \times 24178^6} = 0.4516.$$

3. Let X denote the number of defects on the selected roll of tape. Then for any given value of λ, the p.f. of X is.

$$f(x \mid \lambda) = \frac{\exp(-\lambda)\lambda^x}{x!} \quad \text{for} \quad x = 0, 1, 2, \ldots.$$

Therefore,

$$\xi(1.0 \mid X = 3) = \Pr(\lambda = 1.0 \mid X = 3) = \frac{\xi(1.0)f(3 \mid 1.0)}{\xi(1.0)f(3 \mid 1.0) + \xi(1.5)f(3 \mid 1.5)}.$$

From the table of the Poisson distribution in the back of the book it is found that

$$f(3 \mid 1.0) = 0.0613 \quad \text{and} \quad f(3 \mid 1.5) = 0.1255.$$

Therefore, $\xi(1.0 \mid X = 3) = 0.2456$ and $\xi(1.5 \mid X = 3) = 1 - \xi(1.0 \mid X = 3) = 0.7544$.

5. If α and β denote the parameters of the beta distribution, then we must have

$$\frac{\alpha}{\alpha + \beta} = \frac{1}{3} \quad \text{and} \quad \frac{\alpha\beta}{(\alpha + \beta)^2(\alpha + \beta + 1)} = \frac{2}{90}.$$

Since $\dfrac{\alpha}{\alpha + \beta} = \dfrac{1}{3}$, it follows that $\dfrac{\beta}{(\alpha + \beta)} = \dfrac{2}{3}$. Therefore,

$$\frac{\alpha\beta}{(\alpha + \beta)^2} = \frac{\alpha}{\alpha + \beta} \cdot \frac{\beta}{\alpha + \beta} = \frac{1}{3} \cdot \frac{2}{3} = \frac{2}{9}.$$

It now follows from the second equation that $\dfrac{2}{9(\alpha + \beta + 1)} = \dfrac{2}{90}$ and, hence, that $\alpha + \beta + 1 = 10$. Therefore, $\alpha + \beta = 9$ and it follows from the first equation that $\alpha = 3$ and $\beta = 6$. Hence, the prior p.d.f. of θ is as follows, for $0 < \theta < 1$:

$$\xi(\theta) = \frac{\Gamma(9)}{\Gamma(3)\Gamma(6)} \theta^2 (1 - \theta)^5.$$

7. Since $f_n(\boldsymbol{x} \mid \theta)$ is given by Eq. (7.2.11) with n = 8 and y = 3, then

$$f_n(\boldsymbol{x} \mid \theta)\xi(\theta) = 2\theta^3(1 - \theta)^6.$$

When we compare this expression with Eq. (5.8.3), we see that it has the same form as the p.d.f. of a beta distribution with parameters $\alpha = 4$ and $\beta = 7$. Therefore, this beta distribution is the posterior distribution of θ.

9. It follows from Exercise 8 that if the experiment yields a total of three defectives and five nondefectives, the posterior distribution will be the same regardless of whether the eight items were selected in one batch or one at a time in accordance with some stopping rule. Therefore, the posterior distribution in this exercise will be the same beta distribution as that obtained in Exercise 6.

11. Let y_1 denote the smallest and let y_6 denote the largest of the six observations. Then the joint p.d.f. of the six observations is

$$f_n(\boldsymbol{x} \mid \theta) = \begin{cases} 1 & \text{for } \theta - \dfrac{1}{2} < y_1 < y_6 < \theta + \dfrac{1}{2}, \\ 0 & \text{otherwise.} \end{cases}$$

The condition that $\theta - \dfrac{1}{2} < y_1 < y_6 < \theta + \dfrac{1}{2}$ is the same as the condition that $y_6 - 1/2 < \theta < y_1 + 1/2$. Since $\xi(\theta)$ is again as given in Exercise 10, it follows that $f_n(\boldsymbol{x} \mid \theta)\xi(\theta)$ will be positive only for values of θ which satisfy both the requirement that $10 < \theta < 20$. Since $y_1 = 10.9$ and $y_6 = 11.7$ in this exercise, $f_n(\boldsymbol{x} \mid \theta)\xi(\theta)$ is positive only for $y_6 - 1/2 < \theta < y_1 + 1/2$ and the requirement that $10 < \theta < 20$. Since $y_1 = 10.9$ and $y_6 = 11.7$ in this exercise, $f_n(\boldsymbol{x} \mid \theta)\xi(\theta)$ is positive only for $11.2 < \theta < 11.4$. Furthermore, since $f_n(\boldsymbol{x} \mid \theta)\xi(\theta)$ is constant over this interval, the posterior p.d.f. $\xi(\theta \mid \boldsymbol{x})$ will also be constant over the interval. In other words, the posterior distribution of θ must be a uniform distribution on this interval.

7.3 Conjugate Prior Distributions

1. The posterior mean of θ will be

$$\frac{100 \times 0 + 20v^2 \times 0.125}{100 + 20v^2} = 0.12.$$

We can solve this equation for v^2 by multiplying both sides by $100 + 20v^2$ and collecting terms. The result is $v^2 = 120$.

3. Since the observed number of defective items is 3 and the observed number of nondefective items is 97, it follows from Theorem 7.3.1 that the posterior distribution of θ is a beta distribution with parameters $2 + 3 = 5$ and $200 + 97 = 297$.

5. By Theorem 7.3.2, the posterior distribution will be the gamma distribution for which the parameters are $3 + \sum_{i=1}^{n} x_i = 3 + 13 = 16$ and $1 + n = 1 + 5 = 6$.

7. In the notation of Theorem 7.3.3, we have $\sigma^2 = 4$, $\mu = 68$, $v^2 = 1$, $n = 10$, and $\bar{x}_n = 69.5$. Therefore, the posterior distribution of θ is the normal distribution with mean $\mu_1 = 967/14$ and variance $v_1^2 = 2/7$.

9. Since the posterior distribution of θ is normal, the prior distribution of θ must also have been normal. Furthermore, from Eqs. (7.3.1) and (7.3.2), we obtain the relations:

$$8 = \frac{\mu + (20)(10)v^2}{1 + 20v^2}$$

and

$$\frac{1}{25} = \frac{v^2}{1 + 20v^2}.$$

It follows that $v^2 = 1/5$ and $\mu = 0$.

11. In this exercise, $\sigma^2 = 4$ and $n = 100$. Therefore, by Eq. (7.3.2),

$$v_1^2 = \frac{4v^2}{4 + 100v^2} = \frac{1}{25 + (1/v^2)} < \frac{1}{25}.$$

Since the variance of the posterior distribution is less than 1/25, the standard deviation must be less than 1/5.

13. The mean of the gamma distribution with parameters α and β is α/β and the standard deviation is $\alpha^{1/2}/\beta$. Therefore, the coefficient of variation is $\alpha^{-1/2}$. Since the coefficient of variation of the prior gamma distribution of θ is 2, it follows that $\alpha = 1/4$ in the prior distribution. Furthermore, it now follows from Theorem 7.3.4 that the coefficient of variation of the posterior gamma distribution of θ is $(\alpha + n)^{-1/2} = (n + 1/4)^{-1/2}$. This value will be less than 0.1 if and only if $n \geq 99.75$.Thus, the required sample size is $n \geq 100$.

15. (a) Let $y = 1/\theta$. Then $\theta = 1/y$ and $d\theta = -dy/y^2$. Hence,

$$\int_0^\infty \xi(\theta)d\theta = \int_0^\infty \frac{\beta^\alpha}{\Gamma(\alpha)} y^{\alpha-1} \exp(-\beta y)dy = 1.$$

(b) If an observation X has a normal distribution with a known value of the mean μ and an unknown value of the variance θ, then the p.d.f. of X has the form

$$f(x \mid \theta) \propto \frac{1}{\theta^{1/2} \exp\left[-\dfrac{(x-\mu)^2}{2\theta}\right]}.$$

Also, the prior p.d.f. of θ has the form

$$\xi(\theta) \propto \theta^{-(\alpha+1)} \exp(-\beta/\theta).$$

Therefore, the posterior p.d.f. $\xi(\theta \mid x)$ has the form

$$\xi(\theta \mid x) \propto \xi(\theta) f(x \mid \theta) \propto \theta^{-(\alpha+3/2)} \exp\left\{-\left[\beta + \frac{1}{2}(x-\mu)^2\right] \cdot \frac{1}{\theta}\right\}.$$

Hence, the posterior p.d.f. of θ has the same form as $\xi(\theta)$ with α replaced by $\alpha + 1/2$ and β replaced by $\beta + 1/2(x-\mu)^2$. Since this distribution will be the prior distribution of θ for future observations, it follows that the posterior distribution after any number of observations will also belong to the same family of distributions.

17. The joint p.d.f. of the three observations is

$$f(x_1, x_2, x_3 \mid \theta) = \begin{cases} 1/\theta^3 & \text{for } 0 < x_i < \theta \ (i = 1, 2, 3), \\ 0 & \text{otherwise.} \end{cases}$$

Therefore, the posterior p.d.f. $\xi(\theta \mid x_1, x_2, x_3)$ will be positive only if $\theta \geq 4$, as required by the prior p.d.f., and also $\theta > 8$, the largest of the three observed values. Hence, for $\theta > 8$,

$$\xi(\theta \mid x_1, x_2, x_3) \propto \xi(\theta) f(x_1, x_2, x_3 \mid \theta) \propto 1/\theta^7.$$

Since

$$\int_8^\infty \frac{1}{\theta^7} d\theta = \frac{1}{6(8)^6},$$

it follows that

$$\xi(\theta \mid x_1, x_2, x_3) = \begin{cases} 6(8^6)/\theta^7 & \text{for } \theta > 8 \\ 0 & \text{for } \theta \leq 8. \end{cases}$$

19. The joint p.d.f. of $X_1, \ldots, X_n$ has the following form, for $0 < x_i < 1 (i = 1, \ldots, n)$:

$$\begin{aligned} f_n(\boldsymbol{x} \mid \theta) &= \theta^n \left(\prod_{i=1}^n x_i\right)^{\theta-1} \propto \theta^n \left(\prod_{i=1}^n x_i\right)^{\theta} \\ &= \theta^n \exp\left(\theta \sum_{i=1}^n \log x_i\right). \end{aligned}$$

The prior p.d.f. of θ has the form

$$\xi(\theta) \propto \theta^{\alpha-1} \exp(-\beta\theta).$$

Hence, the posterior p.d.f. of θ has the form

$$\xi(\theta \mid \boldsymbol{x}) \propto \xi(\theta) f_n(\boldsymbol{x} \mid \theta) \propto \theta^{\alpha+n-1} \exp\left[-\left(\beta - \sum_{i=1}^n \log x_i\right)\theta\right].$$

This expression can be recognized as being, except for a constant factor, the p.d.f. of the gamma distribution with parameters $\alpha_1 = \alpha + n$ and $\beta_1 = \beta - \sum_{i=1}^n \log x_i$. Therefore, the mean of the posterior distribution is α_1/β_1 and the variance is α_1/β_1^2.

21. The posterior p.d.f. is proportional to the likelihood $\theta^n \exp\left(-\theta \sum_{i=1}^n x_i\right)$ times $1/\theta$. This product can be written as $\theta^{n-1} \exp(-\theta n \overline{x}_n)$. As a function of θ this is recognizable as the p.d.f. of the gamma distribution with parameters n and $n\overline{x}_n$. The mean of this posterior distribution is then $n/[n\overline{x}_n] = 1/\overline{x}_n$.

23. (a) Let the prior p.d.f. of θ be $\xi_{\alpha,\beta}(\theta)$. Suppose that $X_1, \ldots, X_n$ are i.i.d. with conditional p.d.f. $f(x|\theta)$ given θ, where f is as stated in the exercise. The posterior p.d.f. after observing these data is

$$\xi(\theta|\boldsymbol{x}) = \frac{a(\theta)^{\alpha+n} \exp\left[c(\theta)\left\{\beta + \sum_{i=1}^b d(x_i)\right\}\right]}{\int_\Omega a(\theta)^{\alpha+n} \exp\left[c(\theta)\left\{\beta + \sum_{i=1}^b d(x_i)\right\}\right] d\theta}. \tag{S.7.1}$$

Eq. (S.7.1) is of the form of $\xi_{\alpha',\beta'}(\theta)$ with $\alpha' = \alpha + n$ and $\beta' = \beta + \sum_{i=1}^n d(x_i)$. The integral in the denominator of (S.7.1) must be finite with probability 1 (as a function of $x_1, \ldots, x_n$) because $\prod_{i=1}^n b(x_i)$ times this denominator is the marginal (joint) p.d.f. of $X_1, \ldots, X_n$.

(b) This was essentially the calculation done in part (a).

25. For every θ, the p.d.f. (or p.f.) $f(x|\theta)$ for an exponential family is strictly postive for all x such that $b(x) > 0$. That is, the set of x for which $f(x|\theta) > 0$ is the same for all θ. This is not true for uniform distributions where the set of x such that $f(x|\theta) > 0$ is $[0, \theta]$.

7.4 Bayes Estimators

1. The posterior distribution of θ would be the beta distribution with parameters 2 and 1. The mean of the posterior distribution is 2/3, which would be the Bayes estimate under squared error loss. The median of the posterior distribution would be the Bayes estimate under absolute error loss. To find the median, write the c.d.f. as

$$F(\theta) = \int_0^\theta 2tdt = \theta^2,$$

for $0 < \theta < 1$. The quantile function is then $F^{-1}(p) = p^{1/2}$, so the median is $(1/2)^{1/2} = 0.7071$.

3. If y denotes the number of defective items in the sample, then the posterior distribution of θ will be the beta distribution with parameters $5 + y$ and $10 + 20 - y = 30 - y$. The variance V of this beta distribution is

$$V = \frac{(5+y)(30-y)}{(35)^2(36)}.$$

Since the Bayes estimate of θ is the mean μ of the posterior distribution, the mean squared error of this estimate is $E[(\theta - \mu)^2 \mid \boldsymbol{x}]$, which is the variance V of the posterior distribution.

   (a) V will attain its maximum at a value of y for which $(5+y)(30-y)$ is a maximum. By differentiating with respect to y and setting the derivative equal to 0, we find that the maximum is attained when $y = 12.5$. Since the number of defective items y must be an integer, the maximum of V will be attained for $y = 12$ or $y = 13$. When these values are substituted into $(5 + y)(30 - y)$, it is found that they both yield the same value.

   (b) Since $(5+y)(30-y)$ is a quadratic function of y and the coefficient of y^2 is negative, its minimum value over the interval $0 \leq y \leq 20$ will be attained at one of the endpoints of the interval. It is found that the value for $y = 0$ is smaller than the value for $y = 20$.

5. It was shown in Exercise 5 of Sec. 7.3 that the posterior distribution of θ is the gamma distribution with parameters $\alpha = 16$ and $\beta = 6$. The Bayes estimate of θ is the mean of this distribution and is equal to $16/6 = 8/3$.

7. The Bayes estimator is the mean of the posterior distribution of θ, as given in Exercise 6. Since θ is the mean of the Poisson distribution, it follows from the law of large numbers that $\overline{X}_n$ converges to θ in probability as $n \to \infty$. It now follows from Exercise 6 that, since $\gamma_n \to 1$, the Bayes estimators will also converge to θ in probability as $n \to \infty$. Hence, the Bayes estimators form a consistent sequence of estimators of θ.

9. For any given values in the random sample, the Bayes estimate of θ is the mean of the posterior distribution of θ. Therefore, the mean squared error of the estimate will be the variance of the posterior distribution of θ. It was shown in Exercise 10 of Sec. 7.3 that this variance will be 0.01 or less for $n \geq 396$.

11. Let $X_1, \ldots, X_n$ denote the observations in the random sample, and let α and β denote the parameters of the prior gamma distribution of θ. It was shown in Theorem 7.3.4 that the posterior distribution of θ will be the gamma distribution with parameters $\alpha + n$ and $\beta + n\overline{X}_n$. The Bayes estimator, which is the mean of this posterior distribution is, therefore,

$$\frac{\alpha + n}{\beta + n\overline{X}_n} = \frac{1 + (\alpha/n)}{\overline{X}_n + (\beta/n)}.$$

Since the mean of the exponential distribution is $1/\theta$, it follows from the law of large numbers that $\overline{X}_n$ will converge in probability to $1/\theta$ as $n \to \infty$. It follows, therefore, that the Bayes estimators will converge in probability to θ as $n \to \infty$. Hence, the Bayes estimators form a consistent sequence of estimators of θ.

13. If θ has the Pareto distribution with parameters $\alpha > 1$ and $x_0 > 0$, then

$$E(\theta) = \int_{x_0}^{\infty} \theta \cdot \frac{\alpha x_0^{\alpha}}{\theta^{\alpha+1}}\, d\theta = \frac{\alpha}{\alpha - 1} x_0.$$

It was shown in Exercise 18 of Sec. 7.3 that the posterior distribution of θ will be a Pareto distribution with parameters $\alpha + n$ and $\max\{x_0, X_1, \ldots, X_n\}$. The Bayes estimator is the mean of this posterior distribution and is, therefore, equal to $(\alpha + n)\max\{x_0, X_1, \ldots, X_n\}/(\alpha + n - 1)$.

15. Let a_0 be a $1/(1+c)$ quantile of the posterior distribution, and let a_1 be some other value. Assume that $a_1 < a_0$. The proof for $a_1 > a_0$ is similar. Let $g(\theta|x)$ denote the posterior p.d.f. The posterior mean of the loss for action a is

$$h(a) = c\int_{-\infty}^{a} (a - \theta) g(\theta|x) d\theta + \int_{a}^{\infty} (\theta - a) g(\theta|x) d\theta.$$

We shall now show that $h(a_1) \geq h(a_0)$, with strict inequality if a_1 is not a $1/(1+c)$ quantile.

$$\begin{aligned} h(a_1) - h(a_0) &= c\int_{-\infty}^{a_0} (a_1 - a_0) g(\theta|x) d\theta + \int_{a_0}^{a_1} (ca_1 - (1+c)\theta + a_0) g(\theta|x) d\theta \\ &\quad + \int_{a_1}^{\infty} (a_0 - a_1) g(\theta|x) d\theta \end{aligned} \qquad \text{(S.7.2)}$$

The first integral in (S.7.2) equals $c(a_1 - a_0)/(1+c)$ because a_0 is a $1/(1+c)$ quantile of a the posterior distribution, and the posterior distribution is continuous. The second integral in (S.7.2) is at least as large as $(a_0 - a_1)\Pr(a_0 < \theta \leq a_1|x)$ since $-(1+c)\theta > -(1+c)a_1$ for all θ in that integral. In fact, the integral will be strictly larger than $(a_0 - a_1)\Pr(a_0 < \theta \leq a_1|x)$ if this probability is positive. The last integral in (S.7.2) equals $(a_0 - a_1)\Pr(\theta > a_1|x)$. So

$$h(a_1) - h(a_0) \geq c\frac{a_1 - a_0}{1+c} + (a_0 - a_1)\Pr(\theta > a_0|x) = 0. \qquad \text{(S.7.3)}$$

The equality follows from the fact that $\Pr(\theta > a_0|x) = c/(1+c)$. The inequality in (S.7.3) will be strict if and only if $\Pr(a_0 < \theta \leq a_1|x) > 0$, which occurs if and only if a_1 is not another $1/(1+c)$ quantile.

7.5 Maximum Likelihood Estimators

1. We can easily compute

$$\begin{aligned} E(Y) &= \frac{1}{n}\sum_{i=1}^{n} x_i = \overline{x}_n, \\ E(Y^2) &= \frac{1}{n}\sum_{i=1}^{n} x_i^2. \end{aligned}$$

Then

$$\mathrm{Var}(Y) = \frac{1}{n}\sum_{i=1}^{n} x_i^2 - \overline{x}_n^2 = \frac{1}{n}\sum_{i=1}^{n}(x_i - \overline{x}_n)^2.$$

3. The likelihood function for the given sample is $p^{58}(1-p)^{12}$. Among all values of p in the interval $1/2 \le p \le 2/3$, this function is a maximum when $p = 2/3$.

5. Let y denote the sum of the observed values $x_1, \ldots, x_n$. Then the likelihood function is

$$f_n(\boldsymbol{x} \mid \theta) = \frac{\exp(-n\theta)\theta^y}{\prod_{i=1}^{n}(x_i!)}.$$

(a) If $y > 0$ and we let $L(\theta) = \log f_n(\boldsymbol{x} \mid \theta)$, then

$$\frac{\partial}{\partial\theta}L(\theta) = -n + \frac{y}{\theta}.$$

The maximum of $L(\theta)$ will be attained at the value of θ for which this derivative is equal to 0. In this way, we find that $\hat{\theta} = y/n = \overline{x}_n$.

(b) If $y = 0$, then $f_n(\boldsymbol{x} \mid \theta)$ is a decreasing function of θ. Since $\theta = 0$ is not a value in the parameter space, there is no M.L.E.

7. Let y denote the sum of the observed values $x_1, \ldots, x_n$. Then the likelihood function is

$$f_n(\boldsymbol{x} \mid \beta) = \beta^n \exp(-\beta y).$$

If we let $L(\beta) = \log f_n(\boldsymbol{x} \mid \beta)$, then

$$\frac{\partial L(\beta)}{\partial\beta} = \frac{n}{\beta} - y.$$

The maximum of $L(\beta)$ will be attained at the value of β for which this derivative is equal to 0. Therefore, $\hat{\beta} = n/y = 1/\overline{x}_n$.

9. If $0 < x_i < 1$ for $i = 1, \ldots, n$, then the likelihood function will be as follows:

$$f_n(\boldsymbol{x} \mid \theta) = \theta^n \left(\prod_{i=1}^{n} x_i\right)^{\theta-1}.$$

If we let $L(\theta) = \log f_n(\boldsymbol{x} \mid \theta)$, then

$$\frac{\partial}{\partial\theta}L(\theta) = \frac{n}{\theta} + \sum_{i=1}^{n}\log x_i.$$

Therefore, $\hat{\theta} = -n/\sum_{i=1}^{n}\log x_i$. It should be noted that $\hat{\theta} > 0$.

11. The p.d.f. of each observation can be written as follows:

$$f(x \mid \theta_1, \theta_2) = \begin{cases} \frac{1}{\theta_2-\theta_1} & \text{for } \theta_1 \le x \le \theta_2, \\ 0 & \text{otherwise.} \end{cases}$$

Therefore, the likelihood function is

$$f_n(\boldsymbol{x} \mid \theta_1, \theta_2) = \frac{1}{(\theta_2-\theta_1)^n}$$

for $\theta_1 \le \min\{x_1, \ldots, x_n\} < \max\{x_1, \ldots, x_n\} \le \theta_2$, and $f_n(\boldsymbol{x} \mid \theta_1, \theta_2) = 0$ otherwise. Hence, $f_n(\boldsymbol{x} \mid \theta_1, \theta_2)$ will be a maximum when $\theta_2 - \theta_1$ is made as small as possible. Since the smallest possible value of θ_2 is $\max\{x_1, \ldots, x_n\}$ and the largest possible value of θ_1 is $\min\{x_1, \ldots, x_n\}$, these values are the M.L.E.'s.

13. It follows from Eq. (5.10.2) (with x_1 and x_2 now replaced by x and y) that the likelihood function is

$$f_n(\boldsymbol{x}, \boldsymbol{y} \mid \mu_1, \mu_2) \propto \exp\left\{-\frac{1}{2(1-\rho^2)}\sum_{i=1}^{n}\left[\left(\frac{x_i-\mu_1}{\sigma_1}\right)^2 - 2\rho\left(\frac{x_i-\mu_1}{\sigma_1}\right)\left(\frac{y_i-\mu_2}{\sigma_2}\right) + \left(\frac{y_i-\mu_2}{\sigma_2}\right)^2\right]\right\}.$$

If we let $L(\mu_1, \mu_2) = \log\ f(\boldsymbol{x}, \boldsymbol{y} \mid \mu_1, \mu_2)$, then

$$\frac{\partial L(\mu_1,\mu_2)}{\partial \mu_1} = \frac{1}{1-\rho^2}\left[\frac{1}{\sigma_1^2}\left(\sum_{i=1}^{n} x_i - n\mu_1\right) - \frac{\rho}{\sigma_1\sigma_2}\left(\sum_{i=1}^{n} y_i - n\mu_2\right)\right],$$

$$\frac{\partial L(\mu_1,\mu_2)}{\partial \mu_2} = \frac{1}{1-\rho^2}\left[\frac{1}{\sigma_2^2}\left(\sum_{i=1}^{n} y_i - n\mu_2\right) - \frac{\rho}{\sigma_1\sigma_2}\left(\sum_{i=1}^{n} x_i - n\mu_1\right)\right].$$

When these derivatives are set equal to 0, the unique solution is $\mu_1 = \overline{x}_n$ and $\mu_2 = \overline{y}_n$. Hence, these values are the M.L.E.'s.

7.6 Properties of Maximum Likelihood Estimators

1. The M.L.E. of $\exp(-1/\theta)$ is $\exp(-1/\hat{\theta})$, where $\hat{\theta} = -n/\sum_{i=1}^{n}\log(x_i)$ is the M.L.E. of θ. That is, the M.L.E. of $\exp(-1/\theta)$ is

$$\exp\left(\sum_{i=1}^{n}\log(x_i)/n\right) = \exp\left(\log\left[\prod_{i=1}^{n} x_i\right]^{1/n}\right) = \left(\prod_{i=1}^{n} x_i\right)^{1/n}.$$

3. The median of an exponential distribution with parameter β is the number m such that

$$\int_0^m \beta\exp(-\beta x)dx = \frac{1}{2}.$$

Therefore, $m = (\log\ 2)/\beta$, and it follows that $\hat{m} = (\log 2)/\hat{\beta}$. It was shown in Exercise 7 of Sec. 7.5 that $\hat{\beta} = 1/\overline{X}_n$.

5. Since the mean of the uniform distribution is $\mu = (a+b)/2$, it follows that $\hat{\mu} = (\hat{a}+\hat{b})/2$. It was shown in Exercise 11 of Sec. 7.5 that $\hat{a} = \min\{X_1, \ldots, X_n\}$ and $\hat{b} = \max\{X_1, \ldots, X_n\}$.

7. $\nu = \Pr(X > 2) = \Pr\left(Z > \frac{2-\mu}{\sigma}\right) = 1 - \Phi\left(\frac{2-\mu}{\sigma}\right) = \Phi\left(\frac{\mu-2}{\sigma}\right).$

Therefore, $\hat{\nu} = \Phi((\hat{\mu}-2)/\hat{\sigma})$.

9. If we let $y = \sum_{i=1}^{n} x_i$, then the likelihood function is

$$f_n(\boldsymbol{x} \mid \alpha, \beta) = \frac{\beta^{n\alpha}}{[\Gamma(\alpha)]^n}\left(\prod_{i=1}^{n} x_i\right)^{\alpha-1}\exp(-\beta y).$$

If we now let $L(\alpha,\beta) = \log\ f_n(\boldsymbol{x} \mid \alpha, \beta)$, then

$$L(\alpha,\beta) = n\alpha\ \log\ \beta - n\ \log\ \Gamma(\alpha) + (\alpha-1)\ \log\left(\prod_{i=1}^{n} x_i\right) - \beta y.$$

Hence,

$$\frac{\partial L(\alpha,\beta)}{\partial\beta} = \frac{n\alpha}{\beta} - y.$$

Since $\hat{\alpha}$ and $\hat{\beta}$ must satisfy the equation $\partial L(\alpha,\beta)/\partial\beta = 0$ [as well as the equation $\partial L(\alpha,\beta)/\partial\alpha = 0$], it follows that $\hat{\alpha}/\hat{\beta} = y/n = \overline{x}_n$.

11. Let $Y_n = \max\{X_1,\dots,X_n\}$. It was shown in Example 7.5.7 that $\hat{\theta} = Y_n$. Therefore, for $\varepsilon > 0$,

$$\Pr(|\hat{\theta}-\theta| < \varepsilon) = \Pr(Y_n > \theta - \varepsilon) = 1 - \left(\frac{\theta-\varepsilon}{\theta}\right)^n.$$

It follows that $\lim_{n\to\infty} \Pr(|\hat{\theta}-\theta| < \varepsilon) = 1$. Therefore, $\hat{\theta} \xrightarrow{P} \theta$.

13. Let $Z_i = -\log X_i$ for $i = 1,\dots,n$. Then by Exercise 9 of Sec. 7.5, $\hat{\theta} = 1/\overline{Z}_n$. If X_i has the p.d.f. $f(x \mid \theta)$ specified in that exercise, then the p.d.f. $g(z \mid \theta)$ of Z_i will be as follows, for $z > 0$:

$$g(z \mid \theta) = f(\exp(-z) \mid \theta)\left|\frac{dx}{dz}\right| = \theta(\exp(-z))^{\theta-1}\exp(-z) = \theta\exp(-\theta z).$$

Therefore, Z_i has an exponential distribution with parameter θ. It follows that $E(Z_i) = 1/\theta$. Furthermore, since $X_1,\dots,X_n$ form a random sample from a distribution for which the p.d.f. is $f(x \mid \theta)$, it follows that $Z_1,\dots,Z_n$ will have the same joint distribution as a random sample from an exponential distribution with parameter θ. Therefore, by the law of large numbers, $\overline{Z}_n \xrightarrow{P} 1/\theta$. It follows that $\hat{\theta} \xrightarrow{P} \theta$.

15. As explained in this section, the likelihood function for the 21 observations is equal to the joint p.d.f. of the 20 observations for which the exact value is known, multiplied by the probability $\exp(-15/\mu)$ that the 21st observation is greater than 15. If we let y denote the sum of the first 20 observations, then the likelihood function is

$$\frac{1}{\mu^{20}}\exp(-y/\mu)\exp(-15/\mu).$$

Since $y = (20)(6) = 120$, this likelihood function reduces to

$$\frac{1}{\mu^{20}}\exp(-135/\mu).$$

The value of μ which maximizes this likelihood function is $\hat{\mu} = 6.75$.

17. The likelihood function determined by any observed value x of X is $\binom{10}{x}p^x(1-p)^{10-x}$. By Eq. (5.5.1) the likelihood function determined by any observed value y of Y is $\binom{3+y}{y}p^4(1-p)^y$. Therefore, when $X = 4$ and $Y = 6$, each of these likelihood functions is proportional to $p^4(1-p)^6$. The M.L.E. obtained by either statistician will be the value of p which maximizes this expression. That value is $\hat{p} = 2/5$.

19. The mean of an exponential random variable with parameter β is $1/\beta$, so the method of moments estimator is one over the sample mean, which is also the M.L.E.

21. The M.L.E. of the mean is the sample mean, which is the method of moments estimator. The M.L.E. of σ^2 is the mean of the X_i^2's minus the square of the sample mean, which is also the method of moments estimator of the variance.

23. (a) The means of X_i and X_i^2 are respectively $\alpha/(\alpha+\beta)$ and $\alpha(\alpha+1)/[(\alpha+\beta)(\alpha+\beta+1)$. We set these equal to the sample moments $\overline{x}_n$ and $\overline{x^2}_n$ and solve for α and β. After some tedious algebra, we get

$$\hat{\alpha} = \frac{\overline{x}_n(\overline{x}_n - \overline{x^2}_n)}{\overline{x^2}_n - \overline{x}_n^2},$$

$$\hat{\beta} = \frac{(1-\overline{x}_n)(\overline{x}_n - \overline{x^2}_n)}{\overline{x^2}_n - \overline{x}_n^2}.$$

(b) The M.L.E. involves derivatives of the gamma function and the products $\prod_{i=1}^n x_i$ and $\prod_{i=1}^n (1-x_i)$.

25. When we observe only the first $n-k$ Y_i's, the M.L.E.'s of μ_1 and σ_1^2 are not affected. The M.L.E.'s of α, β and $\sigma_{2.1}^2$ are just as in the previous exercise but with n replaced by $n-k$. The M.L.E.'s of μ_2, σ_2^2 and ρ are obtained by substituting $\hat{\alpha}$, $\hat{\beta}$ and $\widehat{\sigma_{2.1}^2}$ into the three equations Exercise 24:

$$\hat{\mu}_2 = \hat{\alpha} + \hat{\beta}\hat{\mu}_1$$

$$\widehat{\sigma_2^2} = \widehat{\sigma_{2.1}^2} + \hat{\beta}^2\widehat{\sigma_1^2}$$

$$\hat{\rho} = \frac{\hat{\beta}\widehat{\sigma_1}}{\widehat{\sigma_2}}.$$

7.7 Sufficient Statistics

In Exercises 1–11, let t denote the value of the statistic T when the observed values of $X_1, \ldots, X_n$ are $x_1, \ldots, x_n$. In each exercise, we shall show that T is a sufficient statistic by showing that the joint p.f. or joint p.d.f. can be factored as in Eq. (7.7.1).

1. The joint p.f. is

$$f_n(\boldsymbol{x} \mid p) = p^t(1-p)^{n-t}.$$

3. The joint p.f. is

$$f_n(\boldsymbol{x} \mid p) = \left[\prod_{i=1}^n \binom{r+x_i-1}{x_i}\right][p^{nr}(1-p)^t].$$

Since the expression inside the first set of square brackets does not depend on the parameter p, it follows that T is a sufficient statistic for p.

5. The joint p.d.f. is

$$f_n(\boldsymbol{x} \mid \beta) = \left\{\frac{1}{[\Gamma(\alpha)]^n}\left(\prod_{i=1}^n x_i\right)^{\alpha-1}\right\}\{\beta^{n\alpha}\exp(-n\beta t)\}.$$

7. The joint p.d.f. is

$$f_n(\boldsymbol{x} \mid \alpha) = \left\{ \frac{1}{[\Gamma(\beta)]^n} \left[\prod_{i=1}^{n} (1 - x_i) \right]^{\beta-1} \right\} \left\{ \left[\frac{\Gamma(\alpha+\beta)}{\Gamma(\alpha)} \right]^n t^{\alpha-1} \right\}$$

9. The joint p.d.f. is

$$f_n(\boldsymbol{x} \mid a) = \frac{h(a,t)}{(b-a)^n},$$

where h is defined in Example 7.7.5.

11. The joint p.d.f. or joint p.f. is

$$f_n(\boldsymbol{x} \mid \theta) = \left\{ \prod_{i=1}^{n} b(x_i) \right\} \{[a(\theta)]^n \exp[c(\theta)t]\}.$$

13. The statistic T will be a sufficient statistic for θ if and only if $f_n(\boldsymbol{x} \mid \theta)$ can be factored as in Eq. (7.7.1). However, since $r(\boldsymbol{x})$ can be expressed as a function of $r'(\boldsymbol{x})$, and conversely, there will be a factorization of the form given in Eq. (7.7.1) if and only if there is a similar factorization in which the function v is a function of $r'(\boldsymbol{x})$ and θ. Therefore, T will be a sufficient statistic if and only if T' is a sufficient statistic.

15. It follows from Exercise 11 and Exercise 24(i) of Sec. 7.3 that the statistic $T' = \sum_{i=1}^{n} \log(1 - X_i)$ is a sufficient statistic. Since T is a one-to-one function of T', it follows from Exercise 13 that T is also a sufficient statistic.

17. First, suppose that T is sufficient. Then the likelihood function from observing $\boldsymbol{X} = \boldsymbol{x}$ is $u(\boldsymbol{x})v[r(\boldsymbol{x}), \theta]$, which is proportional to $v[r(\boldsymbol{x}), \theta]$. The likelihood from observing $T = t$ (when $t = r(\boldsymbol{x})$) is

$$\sum u(\boldsymbol{x})v[r(\boldsymbol{x}), \theta] = v[t, \theta] \sum u(\boldsymbol{x}), \tag{S.7.4}$$

where the sums in (S.7.4) are over all $\boldsymbol{x}$ such that $t = r(\boldsymbol{x})$. Notice that the right side of (S.7.4) is proportional to $v[t, \theta] = v[r(\boldsymbol{x}), \theta]$. So the two likelihoods are proportional. Next, suppose that the two likelihoods are proportional. That is, let $f(\boldsymbol{x}|\theta)$ be the p.f. of $\boldsymbol{X}$ and let $h(t|\theta)$ be the p.f. of T. If $t = r(\boldsymbol{x})$ then there exists $c(\boldsymbol{x})$ such that

$$f(\boldsymbol{x}|\theta) = u(\boldsymbol{x})h(t|\theta).$$

Let $v[t, \theta] = h(t|\theta)$ and apply the factorization criterion to see that T is sufficient.

7.8 Jointly Sufficient Statistics

In Exercises 1–4, let t_1 and t_2 denote the values of T_1 and T_2 when the observed values of $X_1, \ldots, X_n$ are $x_1, \ldots, x_n$. In each exercise, we shall show that T_1 and T_2 are jointly sufficient statistics by showing that the joint p.d.f. of $X_1, \ldots, X_n$ can be factored as in Eq. (7.8.1).

1. The joint p.d.f. is

$$f_n(\boldsymbol{x} \mid \alpha, \beta) = \frac{\beta^{n\alpha}}{[\Gamma(\alpha)]^n} t_1^{\alpha-1} \exp(-\beta t_2).$$

3. Let the function h be as defined in Example 7.8.4. Then the joint p.d.f. can be written in the following form:

$$f_n(x \mid x_0, \alpha) = \frac{(\alpha x_0^\alpha)^n}{t_2^{\alpha+1}} h(x_0, t_1).$$

5. The joint p.d.f. of the vectors (X_i, Y_i), for $i = 1, \ldots, n$, was given in Eq. (5.10.2). The following relations hold:

$$\sum_{i=1}^n (x_i - \mu_1)^2 = \sum_{i=1}^n x_i^2 - 2\mu_1 \sum_{i=1}^n x_i + n\mu_1^2,$$
$$\sum_{i=1}^n (y_i - \mu_2)^2 = \sum_{i=1}^n y_i^2 - 2\mu_2 \sum_{i=1}^n y_i + n\mu_2^2,$$
$$\sum_{i=1}^n (x_i - \mu_1)(y_i - \mu_2) = \sum_{i=1}^n x_i y_i - \mu_2 \sum_{i=1}^n x_i - \mu_1 \sum_{i=1}^n y_i + n\mu_1\mu_2.$$

Because of these relations, it can be seen that the joint p.d.f. depends on the observed values of the n vectors in the sample only through the values of the five statistics given in this exercise. Therefore, they are jointly sufficient statistics.

7. In each part of this exercise we shall first present the p.d.f. f, and then we shall identify the functions a, b, c_1, d_1, c_2, and d_2 in the form for a two-parameter exponential family given in Exercise 6.

   (a) Let $\theta = (\mu, \sigma^2)$. Then $f(x \mid \theta)$ is as given in the solution to Exercise 24(d) of Sec. 7.3. Therefore, $a(\theta) = \frac{1}{(2\pi\sigma^2)^{1/2}} \exp\left(-\frac{\mu^2}{2\sigma^2}\right)$, $b(x) = 1$, $c_1(\theta) = -\frac{1}{2\sigma^2}$, $d_1(x) = x^2$, $c_2(\theta) = \frac{\mu}{\sigma^2}$, $d_2(x) = x$.

   (b) Let $\theta = (\alpha, \beta)$. Then $f(x \mid \theta)$ is as given in the solution to Exercise 24(f) of Sec. 7.3. Therefore, $a(\theta) = \frac{\beta^\alpha}{\Gamma(\alpha)}$, $b(x) = 1$, $c_1(\theta) = \alpha - 1$, $d_1(x) = \log\ x$, $c_2(\theta) = -\beta$, $d_2(x) = x$.

   (c) Let $\theta = (\alpha, \beta)$. Then $f(x \mid \theta)$ is as given in the solution to Exercise 24(h) of Sec. 7.3. Therefore, $a(\theta) = \frac{\Gamma(\alpha+\beta)}{\Gamma(\alpha)\Gamma(\beta)}$, $b(x) = 1$, $c_1(\theta) = \alpha - 1$, $d_1(x) = \log x$, $c_2(\theta) = \beta - 1$, and $d_2(x) = \log(1 - x)$.

9. By Example 7.5.4, $\hat{p} = \overline{X}_n$. By Exercise 1 of Sec. 7.7, $\hat{p}$ is a sufficient statistic. Therefore, $\hat{p}$ is a minimal sufficient statistic.

11. By Example 7.8.5, the order statistics are minimal jointly sufficient statistics. Therefore, the M.L.E. of θ, all by itself, cannot be a sufficient statistic. (We know from Example 7.6.5 that there is no simple expression for this M.L.E., so we cannot solve this exercise by first deriving the M.L.E. and then checking to see whether it is a sufficient statistic.)

13. By Exercise 11 of Sec. 7.5, $\hat{a} = \min\{X_1, \ldots, X_n\}$ and $\hat{b} = \max\{X_1, \ldots X_n\}$. By Example 7.8.4, $\hat{a}$ and $\hat{b}$ are jointly sufficient statistics. Therefore, $\hat{a}$ and $\hat{b}$ are minimal jointly sufficient statistics.

15. The Bayes estimator of p is given by Eq. (7.4.5). Since $\sum_{i=1}^n x_i$ is a sufficient statistic for p, the Bayes estimator is also a sufficient statistic for p. Hence, this estimator will be a minimal sufficient statistic.

17. The Bayes estimator of μ is given by Eq. (7.4.6). Since $\overline{X}_n$ is a sufficient statistic for μ, the Bayes estimator is also a sufficient statistic. Hence, this estimator will be a minimal sufficient statistic.

7.9 Improving an Estimator

1. The statistic $Y_n = \sum_{i=1}^n X_i^2$ is a sufficient statistic for θ. Since the value of the estimator δ_1 cannot be determined from the value of Y_n alone, δ_1 is inadmissible.

3. The mean of the uniform distribution on the interval $[0, \theta]$ is $\theta/2$ and the variance is $\theta^2/12$. Therefore, $E_\theta(\overline{X}_n) = \theta/2$ and $\text{Var}_\theta(\overline{X}_n) = \theta^2/(12n)$. In turn, it now follows that

$$E_\theta(\delta_1) = \theta \text{ and } \text{Var}_\theta(\delta_1) = \frac{\theta^2}{3n}.$$

   Hence, for $\theta > 0$,

$$R(\theta, \delta_1) = E_\theta[(\delta_1 - \theta)^2] = \text{Var}_\theta(\delta_1) = \frac{\theta^2}{3n}.$$

5. For any constant c,

$$\begin{aligned} R(\theta, cY_n) &= E_\theta[(cY_n - \theta)^2] = c^2 E_\theta(Y_n^2) - 2c\theta E_\theta(Y_n) + \theta^2 \\ &= \left(\frac{n}{n+2}c^2 - \frac{2n}{n+1}c + 1\right)\theta^2. \end{aligned}$$

   Hence, for any given value of n and any given value of $\theta > 0$, $R(\theta, cY_n)$ will be a minimum when c is chosen so that the coefficient of θ^2 in this expression is a minimum. By differentiating with respect to c, we find that the minimizing value of c is $c = (n+2)(n+1)$. Hence, the estimator $(n+2)Y_n/(n+1)$ dominates every other estimator of the form cY_n.

7. (a) Since the value of δ is always 3, $R(\beta, \delta) = (\beta - 3)^2$.

   (b) Since $R(3, \delta) = 0$, no other estimator δ_1 can dominate δ unless it is also true that $R(3, \delta_1) = 0$. But the only way that the M.S.E. of an estimator δ_1 can be 0 is for the estimator δ_1 to be equal to 3 with probability 1. In other words, the estimator δ_1 must be the same as the estimator δ. Therefore, δ is not dominated by any other estimator and it is admissible.

   In other words, the estimator that always estimates the value of β to be 3 is admissible because it is the best estimator to use if β happens to be equal to 3. Of course, it is a poor estimator to use if β happens to be different from 3.

9. Suppose that X has a continuous distribution for which the p.d.f. is f. Then

$$E(X) = \int_{-\infty}^0 xf(x)\,dx + \int_0^\infty xf(x)\,dx.$$

   Suppose first that $E(X) \le 0$. Then

$$\begin{aligned} |E(X)| = -E(X) &= \int_{-\infty}^0 (-x)f(x)\,dx - \int_0^\infty xf(x)\,dx \\ &\le \int_{-\infty}^0 (-x)f(x)\,dx + \int_0^\infty xf(x)\,dx \\ &= \int_{-\infty}^\infty |x|f(x)\,dx = E(|X|). \end{aligned}$$

   A similar proof can be given if X has a discrete distribution or a more general type of distribution, or if $E(X) > 0$.

   Alternatively, the result is immediate from Jensen's inequality, Theorem 4.2.5.

11. Since $\hat{\theta}$ is the M.L.E. of θ, we know from the discussion in Sec. 7.8 that $\hat{\theta}$ is a function of T alone. Since $\hat{\theta}$ is already a function of T, taking the conditional expectation $E(\hat{\theta}|T)$ will not affect $\hat{\theta}$. Hence, $\delta_0 = E(\hat{\theta}|T) = \hat{\theta}$.

13. We shall carry out the analysis for Y_1. The analysis for every other value of i is similar. Since Y_1 must be 0 or 1,

$$\begin{aligned} E(Y_1|T=t) &= \Pr(Y_1 = 1|T=t) = \Pr(X_1 = 0|T=t) \\ &= \frac{\Pr(X_1 = 0 \text{ and } T = t)}{\Pr(T=t)} = \frac{\Pr\left(X_1 = 0 \text{ and } \sum_{i=2}^{n} X_i = t\right)}{\Pr(T=t)}. \end{aligned}$$

The random variables X_1 and $\sum_{i=2}^{n} X_i$ are independent, X_1 has a Poisson distribution with mean θ, and $\sum_{i=2}^{n} X_i$ has a Poisson distribution with mean $(n-1)\theta$. Therefore,

$$\Pr\left(X_1 = 0 \text{ and } \sum_{i=2}^{n} X_i = t\right) = \Pr(X_1 = 0)\Pr\left(\sum_{i=2}^{n} X_i = t\right) = \exp(-\theta) \cdot \frac{\exp(-(n-1)\theta)[(n-1)\theta]^t}{t!}.$$

Also, since T has a Poisson distribution with mean $n\theta$,

$$\Pr(T=t) = \frac{\exp(-n\theta)(n\theta)^t}{t!}$$

It now follows that $E(Y_1|T=t) = ([n-1]/n)^t$.

15. Let $\hat{\theta}$ be the M.L.E. of θ. Then the M.L.E. of $\exp(\theta + 0.125)$ is $\exp(\hat{\theta} + 0.125)$. The M.L.E. of θ is $\overline{X}_n$, so the M.L.E. of $\exp(\theta + 0.125)$ is $\exp(\overline{X}_n + 0.125)$. The M.S.E. of an estimator of the form $\exp(\overline{X}_n + c)$ is

$$\begin{aligned} & E\left[(\exp[\overline{X}_n + c] - \exp[\theta + 0.125])^2\right] \\ &= \text{Var}(\exp[\overline{X}_n + c]) + \left[E(\exp[\overline{X}_n + c]) - \exp(\theta + 0.125)\right]^2 \\ &= \exp(2\theta + 0.25/n + 2c)[\exp(0.25/n) - 1] + [\exp(\theta + 0.125/n + c) - \exp(\theta + 0.125)]^2 \\ &= \exp(2\theta)\{\exp(0.25/n + 2c)[\exp(0.25/n) - 1] + \exp(0.25/n + 2c) - 2\exp(0.125[1 + 1/n] + c) \\ &\quad + \exp(0.5)\} \\ &= \exp(2\theta)\left[\exp(2c)\exp(0.5/n) - 2\exp(c)\exp(0.125[1 + 1/n]) + \exp(0.5)\right]. \end{aligned}$$

Let $a = \exp(c)$ in this last expression. Then we can minimize the M.S.E. simultaneously for all θ by minimizing

$$a^2 \exp(0.5/n) - 2a\exp(0.125[1 + 1/n]) + \exp(0.5).$$

The minimum occurs at $a = \exp(0.125 - 0.375/n)$, so $c = 0.125 - 0.375/n$.

7.10 Supplementary Exercises

1. (a) The prior distribution of θ is the beta distribution with parameters 1 and 1, so the posterior distribution of θ is the beta distribution with parameters $1 + 10 = 11$ and $1 + 25 - 10 = 16$.

   (b) With squared error loss, the estimate to use is the posterior mean, which is 11/27 in this case.

3. The prior distribution of θ is the beta distribution with $\alpha = 3$ and $\beta = 4$, so it follows from Theorem 7.3.1 that the posterior distribution is the beta distribution with $\alpha = 3 + 3 = 6$ and $\beta = 4 + 7 = 11$. The Bayes estimate is the mean of this posterior distribution, namely 6/17.

5. The joint p.d.f. of X_1 and X_2 is

$$\frac{1}{2\pi\sigma_1\sigma_2}\exp\left[-\frac{1}{2\sigma_1^2}(x_1 - b_1\mu)^2 - \frac{1}{2\sigma_2^2}(x_2 - b_2\mu)^2\right].$$

If we let $L(\mu)$ denote the logarithm of this expression, and solve the equation $dL(\mu)/d\mu = 0$, we find that

$$\hat{\mu} = \frac{\sigma_2^2 b_1 x_1 + \sigma_1^2 b_2 x_2}{\sigma_2^2 b_1^2 + \sigma_1^2 b_2^2}.$$

7. The joint p.d.f. of X_1, X_2, X_3 is

$$f(\boldsymbol{x}|\theta) = \frac{1}{\theta}\exp\left(-\frac{1}{\theta}x_1\right)\cdot\frac{1}{2\theta}\exp\left(-\frac{1}{2\theta}x_2\right)\cdot\frac{1}{3\theta}\exp\left(-\frac{1}{3\theta}x_3\right) = \frac{1}{6\theta^3}\exp\left[-\left(x_1 + \frac{x_2}{2} + \frac{x_3}{3}\right)\frac{1}{\theta}\right].$$

   (a) By solving the equation $\partial \log(f)/\partial\theta = 0$, we find that

$$\hat{\theta} = \frac{1}{3}\left(X_1 + \frac{1}{2}X_2 + \frac{1}{3}X_3\right).$$

   (b) In terms of ψ, the joint p.d.f. of X_1, X_2, X_3 is

$$f(\boldsymbol{x} \mid \psi) = \frac{\psi^3}{6}\exp\left[-\left(x_1 + \frac{1}{2}x_2 + \frac{1}{3}x_3\right)\psi\right].$$

Since the prior p.d.f. of ψ is

$$\xi(\psi) \propto \psi^{\alpha-1}\exp(-\beta\psi),$$

it follows that the posterior p.d.f. is

$$\xi(\psi \mid \boldsymbol{x}) \propto f(\boldsymbol{x} \mid \psi)\xi(\psi) \propto \psi^{\alpha+2}\exp\left[-\left(\beta + x_1 + \frac{1}{2}x_2 + \frac{1}{3}x_3\right)\psi\right].$$

Hence, the posterior distribution of ψ is the gamma distribution with parameters $\alpha + 3$ and $\beta + x_1 + x_2/2 + x_3/3$.

9. The posterior p.d.f. of θ given $X = x$ satisfies the relation

$$\xi(\theta \mid \boldsymbol{x}) \propto f(x \mid \theta)\xi(\theta) \propto \exp(-\theta), \text{ for } \theta > x.$$

Hence,

$$\xi(\theta \mid \boldsymbol{x}) = \begin{cases} \exp(x - \theta) & \text{for } \theta > x, \\ 0 & \text{otherwise.} \end{cases}$$

(a) The Bayes estimator is the mean of this posterior distribution, $\hat{\theta} = x + 1$.

(b) The Bayes estimator is the median of this posterior distribution, $\hat{\theta} = x + \log\, 2$.

11. Under these conditions, X has a binomial distribution with parameters n and $\theta = \frac{1}{2} \cdot \frac{1}{2} + \frac{1}{2}p = \frac{1}{4} + \frac{1}{2}p$. Since $0 \le p \le 1$, it follows that $1/4 \le \theta \le 3/4$. Hence, as in Exercise 3 of Sec. 7.5, the M.L.E. of $\hat{\theta}$ is

$$\hat{\theta} = \begin{cases} \dfrac{X}{n} & \text{if } \dfrac{1}{4} \le \dfrac{X}{n} \le \dfrac{3}{4}, \\ \dfrac{1}{4} & \text{if } \dfrac{X}{n} < \dfrac{1}{4}, \\ \dfrac{3}{4} & \text{if } \dfrac{X}{n} > \dfrac{3}{4}. \end{cases}$$

It then follows from Theorem 7.6.1 that the M.L.E. of p is $\hat{p} = 2(\hat{\theta} - 1/4)$.

13. The Bayes estimate of θ will be the median of the posterior Pareto distribution. This will be the value m such that

$$\frac{1}{2} = \int_m^\infty \frac{5}{\theta^6}\, d\theta = \frac{1}{m^5}.$$

Hence, $\hat{\theta} = m = 2^{1/5}$.

15. The joint p.d.f. of the observations is $\alpha^n\, {x_0}^{n\alpha} / \left(\prod_{i=1}^n x_i\right)^{\alpha+1}$ for $\min\{x_1, \ldots, x_n\} \ge x_0$. This p.d.f. is maximized when x_0 is made as large as possible. Thus,

$$\hat{x}_0 = \min\{X_1, \ldots, X_n\}.$$

17. It follows from Exercise 15 that $\hat{x}_0 = \min\{X_1, \ldots, X_n\}$ will again be the M.L.E. of x_0 , since this value of x_0 maximizes the likelihood function regardless of the value of α. If we substitute $\hat{x}_0$ for x_0 and let $L(\alpha)$ denote the logarithm of the resulting likelihood function, which was given in Exercise 15, then

$$L(\alpha) = n \log \alpha + n\ \alpha \log \hat{x}_0 - (\alpha + 1) \sum_{i=1}^n \log x_i$$

and

$$\frac{dL(\alpha)}{d\alpha} = \frac{n}{\alpha} + n \log \hat{x}_0 - \sum_{i=1}^n \log x_i.$$

Hence, by setting this expression equal to 0, we find that

$$\hat{\alpha} = \left(\frac{1}{n} \sum_{i=1}^n \log x_i - \log \hat{x}_0\right)^{-1}.$$

19. The p.f. of X is

$$f(x|n,p) = \binom{n}{x} p^x (1-p)^{n-x}.$$

The M.L.E. of n will be the value that maximizes this expression for given values of x and p. The ratio given in the hint to this exercise reduces to

$$R = \frac{n+1}{n+1-x} \quad (1-p).$$

Since R is a decreasing function of n, it follows that $f(x|n,p)$ will be maximized at the smallest value of n for which $R < 1$. After some algebra, it is found that $R < 1$ if and only if $n > x/p - 1$. Hence, n will be the smallest integer greater than $x/p - 1$. If $x/p - 1$ is itself an integer, then $x/p - 1$ and x/p are both M.L.E.'s.

21. The Bayes estimator of θ is the mean of the posterior distribution of θ, and the expected loss or M.S.E. of this estimator is the variance of the posterior distribution. This variance, as given by Eq. (7.3.2), is

$$\nu_1^2 = \frac{(100)(25)}{100 + 25n} = \frac{100}{n+4}.$$

Hence, n must be chosen to minimize

$$\frac{100}{n+4} + \frac{1}{4}\, n.$$

By setting the first derivative equal to 0, it is found that the minimum occurs when $n = 16$.

Chapter 8

Sampling Distributions of Estimators

8.1 The Sampling Distribution of a Statistic

1. The c.d.f. of $U = \max\{X_1, \ldots, X_n\}$ is

$$F(u) = \begin{cases} 0 & \text{for } u \le 0, \\ (u/\theta)^n & \text{for } 0 < u < \theta, \\ 1 & \text{for } u \ge \theta. \end{cases}$$

Since $U \le \theta$ with probability 1, the event that $|U - \theta| \le 0.1\theta$ is the same as the event that $U \ge 0.9\theta$. The probability of this is $1 - F(0.9\theta) = 1 - 0.9^n$. In order for this to be at least 0.95, we need $0.9^n \le 0.05$ or $n \ge \log(0.05)/\log(0.9) = 28.43$. So $n \ge 29$ is needed.

3. Once again, $\overline{X}_n$ has the normal distribution with mean θ and variance $4/n$. Hence, the random variable $Z = (\overline{X}_n - \theta)/(2/\sqrt{n})$ will has the standard normal distribution. Therefore,

$$\begin{aligned} E_\theta(|\overline{X}_n - \theta|) &= \frac{2}{\sqrt{n}} E_\theta(|Z|) = \frac{2}{\sqrt{n}} \int_{-\infty}^{\infty} |z| \frac{1}{\sqrt{2\pi}} \exp(-z^2/2) dz = 2\sqrt{\frac{2}{n\pi}} \int_0^\infty z \exp(-z^2/2) dz \\ &= 2\sqrt{\frac{2}{n\pi}}. \end{aligned}$$

But $2\sqrt{2/(n\pi)} \le 0.1$ if and only if $n \ge 800/\pi = 254.6$. Hence, n must be least 255.

5. When $p = 0.2$, the random variable $Z_n = n\overline{X}_n$ will have a binomial distribution with parameters n and $p = 0.2$, and

$$\Pr(|\overline{X}_n - p| \le 0.1) = \Pr(0.1n \le Z_n \le 0.3n).$$

The value of n for which this probability will be at least 0.75 must be determined by trial and error from the table of the binomial distribution in the back of the book. For $n = 8$, the probability becomes

$$\Pr(0.8 \le Z_8 \le 2.4) = \Pr(Z_8 = 1) + \Pr(Z_8 = 2) = 0.3355 + 0.2936 = 0.6291.$$

For $n = 9$, we have

$$\Pr(0.9 \le Z_9 \le 2.7) = \Pr(Z_9 = 1) + \Pr(Z_9 = 2) = 0.3020 + 0.3020 = 0.6040.$$

For $n = 10$, we have

$$\Pr(1 \le Z_{10} \le 3) = \Pr(Z_{10} = 1) + \Pr(Z_{10} = 2) + \Pr(Z_{10} = 3) = 0.2684 + 0.3020 + 0.2013 = 0.7717.$$

Hence, $n = 10$ is sufficient.

It should be noted that although a sample size of $n = 10$ will meet the required conditions, a sample size of $n = 11$ will not meet the required conditions. For $n = 11$, we would have

$$\Pr(1.1 \le Z_{11} \le 3.3) = \Pr(Z_{11} = 2) + \Pr(Z_{11} = 3).$$

Thus, only two terms of the binomial distribution for $n = 11$ are included, whereas three terms of binomial distribution for $n = 10$ were included.

7. It follows from the results given in the solution to Exercise 6 that, when $p = 0.2$,

$$E_p(|\overline{X}_n - p|^2) = \mathrm{Var}(\overline{X}_n) = \frac{0.16}{n},$$

and $0.16/n \le 0.01$ if and only if $n \ge 16$.

9. The M.L.E. is $\hat{\theta} = n/T$, where T was shown to have the gamma distribution with parameters n and θ. Let $G(\cdot)$ denote the c.d.f. of the sampling distribution of T. Let $H(\cdot)$ be the c.d.f. of the sampling distribution of $\hat{\theta}$. Then $H(t) = 0$ for $t \le 0$, and for $t > 0$,

$$H(t) = \Pr(\hat{\theta} \le t) = \Pr\left(\frac{n}{T} \le t\right) = \Pr\left(T \ge \frac{n}{t}\right) = 1 - G\left(\frac{n}{t}\right).$$

8.2 The Chi-Square Distributions

1. The distribution of $20T/0.09$ is the χ^2 distribution with 20 degrees of freedom. We can write $\Pr(T \le c) = \Pr(20T/0.09 \le 20c/0.09)$. In order for this probability to be 0.9, we need $20c/0.09$ to equal the 0.9 quantile of the χ^2 distribution with 20 degrees of freedom. That quantile is 28.41. Set $28.41 = 20c/0.09$ and solve for $c = 0.1278$.

3. The median of each distribution is found from the table of the χ^2 distribution given at the end of the book.

5. We must determine $\Pr(X^2 + Y^2 + Z^2 \le 1)$. Since $X^2 + Y^2 + Z^2$ has the χ^2 distribution with three degrees of freedom, it is found from the table at the end of the book that the required probability is slightly less than 0.20.

7. By the probability integral transformation, we know that $T_i = F_i(X_i)$ has a uniform distribution on the interval $[0, 1]$. Now let $Z_i = -\log T_i$. We shall determine the p.d.f. g of Z_i. The p.d.f. of T_i is

$$f(t) = \begin{cases} 1 & \text{for } 0 < t < 1, \\ 0 & \text{otherwise.} \end{cases}$$

Since $T_i = \exp(-Z_i)$, we have $dt/dz = -\exp(-z)$. Therefore, for $z > 0$,

$$g(z) = f(\exp(-z))\left|\frac{dt}{dz}\right| = \exp(-z).$$

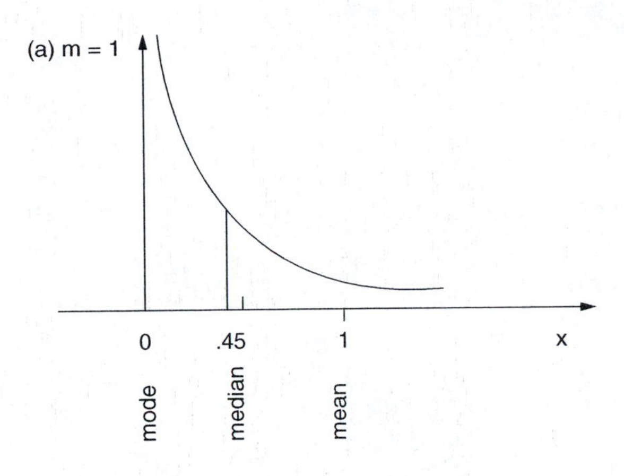

Figure S.8.1: First figure for Exercise 3 of Sec. 8.2.

Thus, it is now seen that Z_i has the exponential distribution with parameter $\beta = 1$ or, in other words, the gamma distribution with parameters $\alpha = 1$ and $\beta = 1$. Therefore, by Exercise 1 of Sec. 5.7, $2Z_i$ has the gamma distribution with parameters $\alpha = 1$ and $\beta = 1/2$. Finally, by Theorem 5.7.7 $\sum_{i=1}^{n} 2Z_i$ will have the gamma distribution with parameters $\alpha = n$ and $\beta = 1/2$ or, equivalently, the χ^2 distribution with 2n degrees of freedom.

9. It is known that $\overline{X}_n$ has the normal distribution with mean μ and variance σ^2/n. Therefore, $(\overline{X}_n - \mu)/(\sigma/\sqrt{n})$ has a standard normal distribution and the square of this variable has the χ^2 distribution with one degree of freedom.

11. The simplest way to determine the mean is to calculate $E(X^{1/2})$ directly, where X has the χ^2 distribution with n degrees of freedom. Thus,

$$\begin{aligned} E(X^{1/2}) &= \int_0^\infty x^{1/2} \frac{1}{2^{n/2}\Gamma(n/2)} x^{(n/2)-1} \exp(-x/2)dx = \frac{1}{2^{n/2}\Gamma(n/2)} \int_0^\infty x^{(n-1)/2} \exp(-x/2)dx \\ &= \frac{1}{2^{n/2}\Gamma(n/2)} \cdot 2^{(n+1)/2}\Gamma[(n+1)/2] = \frac{\sqrt{2}\Gamma[(n+1)/2]}{\Gamma(n/2)}. \end{aligned}$$

13. We already found that the distribution of $W = n\widehat{\sigma^2}/\sigma^2$ is the χ^2 distribution with n degrees of freedom, which is also the gamma distribution with parameters $n/2$ and $1/2$. If we multiply a gamma random variable by a constant, we change its distribution to another gamma distribution with the same first parameter and the second parameter gets divided by the constant. (See Exercise 1 in Sec. 6.3.) Since $\widehat{\sigma^2} = (\sigma^2/n)W$, we see that the distribution of $\widehat{\sigma^2}$ is the gamma distribution with parameters $n/2$ and $n/(2\sigma^2)$.

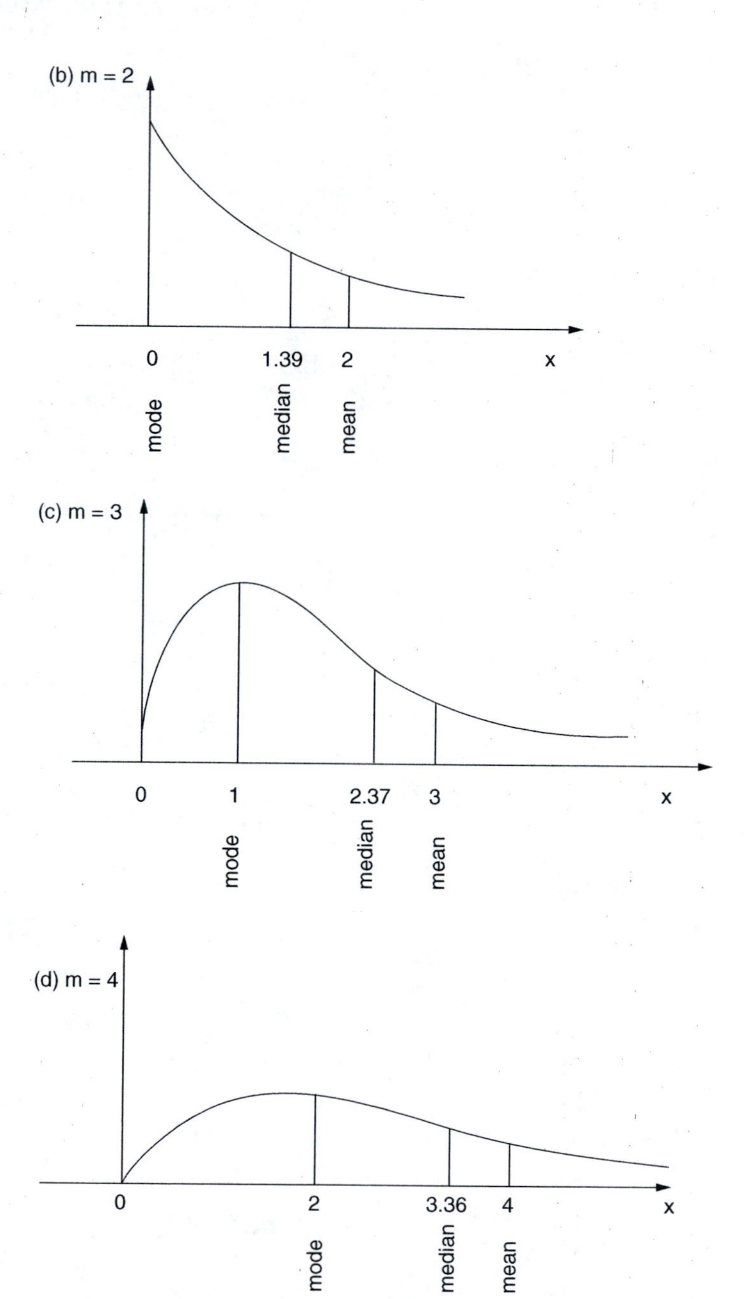

Figure S.8.2: Second figure for Exercise 3 of Sec. 8.2.

8.3 Joint Distribution of the Sample Mean and Sample Variance

1. We found that $U = n\widehat{\sigma^2}/\sigma^2$ has the χ^2 distribution with $n-1$ degrees of freedom, which is also the gamma distribution with parameters $(n-1)/2$ and $1/2$. If we multiply a gamma random variable by a number c, we change the second parameter by dividing it by c. So, with $c = \sigma^2/n$, we find that $cU = \hat{\sigma^2}$ has the gamma distribution with parameters $(n-1)/2$ and $n/(2\sigma^2)$.

3. (a) Consider the matrix

$$\boldsymbol{A} = \begin{bmatrix} 1/\sqrt{2} & 1/\sqrt{2} \\ a_1 & a_2 \end{bmatrix}.$$

For $\boldsymbol{A}$ to be orthogonal, we must have $a_1^2 + a_2^2 = 1$ and $\dfrac{1}{\sqrt{2}}a_1 + \dfrac{1}{\sqrt{2}}a_2 = 0$. It follows from the second equation that $a_1 = -a_2$ and, in turn, from the first equation that $a_1^2 = 1/2$. Hence, either the pair of values $a_1 = 1/\sqrt{2}$ and $a_2 = -1/\sqrt{2}$ or the pair $a_1 = -1/\sqrt{2}$ and $a_2 = 1/\sqrt{2}$ will make $\boldsymbol{A}$ orthogonal.

(b) Consider the matrix

$$\boldsymbol{A} = \begin{bmatrix} 1/\sqrt{3} & 1/\sqrt{3} & 1/\sqrt{3} \\ a_1 & a_2 & a_3 \\ b_1 & b_2 & b_3 \end{bmatrix}$$

For $\boldsymbol{A}$ to be orthogonal, we must have

$$a_1^2 + a_2^2 + a_3^2 = 1$$

and

$$\frac{1}{\sqrt{3}}a_1 + \frac{1}{\sqrt{3}}a_2 + \frac{1}{\sqrt{3}}a_3 = 0.$$

Therefore, $a_3 = -a_1 - a_2$ and it follows from the first equation that

$$a_1^2 + a_2^2 + (a_1 + a_2)^2 = 2a_1^2 + 2a_2^2 + 2a_1a_2 = 1.$$

Any values of a_1 and a_2 satisfying this equation can be chosen. We shall use $a_1 = 2/\sqrt{6}$ and $a_2 = -1/\sqrt{6}$. Then $a_3 = -1/\sqrt{6}$.

Finally, we must have $b_1^2 + b_2^2 + b_3^2 = 1$ as well as

$$\frac{1}{\sqrt{3}}b_1 + \frac{1}{\sqrt{3}}b_2 + \frac{1}{\sqrt{3}}b_3 = 0$$

and

$$\frac{2}{\sqrt{6}}b_1 - \frac{1}{\sqrt{6}}b_2 - \frac{1}{\sqrt{6}}b_3 = 0.$$

This final pair of equations can be rewritten as

$$b_2 + b_3 = -b_1 \quad \text{and} \quad b_2 + b_3 = 2b_1.$$

Therefore, $b_1 = 0$ and $b_2 = -b_3$. Since we must have $b_2^2 + b_3^2 = 1$, it follows that we can use either $b_2 = 1/\sqrt{2}$ and $b_3 = -1/\sqrt{2}$ or $b_2 = -1/\sqrt{2}$ and $b_3 = 1/\sqrt{2}$. Thus, one orthogonal matrix is

$$\boldsymbol{A} = \begin{bmatrix} 1/\sqrt{3} & 1/\sqrt{3} & 1/\sqrt{3} \\ 2/\sqrt{6} & -1/\sqrt{6} & -1/\sqrt{6} \\ 0 & 1/\sqrt{2} & -1/\sqrt{2} \end{bmatrix}$$

5. Let $Z_i = (X_i - \mu)/\sigma$ for $i = 1, 2$. Then Z_1 and Z_2 are independent and each has a standard normal distribution. Next, let $Y_1 = (Z_1 + Z_2)/\sqrt{2}$ and $Y_2 = (Z_1 - Z_2)/\sqrt{2}$. Then the 2×2 matrix of coefficients of this transformation is

$$\boldsymbol{A} = \begin{bmatrix} 1/\sqrt{2} & 1/\sqrt{2} \\ 1/\sqrt{2} & -1/\sqrt{2} \end{bmatrix}.$$

Since the matrix $\boldsymbol{A}$ is orthogonal, it follows from Theorem 8.3.4 that Y_1 and Y_2 are also independent and each has a standard normal distribution. Finally, let $W_1 = X_1 + X_2$ and $W_2 = X_1 - X_2$. Then $W_1 = \sqrt{2}\sigma Y_1 + 2\mu$ and $W_2 = \sqrt{2}\sigma Y_2$. Since Y_1 and Y_2 are independent, it now follows from Exercise 15 of Sec. 3.9 that W_1 and W_2 are also independent.

7. (a) The random variable $V = n\hat{\sigma}^2/\sigma^2$ has a χ^2 distribution with $n - 1$ degrees of freedom. The required probability can be written in the form $\Pr(V \leq 1.5n) \geq 0.95$. By trial and error, it is found that for $n = 20$, V has 19 degrees of freedom and $\Pr(V \leq 30) < 0.95$. However, for $n = 21$, V has 20 degrees of freedom and $\Pr(V \leq 31.5) > 0.95$.

   (b) The required probability can be written in the form

$$\Pr\left(\frac{n}{2} \leq V \leq \frac{3n}{2}\right) = \Pr\left(V \leq \frac{3n}{2}\right) - \Pr\left(V \leq \frac{n}{2}\right),$$

   where V again has the χ^2 distribution with $n - 1$ degrees of freedom. By trial and error, it is found that for $n = 12$, V has 11 degrees of freedom and

$$\Pr(V \leq 18) - \Pr(V \leq 6) = 0.915 - 0.130 < 0.8.$$

   However, for $n = 13$, V has 12 degrees of freedom and

$$\Pr(V \leq 19.5) - \Pr(V \leq 6.5) = 0.919 - 0.113 > 0.8.$$

9. The sample mean and the sample variance are independent. Therefore, the information that the sample variance is closer to σ^2 in one sample than it is in the other sample provides no information about which of the two sample means will be closer to μ. In other words, in either sample, the conditional distribution of $\overline{X}_n$, given the observed value of the sample variance, is still the normal distribution with mean μ and variance σ^2/n.

8.4 The t Distributions

1. $E(X^2) = c\int_{-\infty}^{\infty} x^2\left(1 + \frac{x^2}{n}\right)^{-(n+1)/2} dx = 2c\int_0^{\infty} x^2\left(1 + \frac{x^2}{n}\right)^{-(n+1)/2} dx,$

   where $c = \frac{\Gamma[(n+1)/2]}{(n\pi)^{1/2}\Gamma(n/2)}$. If y is defined as in the hint for this exercise, then $x = \left(\frac{ny}{1-y}\right)^{1/2}$ and $\frac{dx}{dy} = \frac{\sqrt{n}}{2}y^{-1/2}(1-y)^{-3/2}$. Therefore,

$$\begin{aligned}
E(X^2) &= \sqrt{n}(const.)\int_0^1 \frac{ny}{1-y}\left(1 + \frac{y}{1-y}\right)^{-(n+1)/2} y^{-1/2}(1-y)^{-3/2}dy \\
&= n^{3/2}(const.)\int_0^1 y^{1/2}(1-y)^{(n-4)/2}dy \\
&= n^{3/2}(const.)\frac{\Gamma(3/2)\Gamma[(n-2)/2]}{\Gamma[(n+1)/2]} = n\pi^{-1/2}\Gamma\left(\frac{3}{2}\right) \cdot \frac{\Gamma[(n-2)/2]}{\Gamma(n/2)} \\
&= n\pi^{-1/2}\left(\frac{1}{2}\sqrt{\pi}\right)\frac{1}{[(n-2)/2]} = \frac{n}{n-2}.
\end{aligned}$$

Since $E(X) = 0$, it now follows that $\text{Var}(X) = n/(n-2)$.

3. $X_1 + X_2$ has the normal distribution with mean 0 and variance 2. Therefore, $Y = (X_1 + X_2)/\sqrt{2}$ has a standard normal distribution. Also, $Z = X_3^2 + X_4^2 + X_5^2$ has the χ^2 distribution with 3 degrees of freedom, and Y and Z are independent. Therefore, $U = \dfrac{Y}{(Z/3)^{1/2}}$ has the t distribution with 3 degrees of freedom. Thus, if we choose $c = \sqrt{3/2}$, the given random variable-will be equal to U.

5. Let $\overline{X}_2 = (X_1 + X_2)/2$ and $S_2^2 = \sum_{i=1}^{2}(X_i - \overline{X}_2)^2$. Then

$$W = \frac{(X_1 + X_2)^2}{(X_1 - X_2)^2} = \frac{2\overline{X}_2^2}{S_2^2}.$$

It follows from Eq. (8.4.4) that $U = \sqrt{2}\,\overline{X}_2/\sqrt{S_2^2}$ has the t distribution with one degree of freedom. Since $W = U^2$, we have

$$\Pr(W < 4) = \Pr(-2 < U < 2) = 2\Pr(U < 2) - 1.$$

It can be found from a table of the t distribution with one degree of freedom that $\Pr(U < 2)$ is just slightly greater than 0.85. Hence, $\Pr(W < 4) = 0.70$.

7. According to Theorem 5.7.4,

$$\lim_{m\to\infty} \frac{(2\pi)^{1/2}(m+1/2)^m \exp(-m-1/2)}{\Gamma(m+1/2)} = 1,$$
$$\lim_{m\to\infty} \frac{(2\pi)^{1/2}(m)^{m-1/2} \exp(-m)}{\Gamma(m)} = 1.$$

Taking the ratio of the above and dividing by $m^{1/2}$, we get

$$\begin{aligned}
\lim_{m\to\infty} \frac{\Gamma(m+1/2)}{\Gamma(m)m^{1/2}} &= \lim_{m\to\infty} \frac{(2\pi)^{1/2}(m+1/2)^m \exp(-m-1/2)}{(2\pi)^{1/2}(m)^{m-1/2}\exp(-m)m^{1/2}} \\
&= \lim_{m\to\infty} \left(\frac{m+1/2}{m}\right)^m \exp(-1/2) \\
&= 1,
\end{aligned}$$

where the last equality follows from Theorem 5.3.3 applied to $(1 + 1/(2m))^m$.

8.5 Confidence Intervals

1. We need to show that

$$\Pr\left[\overline{X}_n - \Phi^{-1}\left(\frac{1+\gamma}{2}\right)\frac{\sigma}{n^{1/2}} < \mu < \overline{X}_n + \Phi^{-1}\left(\frac{1+\gamma}{2}\right)\frac{\sigma}{n^{1/2}}\right] = \gamma. \tag{S.8.1}$$

By subtracting $\overline{X}_n$ from all three sides of the above inequalities and then dividing all three sides by $\sigma/n^{1/2} > 0$, we can rewrite the probability in (S.8.1) as

$$\Pr\left[-\Phi^{-1}\left(\frac{1+\gamma}{2}\right) < \frac{\mu - \overline{X}_n}{\sigma/n^{1/2}} < \Phi^{-1}\left(\frac{1+\gamma}{2}\right)\right].$$

The random variable $(\mu - \overline{X}_n)/(\sigma/n^{1/2})$ has a standard normal distribution no matter what μ and σ^2 are. And the probability that a standard normal random variable is between $-\Phi^{-1}([1+\gamma]/2)$ and $\Phi^{-1}([1+\gamma]/2)$ is $(1+\gamma)/2 - [1-(1+\gamma)/2] = \gamma$.

3. The endpoints of the confidence interval are $\overline{X}_n - c\sigma'/n^{1/2}$ and $\overline{X}_n + c\sigma'/n^{1/2}$. Therefore, $L = 2\sigma'/n^{1/2}$ and $L^2 = 4c^2\sigma'^2/n$. Since

$$W = \frac{\sum_{i=1}^{n}(X_i - \overline{X}_n)^2}{\sigma^2}$$

has the χ^2 distribution with $n-1$ degrees of freedom, $E(W) = n-1$. Therefore, $E(\sigma'^2) = E(\sigma^2 W/[n-1]) = \sigma^2$. It follows that $E(L^2) = 4c^2\sigma^2/n$. As in the text, c must be the $(1+\gamma)/2$ quantile of the t distribution with $n-1$ degrees of freedom.

(a) Here, $(1+\gamma)/2 = 0.975$. Therefore, from a table of the t distribution with $n-1 = 4$ degrees of freedom it is found that $c = 2.776$. Hence, $c^2 = 7.706$ and $E(L^2) = 4(7.706)\sigma^2/5 = 6.16\sigma^2$.

(b) For the t distribution with 9 degrees of freedom, $c = 2.262$. Hence, $E(L^2) = 2.05\sigma^2$.

(c) Here, $c = 2.045$ and $E(L^2) = 0.56\sigma^2$.

It should be noted from parts (a), (b), and (c) that for a fixed value of γ, $E(L^2)$ decreases as the sample size n increases.

(d) Here, $\gamma = 0.90$, so $(1+\gamma)/2 = 0.95$. It is found that $c = 1.895$. Hence, $E(L^2) = 4(1.895)^2\sigma^2/8 = 1.80\sigma^2$.

(e) Here, $\gamma = 0.95$, so $(1+\gamma)/2 = 0.975$ and $c = 2.365$. Hence, $E(L^2) = 2.80\sigma^2$.

(f) Here, $\gamma = 0.99$, so $(1+\gamma)/2 = 0.995$ and $c = 3.499$. Hence, $E(L^2) = 6.12\sigma^2$.

It should be noted from parts (d), (e), and (f) that for a fixed sample size n, $E(L^2)$ increases as γ increases.

5. Since $\sum_{i=1}^{n}(X_i - \overline{X}_n)^2/\sigma^2$ has a χ^2 distribution with $n-1$ degrees of freedom, it is possible to find constants c_1 and c_2 which satisfy the relation given in the hint for this exercise. (As explained in this section, there are an infinite number of different pairs of values of c_1 and c_2 that might be used.) The relation given in the hint can be rewritten in the form

$$\Pr\left[\frac{1}{c_2}\sum_{i=1}^{n}(X_i - \overline{X}_n)^2 < \sigma^2 < \frac{1}{c_1}\sum_{i=1}^{n}(X_i - \overline{X}_n)^2\right] = \gamma.$$

Therefore, the interval with endpoints equal to the observed values of $\sum_{i=1}^{n}(X_i - \overline{X}_n)^2/c_2$ and $\sum_{i=1}^{n}(X_i - \overline{X}_n)^2/c_1$ will be a confidence interval for σ^2 with confidence coefficient γ.

7. The average of the $n = 20$ values is $\overline{x}_n = 156.85$, and $\sigma' = 22.64$. The appropriate t distribution quantile is $T_{19}^{-1}(0.95) = 1.729$. The endpoints of the confidence interval are then $156.85 \pm 22.64 \times 1.729/20^{1/2}$. Completing the calculation, we get the interval $(148.1, 165.6)$.

9. (a) The interval between the smaller an larger values is $(4.7, 5.3)$.

(b) The values of θ consistent with the observed data are those between $5.3 - 0.5 = 4.8$ and $4.7 + 0.5 = 5.2$.

(c) The interval in part (a) contains the set of all possible θ values, hence it is larger than the set of possible θ values.

(d) The value of Z is $5.3 - 4.7 = 0.6$.

(e) According to (8.5.15),

$$\Pr(|\overline{X}_2 - \theta| < 0.1 | Z = 0.6) = \frac{2 \times 0.1}{1 - 0.6} = 0.5.$$

11. The variance stabilizing transformation is $\alpha(x) = \arcsin(x^{1/2})$, and the approximate distribution of $\alpha(\overline{X}_n)$ is the normal distribution with mean $\alpha(p)$ and variance $1/n$. So,

$$\Pr\left(\arcsin(\overline{X}_n^{1/2}) - \Phi^{-1}([1+\gamma]/2)n^{-1/2} < \arcsin p^{1/2} < \arcsin(\overline{X}_n^{1/2}) + \Phi^{-1}([1+\gamma]/2)n^{-1/2}\right) \approx \gamma.$$

This would make the interval with endpoints

$$\arcsin(\overline{x}_n^{1/2}) \pm \Phi^{-1}([1+\gamma]/2)n^{-1/2} \tag{S.8.2}$$

an approximate coefficient γ confidence interval for $\arcsin(p^{1/2})$. The transformation $\alpha(x)$ has an inverse $\alpha^{-1}(y) = \sin^2(y)$ for $0 \le y \le \pi/2$. If the endpoints in (S.8.2) are between 0 and $\pi/2$, then the interval with endpoints

$$\sin^2\left(\arcsin(\overline{x}_n^{1/2}) \pm \Phi^{-1}([1+\gamma]/2)n^{-1/2}\right) \tag{S.8.3}$$

will be an approximate coefficient γ confidence interval for p. If the lower endpoint in (S.8.2) is negative replace the lower endpoint in (S.8.3) by 0. If the upper endpoint in (S.8.2) is greater than $\pi/2$, replace the upper endpoint in (S.8.3) by 1. With these modifications, the interval with the endpoints in (S.8.3) becomes an approximate coefficient γ confidence interval for p.

8.6 Bayesian Analysis of Samples from a Normal Distribution

1. Since X has the normal distribution with mean μ and variance $1/\tau$, we know that Y has the normal distribution with mean $a\mu + b$ and variance a^2/τ. Therefore, the precision of Y is τ/a^2.

3. The joint p.d.f. $f_n(\boldsymbol{x}|\tau)$ of $X_1, \ldots, X_n$ is given shortly after Definition 8.6.1 in the text, and the prior p.d.f. $\xi(\tau)$ is proportional to the expression $\xi_2(\tau)$ in the proof of Theorem 8.6.1. Therefore, the posterior p.d.f. of τ satisfies the following relation:

$$\begin{aligned}\xi(\tau \mid \boldsymbol{x}) &\propto f_n(\boldsymbol{x}|\,\tau)\xi(\tau) \propto \tau^{n/2}\exp\left\{\left[-\frac{1}{2}\sum_{i=1}^{n}(x_i-\mu)^2\right]\tau\right\}\tau^{\alpha_0-1}\exp(-\beta_0\tau)\\ &= \tau^{\alpha_0+(n/2)-1}\exp\left\{-\left[\beta_0+\frac{1}{2}\sum_{i=1}^{n}(x_i-\mu)^2\right]\tau\right\}.\end{aligned}$$

It can now be seen that this posterior p.d.f. is, except for a constant factor, the p.d.f. of the gamma distribution specified in the exercise.

5. Since $E(\tau) = \alpha_0/\beta_0 = 1/2$ and $\text{Var}(\tau) = \alpha_0/\beta_0^2 = 1/3$, then $\alpha_0 = 2$ and $\beta_0 = 4$. Also, $\mu_0 = E(\mu) = -5$. Finally, $\text{Var}(\mu) = \beta_0/[\lambda_0(\alpha_0 - 1)] = 1$. Therefore, $\lambda_0 = 4$.

7. Since $E(\tau) = \alpha_0/\beta_0 = 1$ and $\text{Var}(\tau) = \alpha_0/\beta_0^2 = 4$, then $\alpha_0 = \beta_0 = 1/4$. But $E(\mu)$ exists only if $\alpha_0 > 1/2$.

9. (a) The posterior hyperparameters are computed in the example. The degrees of freedom are $2\alpha_1 = 22$, so the quantile from the t distribution is $T_{22}^{-1}([1+.9]/2) = 1.717$, and the interval is

$$\mu_1 \pm 1.717\left(\frac{\beta_1}{\lambda_1\alpha_1}\right)^{1/2} = 183.95 \pm 1.717\left(\frac{50925.37}{20\times 11}\right)^{1/2} = (157.83, 210.07).$$

(b) This interval has endpoints $182.17 \pm (88678.5/[17\times 18])^{1/2} T_{17}^{-1}(0.95)$. With $T_{17}^{-1}(0.95) = 1.740$, we get the interval $(152.55, 211.79)$.

11. It follows from Theorem 8.6.1 that $\mu_1 = 80/81$, $\lambda_1 = 81/8$, $\alpha_1 = 9$, and $\beta_1 = 491/81$. Therefore, if Eq. (8.6.9) is applied to this posterior distribution, it is seen that the random variable $U = (3.877)(\mu - 0.988)$ has the t distribution with 18 degrees of freedom. Therefore, it is found from a table $\Pr(-2.101 < Y < 2.101) = 0.95$. An equivalent statement is $\Pr(0.446 < \mu < 1.530) = 0.95$. This interval will be the shortest one having probability 0.95 because the center of the interval is μ_1, the point where the p.d.f. of μ is a maximum. Since the p.d.f. of μ decreases as we move away from μ_1 in either direction, it follows that an interval having given length will have the maximum probability when it is centered at μ_1.

13. It follows from Theorem 8.6.1 that $\mu_1 = 67/33$, $\lambda_1 = 33/4$, $\alpha_1 = 7$, and $\beta_1 = 367/33$. In calculating the value of β_1, we have used the relation

$$\sum_{i=1}^{n}(x_i - \overline{x}_n)^2 = \sum_{i=1}^{n} x_i^2 - n\overline{x}_n^2.$$

If Theorem 8.6.2 is now applied to this posterior distribution, it is seen that the random variable $U = (2.279)(\mu - 2.030)$ has the t distribution with 14 degrees of freedom. Therefore, it is found from a table that $\Pr(-2.977 < Y < 2.977) = 0.99$. An equivalent statement is $\Pr(0.724 < \mu < 3.336) = 0.99$.

15. (a) The posterior hyperparameters are

$$\begin{aligned}
\mu_1 &= \frac{2\times 3.5 + 11\times 7.2}{2+11} = 6.63,\\
\lambda_1 &= 2 + 11 = 13,\\
\alpha_1 &= 2 + \frac{11}{2} = 7.5,\\
\beta_1 &= 1 + \frac{1}{2}\left(20.3 + \frac{2\times 11}{2+11}(7.2-3.5)^2\right) = 22.73.
\end{aligned}$$

(b) The interval should run between the values $\mu_1 \pm (\beta_1/[\lambda_1\alpha_1])^{1/2} T_{2\alpha_1}^{-1}(0.975)$. From the table of the t distribution in the book, we obtain $T_{15}^{-1}(0.975) = 2.131$. The interval is then $(5.601, 7.659)$.

17. Using just the first ten observations, we have $\overline{x}_n = 1.379$ and $s_n^2 = 0.9663$. This makes $\mu_1 = 1.379$, $\lambda_1 = 10$, $\alpha_1 = 4.5$, and $\beta_1 = 0.4831$. The posterior distribution of μ and τ is the normal-gamma distribution with these hyperparameters

19. (a) For the 20 observations given in Exercise 7 of Sec. 8.5, the data summaries are $\overline{x}_n = 156.85$ and $s_n^2 = 9740.55$. So, the posterior hyperparameters are

$$\begin{aligned}
\mu_1 &= \frac{0.5\times 150 + 20\times 156.85}{0.5+20} = 156.68,\\
\lambda_1 &= 0.5 + 20 = 20.5,\\
\alpha_1 &= 1 + \frac{20}{2} = 11,\\
\beta_1 &= 4 + \frac{1}{2}\left(9740.55 + \frac{0.5\times 20}{0.5+20}(156.85-150)^2\right) = 4885.7.
\end{aligned}$$

The joint posterior of μ and τ is the normal-gamma distribution with these hyperparameters.

(b) The interval we want has endpoints $\mu_1 \pm (\beta_1/[\alpha_1\lambda_1])^{1/2}T^{-1}_{2\alpha_1}(0.95)$. The quantile we want is $T^{-1}_{22}(0.95) = 1.717$. Substituting the posterior hyperparameters gives the endpoints to be $a = 148.69$ and $b = 164.7$.

8.7 Unbiased Estimators

1. (a) The variance of a Poisson random variable with mean θ is also θ. So the variance is $\sigma^2 = g(\theta) = \theta$.

   (b) The M.L.E. of $g(\theta) = \theta$ was found in Exercise 5 of Sec. 7.5, and it equals $\overline{X}_n$. The mean of $\overline{X}_n$ is the same as the mean of each X_i, namely θ, hence the M.L.E. is unbiased.

3. By Exercise 2, $\delta_1 = \frac{1}{n}\sum_{i=1}^{n} X_i^2$ is an unbiased estimator of $E(X^2)$. Also, we know that $\delta_2 = \frac{1}{n-1}\sum_{i=1}^{n}(X_i - \overline{X}_n)^2$ is an unbiased estimator of $\text{Var}(X)$. Therefore, it follows from the hint for this exercise that $\delta_1 - \delta_2$ will be an unbiased estimator of $[E(X)]^2$.

5. We shall follow the hint for this exercise. If $E[\delta(X)] = \exp(\lambda)$, then

$$\exp(\lambda) = E[\delta(X)] = \sum_{x=0}^{\infty} \delta(x) f(x \mid \lambda) = \sum_{x=0}^{\infty} \frac{\delta(x)\exp(-\lambda)\lambda^x}{x!}.$$

Therefore,

$$\sum_{x=0}^{\infty} \frac{\delta(x)\lambda^x}{x!} = \exp(2\lambda) = \sum_{x=0}^{\infty} \frac{(2\lambda)^x}{x!} = \sum_{x=0}^{\infty} \frac{2^x\lambda^x}{x!}.$$

Since two power series in λ can be equal only if the coefficients of λ^x are equal for $x = 0, 1, 2, \ldots$, if follows that $\delta(x) = 2^x$ for $x = 0, 1, 2, \ldots$. This argument also shows that this estimator $\delta(X)$ is the unique unbiased estimator of $\exp(\lambda)$ in this problem.

7. For any possible values $x_1, \ldots, x_n$ of $X_1, \ldots, X_n$, let $y = \sum_{i=1}^{n} x_i$. Then

$$E[\delta(X_1, \ldots, X_n)] = \sum \delta(x_1, \ldots, x_n) p^y (1-p)^{n-y},$$

where the summation extends over all possible values of $x_1, \ldots, x_n$. Since $p^y(1-p)^{n-y}$ is a polynomial in p of degree n, it follows that $E[\delta(X_1, \ldots, X_n)]$ is the sum of a finite number of terms, each of which is equal to a constant $\delta(x_1, \ldots, x_n)$ times a polynomial in p of degree n. Therefore, $E[\delta(X_1, \ldots, X_n)]$ must itself be a polynomial in p of degree n or less. The degree would actually be less than n if the sum of the terms of order p^n is 0.

9. If $E[\delta(X)] = \exp(-2\lambda)$, then

$$\sum_{x=0}^{\infty} \delta(x)\frac{\exp(-\lambda)\lambda^x}{x!} = \exp(-2\lambda).$$

Therefore,

$$\sum_{x=0}^{\infty} \frac{\delta(x)\lambda^x}{x!} = \exp(-\lambda) = \sum_{x=0}^{\infty} \frac{(-1)^x\lambda^x}{x!}.$$

Therefore, $\delta(X) = (-1)^x$ or, in other world, $\delta(X) = 1$ if x is even and $\delta(x) = -1$ if x is odd.

11. (a) $E(\hat{\theta}) = \alpha E(\overline{X}_m) + (1-\alpha)E(\overline{Y}_n) = \alpha\theta + (1-\alpha)\theta = \theta$. Hence, $\hat{\theta}$ is an unbiased estimator of θ for all values of α, m and n.

(b) Since the two samples are taken independently, $\overline{X}_m$ and $\overline{Y}_n$ are independent. Hence,

$$\text{Var}(\hat{\theta}) = \alpha^2 \text{Var}(\overline{X}_m) + (1-\alpha)^2 \text{Var}(\overline{Y}_n) = \alpha^2\left(\frac{\sigma_A^2}{m}\right) + (1-\alpha)^2\left(\frac{\sigma_B^2}{n}\right).$$

Since $\sigma_A^2 = 4\sigma_B^2$, it follows that

$$\text{Var}(\hat{\theta}) = \left[\frac{4\alpha^2}{m} + \frac{(1-\alpha)^2}{n}\right]\sigma_B^2.$$

By differentiating the coefficient of σ_B^2, it is found that $\text{Var}(\hat{\theta})$ is a minimum when $\alpha = m/(m+4n)$.

13. (a) By Theorem 4.7.1,

$$E(\delta) = E[E(\delta|T)] = E(\delta_0).$$

Therefore, δ and δ_0 have the same expectation. Since δ is unbiased, $E(\delta) = \theta$. Hence, $E(\delta_0) = \theta$ also. In other words, δ_0 is also unbiased.

(b) Let $Y = \delta(\boldsymbol{X})$ and $X = T$ in Theorem 4.7.4. The result there implies that

$$\text{Var}_\theta(\delta(\boldsymbol{X})) = \text{Var}_\theta(\delta_0(\boldsymbol{X})) + E_\theta \text{Var}(\delta(\boldsymbol{X})|T).$$

Since $\text{Var}(\delta(\boldsymbol{X})|T) \geq 0$, so too is $E_\theta \text{Var}(\delta(\boldsymbol{X})|T)$, so $\text{Var}_\theta(\delta(\boldsymbol{X})) \geq \text{Var}_\theta(\delta_0(\boldsymbol{X}))$.

15. (a) $f(1 \mid \theta) + f(2 \mid \theta) = \theta^2[\theta + (1-\theta)] = \theta^2$,
$f(4 \mid \theta) + f(5 \mid \theta) = (1-\theta)^2[\theta + (1-\theta)] = (1-\theta)^2$,
$f(3 \mid \theta) = 2\theta(1-\theta)$.

The sum of the five probabilities on the left sides of these equations is equal to the sum the right sides, which is

$$\theta^2 + (1-\theta)^2 + 2\theta(1-\theta) = [\theta + (1-\theta)]^2 = 1.$$

(b) $E_\theta[\delta_c(X)] = \sum_{x=1}^{5} \delta_c(x) f(x \mid \theta) = 1 \cdot \theta^3 + (2-2c)\theta^2(1-\theta) + (c)2\theta(1-\theta) + (1-2c)\theta(1-\theta)^2 + 0.$

It will be found that the sum of the coefficients of θ^3 is 0, the sum of the coefficients of θ^2 is 0, the sum of the coefficients of θ is 1, and the constant term is 0. Hence, $E_\theta[\delta_c(X)] = \theta$.

(c) For every value of c,

$$\text{Var}_{\theta_0}(\delta_c) = E_{\theta_0}(\delta_c^2) - [E_{\theta_0}(\delta_c)]^2 = E_{\theta_0}(\delta_c^2) - \theta^2.$$

Hence, the value of c for which $\text{Var}_{\theta_0}(\delta_c)$ is a minimum will be the value of c for which $E_{\theta_0}(\delta_c^2)$ is a minimum. Now

$$\begin{aligned} E_{\theta_0}(\delta_c^2) &= (1)^2\theta_0^3 + (2-2c)^2\theta_0^2(1-\theta_0) + (c)^2 2\theta_0(1-\theta_0) \\ &\quad +(1-2c)^2\theta_0(1-\theta_0)^2 + 0 \\ &= 2c^2[2\theta_0^2(1-\theta_0) + \theta_0(1-\theta_0) + 2\theta_0(1-\theta_0)^2] \\ &\quad -4c[2\theta_0^2(1-\theta_0) + \theta_0(1-\theta_0)^2] + \text{ terms not involving } c. \end{aligned}$$

After further simplification of the coefficients of c^2 and c, we obtain the relation

$$E_{\theta_0}(\delta_c^2) = 6\theta_0(1-\theta_0)c^2 + 4\theta_0(1-\theta_0^2)c + \text{ terms not involving } c.$$

By differentiating with respect to c and setting the derivative equal to 0, it is found that the value of c for which $E_{\theta_0}(\delta_c^2)$ is a minimum is $c = (1+\theta_0)/3$.

8.8 Fisher Information

1.

$$\begin{aligned} f(x \mid \mu) &= \frac{1}{\sqrt{2\pi}\sigma} \exp\left\{-\frac{1}{2\sigma^2}(x-\mu)^2\right\}, \\ f'(x \mid \mu) &= \frac{1}{\sqrt{2\pi}\sigma}\frac{(x-\mu)}{\sigma^2} \exp\left\{-\frac{1}{2\sigma^2}(x-\mu)^2\right\} = \frac{x-\mu}{\sigma^2} f(x \mid \mu), \\ f''(x \mid \mu) &= \left[\frac{(x-\mu)^2}{\sigma^4} - \frac{1}{\sigma^2}\right] f(x \mid \mu). \end{aligned}$$

Therefore,

$$\int_{-\infty}^{\infty} f'(x \mid \mu) dx = \frac{1}{\sigma^2}\int_{-\infty}^{\infty} (x-\mu) f(x \mid \mu) d\mu = \frac{1}{\sigma^2} E(X-\mu) = 0,$$

and

$$\int_{-\infty}^{\infty} f''(x \mid \mu) dx = \frac{E[(X-\mu)^2]}{\sigma^4} - \frac{1}{\sigma^2} = \frac{\sigma^2}{\sigma^4} - \frac{1}{\sigma^2} = 0.$$

3.

$$\begin{aligned} f(x \mid \theta) &= \frac{\exp(-\theta)\theta^x}{x!}, \\ \lambda(x \mid \theta) &= -\theta + x \log\theta - \log(x!), \\ \lambda'(x \mid \theta) &= -1 + \frac{x}{\theta}, \\ \lambda''(x \mid \theta) &= -\frac{x}{\theta^2}. \end{aligned}$$

Therefore, by Eq. (8.8.3),

$$I(\theta) = -E_\theta[\lambda''(X \mid \theta)] = \frac{E(X)}{\theta^2} = \frac{1}{\theta}.$$

5. Let $\nu = \sigma^2$. Then

$$\begin{aligned} f(x \mid \nu) &= \frac{1}{\sqrt{2\pi\nu}} \exp\left\{-\frac{x^2}{2\nu}\right\}, \\ \lambda(x \mid \nu) &= -\frac{1}{2}\log\nu - \frac{x^2}{2\nu} + const., \\ \lambda'(x \mid \nu) &= -\frac{1}{2\nu} + \frac{x^2}{2\nu^2}, \\ \lambda''(x \mid \nu) &= \frac{1}{2\nu^2} - \frac{x^2}{\nu^3}. \end{aligned}$$

Therefore,

$$I(\sigma^2) = I(\nu) = -E_\nu[\lambda''(X \mid \nu)] = -\frac{1}{2\nu^2} + \frac{\nu}{\nu^3} = \frac{1}{2\nu^2} = \frac{1}{2\sigma^4}.$$

7. We know that $E(\overline{X}_n) = p$ and $\text{Var}(\overline{X}_n) = p(1-p)/n$. It was shown in Example 8.8.2 that $I(P) = 1/[p(1-p)]$. Therefore, $\text{Var}(\overline{X}_n)$ is equal to the lower bound $1/[nI(p)]$ provided by the information inequality.

9. We shall attack this exercise by trying to find an estimator of the form $c|X|$ that is unbiased. One approach is as follows: We know that X^2/σ^2 has the χ^2 distribution with one degree of freedom. Therefore, by Exercise 11 of Sec. 8.2, $|X|/\sigma$ has the χ distribution with one degree of freedom, and it was shown in that exercise that

$$E\left(\frac{|X|}{\sigma}\right) = \frac{\sqrt{2}\Gamma(1)}{\Gamma(1/2)} = \sqrt{\frac{2}{\pi}}.$$

Hence, $E(|X|) = \sigma\sqrt{2/\pi}$. It follows that $E(|X|\sqrt{\pi/2}) = \sigma$. Let $\delta = |X|\sqrt{\pi/2}$. Then

$$E(\delta^2) = \frac{\pi}{2}E(|X|^2) = \frac{\pi}{2}\sigma^2.$$

Hence,

$$\text{Var}\,\delta = E(\delta^2) - [E(\delta)]^2 = \frac{\pi}{2}\sigma^2 - \sigma^2 = \left(\frac{\pi}{2} - 1\right)\sigma^2.$$

Since $1/I(\sigma) = \sigma^2/2$, it follows that $\text{Var}(\delta) > 1/I(\sigma)$.

Another unbiased estimator is $\delta_1(X) = \sqrt{2\pi}\,X$ if $X \geq 0$ and $\delta_1(X) = 0$ if $X < 0$. However, it can be shown, using advanced methods, that the estimator δ found in this exercise is the *only* unbiased estimator of σ that depends on X only through $|X|$.

11. If $f(x \mid \theta) = a(\theta)b(x)\exp[c(\theta)d(x)]$, then

$$\lambda(x \mid \theta) = \log a(\theta) + \log b(x) + c(\theta)d(x)$$

and

$$\lambda'(x \mid \theta) = \frac{a'(\theta)}{a(\theta)} + c'(\theta)d(x).$$

Therefore,

$$\lambda'_n(\mathbf{X} \mid \theta) = \sum_{i=1}^{n} \lambda'(X_i \mid \theta) = n\frac{a'(\theta)}{a(\theta)} + c'(\theta)\sum_{i=1}^{n} d(X_i).$$

If we choose

$$u(\theta) = \frac{1}{c'(\theta)} \quad \text{and} \quad v(\theta) = -\frac{na'(\theta)}{a(\theta)c'(\theta)},$$

then Eq. (8.8.14) will be satisfied with $T = \sum_{i=1}^{n} d(X_i)$. Hence, this statistic is an efficient estimator of its expectation.

13. The incorrect part of the argument is at the beginning, because the information inequality cannot be applied to the uniform distribution. For each different value of θ, there is a different set of values of x for which $f(x \mid \theta) \geq 0$.

15. We know that the M.L.E. of μ is $\hat{\mu} = \overline{x}_n$ and, from Example 8.8.3, that $I(\mu) = 1/\sigma^2$. The posterior distribution of μ will be approximately a normal distribution with mean $\hat{\mu}$ and variance $1/[nI(\hat{\mu})] = \sigma^2/n$.

17. The derivative of the log-likelihood with respect to p is

$$\lambda'(x|p) = \frac{\partial}{\partial p}\left[\log\binom{n}{x} + x\log(p) + (n-x)\log(1-p)\right] = \frac{x}{p} - \frac{n-x}{1-p} = \frac{x-np}{p(1-p)}.$$

The mean of $\lambda'(X|p)$ is clearly 0, so its variance is

$$I(p) = \frac{\text{Var}(X)}{p^2(1-p)^2} = \frac{n}{p(1-p)}.$$

8.9 Supplementary Exercises

1. According to Exercise 5 in Sec. 8.8, the Fisher information $I(\sigma^2)$ based on a sample of size 1 is $1/[2\sigma^4]$. According to the information inequality, the variance of an unbiased estimator of σ^2 must be at least $2\sigma^4/n$. The variance of $V = \sum_{i=1}^{n} X_i^2/n$ is $\text{Var}(X_1^2)/n$. Since X_1^2/σ^2 has a χ^2 distribution with 1 degree of freedom, its variance is 2. Hence $\text{Var}(X_1^2) = 2\sigma^4$ and $\text{Var}(V)$ equals the lower bound from the information inequality. $E(V) = E(X_1^2) = \sigma^2$, so V is unbiased.

3. It is known from Exercise 18 of Sec. 5.6 that U/V has the Cauchy distribution, which is the t distribution with one degree of freedom. Next, since $|V| = (V^2)^{1/2}$, it follows from Definition 8.4.1 that $U/|V|$ has the required t distribution. Hence, by the previous exercise in this section, $|V|/U$ will also have this t distribution. Since U and V are i.i.d., it now follows that $|U|/V$ must have the same distribution as $|V|/U$.

5. Since X_i has the exponential distribution with parameter β, it follows that $2\beta\ X_i$ has the exponential distribution with parameter 1/2. But this exponential distribution is the χ^2 distribution with 2 degrees of freedom. Therefore, the sum of the i.i.d. random variables $2\beta X_i$ $(i = 1, \ldots, n)$ will have a χ^2 distribution with $2n$ degrees of freedom.

7. (a) $E(\alpha S_X^2 + \beta S_Y^2) = \alpha(m-1)\sigma^2 + \beta(n-1)2\sigma^2$.
Hence, this estimator will be unbiased if $\alpha(m-1) + 2\beta(n-1) = 1$.
(b) Since S_X^2 and S_Y^2 are independent,

$$\begin{aligned}\text{Var}\,(\alpha S_X^2 + \beta S_Y^2) &= \alpha^2\,\text{Var}\,(S_X^2) + \beta^2 var(S_Y^2)\\ &= \alpha^2[2(m-1)\sigma^4] + \beta^2[2(n-1)\cdot 4\sigma^4]\\ &= 2\sigma^4[(m-1)\alpha^2 + 4(n-1)\beta^2].\end{aligned}$$

Therefore, we must minimize

$$A = (m-1)\alpha^2 + 4(n-1)\beta^2$$

subject to the constraint $(m-1)\alpha + 2(n-1)\beta = 1$. If we solve this constraint for β in terms of α, and make this substitution for β in A, we can then minimize A over all values of α. The result is $\alpha = \dfrac{1}{m+n-2}$ and, hence, $\beta = \dfrac{1}{2(m+n-2)}$.

9. Under the given conditions, $Y/(2\sigma)$ has a standard normal distribution and S_n^2/σ^2 has an independent χ^2 distribution with $n-1$ degrees of freedom. Thus, the following random variable will have a t distribution with $n-1$ degrees of freedom:

$$\frac{Y/(2\sigma)}{\{S_n^2/[\sigma^2(n-1)]\}^{1/2}} = \frac{Y/2}{\sigma'},$$

where $\sigma' = [S_n^2/(n-1)]^{1/2}$.

11. Let c denote the .99 quantile of the t distribution with $n-1$ degrees of freedom; i.e., $\Pr(U < c) = .99$ if U has the specified t distribution. Therefore, $\Pr\left[\frac{n^{1/2}(\overline{X}_n - \mu)}{\sigma'} < c\right] = .99$ or, equivalently, $\Pr\left[\mu > \overline{X}_n - \frac{c\sigma'}{n^{1/2}}\right] = .99$. Hence, $L = \overline{X}_n - c\sigma'/n^{1/2}$.

13. (a) The posterior distribution of θ is the normal distribution with mean μ_1 and variance ν_1^2, as given by (7.3.1) and (7.3.2). Therefore, under this distribution,

$$\Pr(\mu_1 - 1.96\nu_1 < \theta < \mu_1 + 1.96\nu_1) = .95.$$

This interval I is the shortest one that has the required probability because it is symmetrically placed around the mean μ_1 of the normal distribution.

(b) It follows from (7.3.1) that $\mu_1 \to \overline{x}_n$ as $\nu^2 \to \infty$ and from (7.3.2) that $\nu_1^2 \to \sigma^2/n$. Hence, the interval I converges to the interval

$$\overline{x}_n - \frac{1.96\sigma}{n^{1/2}} < \theta < \overline{x}_n + \frac{1.96\sigma}{n^{1/2}}.$$

It was shown in Exercise 4 of Sec. 8.5 that this interval is a confidence interval for θ with confidence coefficient .95.

15. In the notation of Sec. 8.8,

$$\begin{aligned}
\lambda(x \mid \theta) &= \log\theta + (\theta - 1)\log x, \\
\lambda'(x \mid \theta) &= \frac{1}{\theta} + \log x, \\
\lambda''(x \mid \theta) &= -1/\theta^2.
\end{aligned}$$

Hence, by Eq. (8.8.3), $I(\theta) = 1/\theta^2$ and it follows that the asymptotic distribution of

$$\frac{n^{1/2}}{\theta}(\hat{\theta}_n - \theta)$$

is standard normal.

17. If $m(p) = (1-p)^2$, then $m'(p) = -2(1-p)$ and $[m'(p)]^2 = 4(1-p)^2$. It was shown in Example 8.8.2 that $I(p) = 1/[p(1-p)]$. Therefore, if T is an unbiased estimator of $m(p)$, it follows from the relation (8.8.14) that

$$\text{Var}(T) \geq \frac{4(1-p)^2 p(1-p)}{n} = \frac{4p(1-p)^3}{n}.$$

19. It was shown in Example 8.8.6 that $I(\beta) = 1/\beta^2$. The distribution of the M.L.E. of β will be approximately the normal distribution with mean β and variance $1/[nI(\beta)]$.

21. (a) The distribution of Y is the Poisson distribution with mean $n\theta$. In order for $r(Y)$ to be an unbiased estimator of $1/\theta$, we need

$$\frac{1}{\theta} = E_\theta(r(Y)) = \sum_{y=0}^{\infty} r(y)\exp(-n\theta)\frac{(n\theta)^y}{y!}.$$

This equation can be rewritten as

$$\exp(n\theta) = \sum_{y=0}^{\infty} \frac{r(y)n^y}{y!}\theta^{y+1}. \quad \text{(S.8.4)}$$

The function on the left side of (S.8.4) has a unique power series representation, hence the right side of (S.8.4) must equal that power series. However, the limit as $\theta \to 0$ of the left side of (S.8.4) is 1, while the limit of the right side is 0, hence the power series on the right cannot represent the function on the left.

(b) $E(n/[Y+1]) = \sum_{y=0}^{\infty} n\exp(-n\theta)[n\theta]^y/(y+1)!$. By letting $u = y+1$ in this sum, we get $n[1 - \exp(-n\theta)]/[n\theta] = 1/\theta - \exp(-n\theta)/\theta$. So the bias is $\exp(-n\theta)/\theta$. Clearly $\exp(-n\theta)$ goes to 0 as $n \to \infty$.

(c) $n/(1+Y) = 1/(\overline{X}_n + 1/n)$. We know that $\overline{X}_n + 1/n$ has approximately the normal distribution with mean $\theta + 1/n$ and variance θ/n. We can ignore the $1/n$ added to θ in the mean since this will eventually be small relative to θ. Using the delta method, we find that $1/(\overline{X}_n + 1/n)$ has approximately the normal distribution with mean $1/\theta$ and variance $(1/\theta^2)^2\theta/n = (n\theta^3)^{-1}$.

Chapter 9

Testing Hypotheses

9.1 Problems of Testing Hypotheses

1. (a) Let δ be the test that rejects H_0 when $X \geq 1$.The power function of δ is

$$\pi(\beta|\delta) = \Pr(X \geq 1|\beta) = \exp(-\beta),$$

for $\beta > 0$.

(b) The size of the test δ is $\sup_{\beta \geq 1} \pi(\beta|\delta)$. Using the answer to part (a), we see that $\pi(\beta|\delta)$ is a decreasing function of β, hence the size of the test is $\pi(1|\delta) = \exp(-1)$.

3. (a) For any given value of p, $\pi(p) = \Pr(Y \geq 7) + \Pr(Y \leq 1)$, where Y has a binomial distribution with parameters $n = 20$ and p. For $p = 0, \Pr(Y \geq 7) = 0$ and $\Pr(Y \leq 1) = 1$. Therefore, $\pi(0) = 1$. For $p = 0.1$, it is found from the table of the binomial distribution that

$$\Pr(Y \geq 7) = .0020 + .0003 + .0001 + .0000 = .0024$$

and $\Pr(Y \leq 1) = .1216 + .2701 = .3917$. Hence, $\pi(0.1) = 0.3941$. Similarly, for $p = 0.2$, it is found that

$$\Pr(Y \geq 7) = .0545 + .0222 + .0074 + .0020 + .0005 + .0001 = .0867$$

and $\Pr(Y \leq 1) = .0115 + .0576 = .0691$. Hence, $\pi(0.2) = 0.1558$. By continuing to use the tables in this way, we can find the values of $\pi(0.3)$, $\pi(0.4)$, and $\pi(0.5)$. For $p = 0.6$, we must use the fact that if Y has a binomial distribution with parameters 20 and 0.6, then $Z = 20 - Y$ has a binomial distribution with parameters 20 and 0.4. Also, $\Pr(Y \geq 7) = \Pr(Z \leq 13)$ and $\Pr(Y \leq 1) = \Pr(Z \geq 19)$. It is found from the tables that $\Pr(Z \leq 13) = .9935$ and $\Pr(Z \geq 19) = .0000$. Hence, $\pi(0.6) = .9935$. Similarly, if $p = 0.7$, then $Z = 20 - Y$ will have a binomial distribution with parameters 20 and 0.3. In this case it is found that $\Pr(Z \leq 13) = .9998$ and $\Pr(Z \geq 19) = .0000$. Hence, $\pi(0.7) = 0.9998$. By continuing in this way, the values of $\pi(0.8)$, $\pi(0.9)$, and $\pi(1.0) = 1$ can be obtained.

(b) Since H_0 is a simple hypothesis, the size α of the test is just the value of the power function at the point specified by H_0. Thus, $\alpha = \pi(0.2) = 0.1558$.

5. A hypothesis is simple if and only if it specifies a single value of both μ and σ. Therefore, only the hypothesis in (a) is simple. All the others are composite. In particular, although the hypothesis in (d) specifies the value of μ, it leaves the value of σ arbitrary.

7. Let C be the critical region of Y_n values for the test δ, and let C^* be the critical region for δ^*. It is easy to see that $C^* \subset C$. Hence

$$\pi(\theta|\delta) - \pi(\theta|\delta^*) = \Pr\left(Y_n \in C \cap (C^*)^C \middle| \theta\right).$$

Here $C \cap (C^*)^C = [4, 4.5]$, so

$$\pi(\theta|\delta) - \pi(\theta|\delta^*) = \Pr(4 \le Y_n \le 4.5|\theta). \tag{S.9.1}$$

(a) For $\theta \le 4$ $\Pr(4 \le Y_n|\theta) = 0$, so the two power functions must be equal by (S.9.1).

(b) For $\theta > 4$,

$$\Pr(4 \le Y_n \le 4.5|\theta) = \frac{(\min\{\theta, 4.5\})^n - 4^n}{\theta^n} > 0.$$

Hence, $\pi(\theta|\delta) > \pi(\theta|\delta^*)$ by (S.9.1).

(c) The only places where the power functions differ are for $\theta > 4$. Since these values are all in Ω_1, it is better for a test to have higher power function for these values. Since δ has higher power function than δ^* for all of these values, δ is the better test.

9. A sensible test would be to reject H_0 if $\overline{X}_n < c'$. So, let $T = \mu_0 - \overline{X}_n$. Then the power function of the test δ that rejects H_0 when $T \ge c$ is

$$\begin{aligned}\pi(\mu|\delta) &= \Pr(T \ge c|\mu)\\ &= \Pr(\overline{X}_n \le \mu_0 - c|\mu)\\ &= \Phi(\sqrt{n}[\mu_0 - c - \mu]).\end{aligned}$$

Since Φ is an increasing function and $\sqrt{n}[\mu_0 - c - \mu]$ is a decreasing function of μ, it follows that $\Phi(\sqrt{n}[\mu_0 - c - \mu])$ is a decreasing function of μ.

11. (a) For $c_1 \ge 2$, $\Pr(Y \le c_1|p = 0.4) \ge 0.23$, hence $c_1 \le 1$. Also, for $c_2 \le 5$, $\Pr(Y \ge c_2|p = 0.4) \ge 0.26$, hence $c_2 \ge 6$. Here are some values of the desired probability for various (c_1, c_2) pairs

c_1	c_2	$\Pr(Y \le c_1\|p = 0.4) + \Pr(Y \ge c_2\|p = 0.4)$
1	6	0.1699
1	7	0.0956
0	6	0.1094
−1	6	0.0994

So, the closest we can get to 0.1 without going over is 0.0994, which is achieved when $c_1 < 0$ and $c_2 = 6$.

(b) The size of the test is 0.0994, as we calculated in part (a).

(c) The power function is plotted in Fig. S.9.1. Notice that the power function is too low for values of $p < 0.4$. This is due to the fact that the test only rejects H_0 when $Y \ge 6$ A better test might be one with $c_1 = 1$ and $c_2 = 7$. Even though the size is slightly smaller (as is the power for $p > 0.4$), its power is much greater for $p < 0.4$.

13. For $c = 3$, $\Pr(X \ge c|\theta = 1) = 0.0803$, while for $c = 2$, the probability is 0.2642. Hence, we must use $c = 3$.

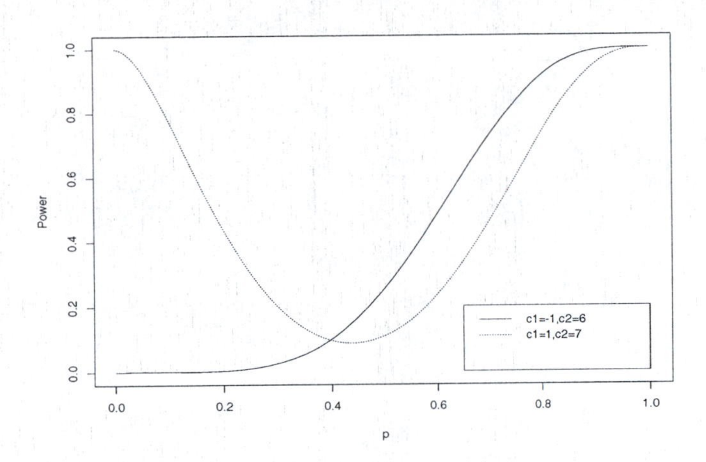

Figure S.9.1: Power function of test in Exercise 11c of Sec. 9.1.

15. The p-value when $X = x$ is observed is the size of the test that rejects H_0 when $X \geq x$, namely

$$\Pr(X \geq x|\theta = 1) = \begin{cases} 0 & \text{if } x \geq 1, \\ 1 - x & \text{if } 0 < x < 1. \end{cases}$$

17. We need $q(y)$ to have the property that $\Pr(q(Y) < p|p) \geq \gamma$ for all p. We shall prove that $q(y)$ equal to the smallest p_0 such that $\Pr(Y \geq y|p = p_0) \geq 1 - \gamma$ satisfies this property. For each p, let $A_p = \{y : q(y) < p\}$. We need to show that $\Pr(Y \in A_p|p) \geq \gamma$. First, notice that $q(y)$ is an increasing function of y. This means that for each p there is y_p such that $A_p = \{0, \ldots, y_p\}$. So, we need to show that $\Pr(Y \leq y_p|p) \geq \gamma$ for all p. Equivalently, we need to show that $\Pr(Y > y_p|p) \leq 1 - \gamma$. Notice that y_p is the largest value of y such that $q(y) < p$. That is, y_p is the largest value of y such that there exists $p_0 < p$ with $\Pr(Y \geq y|p_0) \geq 1 - \gamma$. For each y, $\Pr(Y > y|p)$ is a continuous nondecreasing function of p. If $\Pr(Y > y_p|p) > 1 - \gamma$, then there exists $p_0 < p$ such that

$$1 - \gamma < \Pr(Y > y_p|p_0) = \Pr(Y \geq y_p + 1|p_0).$$

This contradicts the fact that y_p is the largest y such that there is $p_0 < p$ with $\Pr(Y \geq y|p_0) \geq 1 - \gamma$. Hence $\Pr(Y > y_p|p) \leq 1 - \gamma$ and the proof is complete.

19. We want our test to reject H_0 if $\overline{X}_n \leq Y$, where Y might be a random variable. We can write this as not rejecting H_0 if $\overline{X}_n > Y$. We want $\overline{X}_n > Y$ to be equivalent to μ_0 being inside of our interval. We need the test to have level α_0, so

$$\Pr(\overline{X}_n \leq Y|\mu = \mu_0, \sigma^2) = \alpha_0 \tag{S.9.2}$$

is necessary. We know that $n^{1/2}(\overline{X}_n - \mu_0)/\sigma'$ has the t distribution with $n - 1$ degrees of freedom if $\mu = \mu_0$, hence Eq. (S.9.2) will hold if $Y = \mu_0 - n^{-1/2}\sigma' T_{n-1}^{-1}(1 - \alpha_0)$. Now, $\overline{X}_n > Y$ if and only if $\mu_0 < \overline{X}_n + n^{-1/2}\sigma' T_{n-1}^{-1}(1 - \alpha_0)$. This is equivalent to μ_0 in our interval if our interval is

$$\left(-\infty, \overline{X}_n + n^{-1/2}\sigma' T_{n-1}^{-1}(1 - \alpha_0)\right).$$

21. Let $U = n^{1/2}(\overline{X}_n - \mu_0)/\sigma'$.

(a) We reject the null hypothesis in (9.1.22) if and only if

$$U \geq T_{n-1}^{-1}(1-\alpha_0). \tag{S.9.3}$$

We reject the null hypothesis in (9.1.27) if and only if

$$U \leq -T_{n-1}^{-1}(1-\alpha_0). \tag{S.9.4}$$

With $\alpha_0 < 0.5$, $T_{n-1}^{-1}(1-\alpha_0) > 0$. So, (S.9.3) requires $U > 0$ while (S.9.4) requires $U < 0$. These cannot both occur.

(b) Both (S.9.3) and (S.9.4) fail if and only if U is strictly between $-T_{n-1}^{-1}(1-\alpha_0)$ and $T_{n-1}^{-1}(1-\alpha_0)$. This can happen if $\overline{X}_n$ is sufficiently close to μ_0. This has probability $1-2\alpha_0 > 0$.

(c) If $\alpha_0 > 0.5$, then $T_{n-1}^{-1}(1-\alpha_0) < 0$, and both null hypotheses would be rejected if U is between the numbers $T_{n-1}^{-1}(1-\alpha_0) < 0$ and $-T_{n-1}^{-1}(1-\alpha_0) > 0$. This has probability $2\alpha_0 - 1 > 0$.

9.2 Testing Simple Hypotheses

1. According to Theorem 9.2.1, we should reject H_0 if $f_1(x) > f_0(x)$, not reject H_0 if $f_1(x) < f_0(x)$ and do whatever we wish if $f_1(x) = f_0(x)$. Here

$$f_0(x) = \begin{cases} 0.3 & \text{if } x = 1, \\ 0.7 & \text{if } x = 0, \end{cases}$$

$$f_1(x) = \begin{cases} 0.6 & \text{if } x = 1, \\ 0.4 & \text{if } x = 0. \end{cases}$$

We have $f_1(x) > f_0(x)$ if $x = 1$ and $f_1(x) < f_0(x)$ if $x = 0$. We never have $f_1(x) = f_0(x)$. So, the test is to reject H_0 if $X = 1$ and not reject H_0 if $X = 0$.

3. (a) Theorem 9.2.1 can be applied with $a = 3$ and $b = 1$. Therefore, H_0 should not be rejected if $f_1(x)/f_0(x) = 2x < 3$. Since all possible values of X lie in the interval (0,1), and since $2x < 3$ for all values in this interval, the optimal procedure is to not reject H_0 for every possible observed value.

   (b) Since H_0 is never rejecte, $\alpha(\delta) = 0$ and $\beta(\delta) = 1$. Therefore, $3\alpha(\delta) + \beta(\delta) = 1$.

5. (a) The conditions here are different from those of the Neyman-Pearson lemma. Rather than fixing the value of $\alpha(\delta)$ and minimizing $\beta(\delta)$, we must here fix the value of $\beta(\delta)$ and minimize $\alpha(\delta)$. Nevertheless, the same proof as that given for the Neyman-Pearson lemma shows that the optimal procedure is again to reject H_0 if $f_1(\boldsymbol{X})/f_0(\boldsymbol{X}) > k$, where k is now chosen so that

$$\beta(\delta) = \Pr(\text{Acc. } H_0 \,|\, H_1) = \Pr\left[\frac{f_1(\boldsymbol{X})}{f_0(\boldsymbol{X})} < k \,\middle|\, H_1\right] = 0.05.$$

   In this exercise,

$$f_0(\boldsymbol{X}) = \frac{1}{(2\pi)^{n/2}} \exp\left[-\frac{1}{2}\sum_{i=1}^{n}(x_i - 3.5)^2\right]$$

   and

$$f_1(\boldsymbol{X}) = \frac{1}{(2\pi)^{n/2}} \exp\left[-\frac{1}{2}\sum_{i=1}^{n}(x_i - 5.0)^2\right].$$

Therefore,

$$\log\frac{f_1(\boldsymbol{X})}{f_0(\boldsymbol{X})} = \frac{1}{2}\left[\sum_{i=1}^{n}(x_i-3.5)^2-\sum_{i=1}^{n}(x_i-5.0)^2\right]$$
$$= \frac{1}{2}\left[\sum_{i=1}^{n}x_i^2-7\sum_{i=1}^{n}x_i+12.25n-\sum_{i=1}^{n}x_i^2+10\sum_{i=1}^{n}x_i-25n\right]$$
$$= \frac{3}{2}n\overline{x}_n-(const.).$$

It follows that the likelihood ratio $f_1(\boldsymbol{X})/f_0(\boldsymbol{X})$ will be greater than some specified constant k if and only if $\overline{x}_n$ is greater than some other constant k'. Therefore, the optimal procedure is to reject H_0 if $\overline{x}_n > k'$, where k' is chosen so that

$$\Pr(\overline{X}_n < k' \mid H_1) = 0.05.$$

We shall now determine the value of k'. If H_1 is true, then $\overline{X}_n$ will have a normal distribution with mean 5.0 and variance $1/n$. Therefore, $Z = \sqrt{n}(\overline{X}_n - 5.0)$ will have the standard normal distribution, and it follows that

$$\Pr(\overline{X}_n < k' \mid H_1) = \Pr[Z < \sqrt{n}(k'-5.0)] = \Phi[\sqrt{n}(k'-5.0)].$$

If this probability is to be equal to 0.05, then it can be found from a table of values of Φ that $\sqrt{n}(k'-5.0) = -1.645$. Hence, $k' = 5.0 - 1.645n^{-1/2}$.

(b) For $n = 4$, the test procedure is to reject H_0 if $\overline{X}_n > 5.0 - 1.645/2 = 4.1775$. Therefore,

$$\alpha(\delta) = \Pr(\text{Rej. } H_0 \mid H_0) = \Pr(\overline{X}_n > 4.1775 \mid H_0).$$

When H_0 is true, $\overline{X}_n$has a normal distribution with mean 3.5 and variance $1/n = 1/4$. Therefore, $Z = 2(\overline{X}_n - 3.5)$ will have the standard normal distribution, and

$$\alpha(\delta) = \Pr[Z > 2(4.1775-3.5)] = \Pr(Z > 1.355)$$
$$= 1-\Phi(1.355) = 0.0877.$$

7. (a) By the Neyman-Pearson lemma, H_0 should be rejected if $f_1(\boldsymbol{X})/f_0(\boldsymbol{X}) > k$. Here,

$$f_0(\boldsymbol{X}) = \frac{1}{(2\pi)^{n/2}2^{n/2}}\exp\left[-\frac{1}{4}\sum_{i=1}^{n}(x_i-\mu)^2\right]$$

and

$$f_1(\boldsymbol{X}) = \frac{1}{(2\pi)^{n/2}3^{n/2}}\exp\left[-\frac{1}{6}\sum_{i=1}^{n}(x_i-\mu)^2\right].$$

Therefore,

$$\log\frac{f_1(\boldsymbol{X})}{f_0(\boldsymbol{X})} = \frac{1}{12}\sum_{i=1}^{n}(x_i-\mu)^2+(const.).$$

It follows that the likelihood ratio will be greater than a specified constant k if and only if $\sum_{i=1}^{n}(x_i-\mu)^2$ is greater than some other constant c. The constant c is to be chosen so that

$$\Pr\left[\sum_{i=1}^{n}(X_i-\mu)^2 > c \,\middle|\, H_0\right] = 0.05.$$

The value of c can be determined as follows. When H_0 is true, $W = \sum_{i=1}^{n}(X_i - \mu)^2/2$ will have χ^2 distribution with n degrees of freedom. Therefore,

$$\Pr\left[\sum_{i=1}^{n}(X_i - \mu)^2 > c \,\middle|\, H_0\right] = \Pr\left(W > \frac{c}{2}\right).$$

If this probability is to be equal to 0.05, then the value of $c/2$ can be determined from a table of the χ^2 distribution.

(b) For $n = 8$, it is found from a table of the χ^2 distribution with 8 degrees of freedom that $c/2 = 15.51$ and $c = 31.02$.

9. As in Exercise 8, we should reject H_0 if at least one of the n observations is greater than 1. For this test, $\alpha(\delta) = 0$ and

$$\beta(\delta) = \Pr(\text{Acc. } H_0 \,|\, H_1) = \Pr(X_1 < 1, \dots, X_n < 1 \,|\, H_1) = \left(\frac{1}{2}\right)^n.$$

11. Theorem 9.2.1 can be applied with $a = b = 1$. The optimal procedure is to reject H_0 if $f_1(\boldsymbol{X})/f_0(\boldsymbol{X}) > 1$. Here,

$$f_0(\boldsymbol{X}) = \frac{1}{(2\pi)^{n/2}2^n}\exp\left[-\frac{1}{8}\sum_{i=1}^{n}(x_i + 1)^2\right]$$

and

$$f_1(\boldsymbol{X}) = \frac{1}{(2\pi)^{n/2}2^n}\exp\left[-\frac{1}{8}\sum_{i=1}^{n}(x_i - 1)^2\right].$$

After some algebraic reduction, it can be shown that $f_1(\boldsymbol{X})/f_0(\boldsymbol{X}) > 1$ if and only if $\overline{x}_n > 0$. If H_0 is true, $\overline{X}_n$ will have the normal distribution with mean -1 and variance $4/n$. Therefore, $Z = \sqrt{n}(\overline{X}_n + 1)/2$ will have the standard normal distribution, and

$$\alpha(\delta) = \Pr(\overline{X}_n > 0 \,|\, H_0) = \Pr\left(Z > \frac{1}{2}\sqrt{n}\right) = 1 - \Phi\left(\frac{1}{2}\sqrt{n}\right).$$

Similarly, if H_1 is true, $\overline{X}_n$will have the normal distribution with mean 1 and variance $4/n$. Therefore, $Z' = \sqrt{n}(\overline{X}_n - 1)/2$ will have the standard normal distribution, and

$$\beta(\delta) = \Pr(\overline{X}_n < 0 \,|\, H_1) = \Pr\left(Z' < -\frac{1}{2}\sqrt{n}\right) = 1 - \Phi\left(\frac{1}{2}\sqrt{n}\right).$$

Hence, $\alpha(\delta) + \beta(\delta) = 2[1 - \Phi(\sqrt{n}/2)]$. We can now use a program that computes Φ to obtain the following results:

(a) If $n = 1$, $\alpha(\delta) + \beta(\delta) = 2(0.3085) = 0.6170$.

(b) If $n = 4$, $\alpha(\delta) + \beta(\delta) = 2(0.1587) = 0.3173$.

(c) If $n = 16$, $\alpha(\delta) + \beta(\delta) = 2(0.0228) = 0.0455$.

(d) If $n = 36$, $\alpha(\delta) + \beta(\delta) = 2(0.0013) = 0.0027$.

Slight discrepancies appear above due to rounding *after* multiplying by 2 rather than before.

13. (a) The test rejects H_0 if $f_0(\boldsymbol{X}) < f_12(\boldsymbol{X})$. In this case, $f_0(\boldsymbol{x}) = \exp(-[x_1 + x_2]/2)/4$, and $f_1(\boldsymbol{x}) = 4/(2 + x_1 + x_2)^3$ for both $x_1 > 0$ and $x_2 > 0$. Let $T = X_1 + X_2$. Then we reject H_0 if
$$\exp(-T/2)/4 < 4/(2 + T)^3. \tag{S.9.5}$$
(b) If $X_1 = 4$ and $X_2 = 3$ are observed, then $T = 7$. The inequality in (S.9.5) is $\exp(-7/2)/4 < 4/9^3$ or $0.007549 < 0.00549$, which is false, so we do not reject H_0.
(c) If H_0 is true, then T is the sum of two independent exponential random variables with parameter 1/2. Hence, it has the gamma distribution with parameters 2 and 1/2 by Theorem 5.7.7.
(d) The test is to reject H_0 if $f_1(\boldsymbol{X})/f_0(\boldsymbol{X}) > c$, where c is chosen so that the probability is 0.1 that we reject H_0 given $\theta = \theta_0$. We can write
$$\frac{f_1(\boldsymbol{X})}{f_0(\boldsymbol{X})} = \frac{16\exp(T/2)}{(2+T)^3}. \tag{S.9.6}$$
The function on the right side of (S.9.6) takes the value 2 at $T = 0$, decreases to the value 0.5473 at $T = 4$, and increases for $T > 4$. Let G be the c.d.f. of the gamma distribution with parameters 2 and 1/2 (also the χ^2 distribution with 4 degrees of freedom). The level 0.01 test will reject H_0 if $T < c_1$ or $T > c_2$ where c_1 and c_2 satisfy $G(c_1) + 1 - G(c_2) = 0.01$, and either $16\exp(c_1/2)/(2+c_1)^3 = 16\exp(c_2/2)/(2+c_2)^3$ or $c_1 = 0$ and $16\exp(c_2/2)/(2+c_2)^3 > 2$. It follows that $1 - G(c_2) \le 0.01$, that is, $c_2 \ge G^{-1}(0.99) = 13.28$. But
$$16\exp(13.28)/(2 + 13.28)^3 = 3.4 > 2.$$
It follows that $c_1 = 0$ and the test is to reject H_0 if $T > 13.28$.
(e) If $X_1 = 4$ and $X_2 = 3$, then $T = 7$ and we do not reject H_0.

9.3 Uniformly Most Powerful Tests

1. Let $y = \sum_{i=1}^{n} x_i$. Then the joint p.f. is
$$f_n(\boldsymbol{X} \mid \lambda) = \frac{\exp(-n\lambda)\lambda^y}{\prod_{i=1}^n (x_i!)}$$
Therefore, for $0 < \lambda_1 < \lambda_2$,
$$\frac{f_n(\boldsymbol{X} \mid \lambda_2)}{f_n(\boldsymbol{X} \mid \lambda_1)} = \exp(-n(\lambda_2 - \lambda_1))\left(\frac{\lambda_2}{\lambda_1}\right)^y,$$
which is an increasing function of y.

3. Let $y = \prod_{i=1}^{n} x_i$ and let $z = \sum_{i=1}^{n} x_i$. Then the joint p.d.f. is
$$f_n(\boldsymbol{X} \mid \alpha) = \frac{\beta^{n\alpha}}{[\Gamma(\alpha)]^n} y^{\alpha-1}\exp(-\beta z).$$
Therefore, for $0 < \alpha_1 < \alpha_2$,
$$\frac{f_n(\boldsymbol{X} \mid \alpha_2)}{f_n(\boldsymbol{X} \mid \alpha_1)} = (const.) y^{\alpha_2 - \alpha_1},$$
which is an increasing function of y.

5. Let $y = \sum_{i=1}^{n} d(x_i)$. Then the joint p.d.f. or the joint p.f. is

$$f_n(\mathbf{X} \mid \theta) = [a(\theta)]^n \left[\prod_{i=1}^{n} b(x_i)\right] \exp[c(\theta)y].$$

Therefore, for $\theta_1 < \theta_2$,

$$\frac{f_n(\mathbf{X} \mid \theta_2)}{f_n(\mathbf{X} \mid \theta_1)} = \left[\frac{a(\theta_2)}{a(\theta_1)}\right]^n \exp\{[c(\theta_2) - c(\theta_1)]y\}.$$

Since $c(\theta_2) - c(\theta_1) > 0$, this expression is an increasing function of y.

7. No matter what the true value of θ is, the probability that H_0 will be rejected is 0.05. Therefore, the value of the power function at every value of θ is 0.05.

9. The first part of this exercise was answered in Exercise 8. When n = 10 and $\sigma^2 = 2$, the distribution of $Y = \sum_{i=1}^{n} X_i^2/2$ will be the χ^2 distribution with 10 degrees of freedom, and it is found from a table of this distribution that $\Pr(Y \geq 18.31) = 0.05$. Also,

$$\Pr\left(\sum_{i=1}^{n} X_i^2 \geq c \mid \sigma^2 = 2\right) = \Pr\left(Y \geq \frac{c}{2}\right).$$

Therefore, if this probability is to be equal to 0.05, then $c/2 = 18.31$ or $c = 36.62$.

11. It is known from Exercise 1 that the joint p.f. of $X_1, \ldots, X_n$ has a monotone likelihood ratio in he statistic $Y = \sum_{i=1}^{n} X_i$. Therefore, by Theorem 9.3.1, a test which rejects H_0 when $Y \geq c$ will be a UMP test. When $\lambda = 1$ and $n = 10$, Y will have a Poisson distribution with mean 10, and it is found from the table of the Poisson distribution given at the end of this book that

$$\Pr(Y \geq 18) = .0071 + .0037 + .0019 + .0009 + .0004 + .0002 + .0001 = .0143.$$

Therefore, the level of significance of the UMP test which rejects H_0 when $Y \geq 18$ will be 0.0143.

13. (a) By Exercise 12, the test which rejects H_0 when $\overline{X}_n \leq c$ will be a UMP test. For the level of significance to be 0.1, c should be chosen so that $\Pr(\overline{X}_n \leq c \mid \mu = 10) = 0.1$. In this exercise, $n = 4$. When $\mu = 10$, the random variable $z = 2(\overline{X}_n - 10)$ has a standard normal distribution and $\Pr(\overline{X}_n \leq c \mid \mu = 10) = \Pr[Z \leq 2(c - 10)]$. It is found from a table of the standard normal distribution that $\Pr(Z \leq -1.282) = 0.1$. Therefore, $2(c - 10) = -1.282$ or $c = 9.359$.

    (b) When $\mu = 9$, the random variable $2(\overline{X}_n - 9)$ has the standard normal distribution. Therefore, the power of the test is

    $$\Pr(\overline{X}_n \leq 9.359 \mid \mu = 9) = \Pr(Z \leq 0.718) = \Phi(0.718) = 0.7636,$$

    where we have interpolated in the table of the normal distribution between 0.71 and 0.72.

    (c) When $\mu = 11$, the random variable $Z = 2(\overline{X}_n - 11)$ has the standard normal distribution. Therefore, the probability of rejecting H_0 is

    $$\Pr(\overline{X}_n \geq 93359 \mid \mu = 11) = \Pr(Z \geq -3.282) = \Pr(Z \leq 3.282) = \Phi(3.282) = 0.9995.$$

15. By Exercise 4, the joint p.d.f. of $X_1, \ldots, X_n$ has a monotone likelihood ratio in the statistic $-\overline{X}_n$. Therefore, by Exercise 12, a test which rejects H_0 when $-\overline{X}_n \le c'$, for some constant c', will be a UMP test. But this test is equivalent to a test which rejects H_0 when $\overline{X}_n \ge c$, where $c = -c'$. Since $\overline{X}_n$ has a continuous distribution, for any specified value of $\alpha_0 (0 < \alpha_0 < 1)$ there exists a value of c such that $\Pr(\overline{X}_n \ge c \,|\, \beta = 1/2) = \alpha_0$.

17. In this exercise, H_0 is a simple hypothesis. By the Neyman-Pearson lemma, the test which has maximum power at a particular alternative value $\theta_1 > 0$ will reject H_0 if $f(x \,|\, \theta = \theta_1)/f(x \,|\, \theta = 0) > c$, where c is chosen so that the probability that this inequality will be satisfied when $\theta = 0$ is α_0. Here,

$$\frac{f(x \,|\, \theta = \theta_1)}{f(x \,|\, \theta = 0)} > c$$

if and only if $(1-c)x^2 + 2c\theta_1 x > c\theta_1^2 - (1-c)$. For each value of θ_1, the value of c is to be chosen so that the set of points satisfying this inequality has probability α_0 when $\theta = 0$. For two different values of θ_1, these two sets will be different. Therefore, different test procedures will maximize the power at the two different values of θ_1. Hence, no single test is a UMP test.

19. (a) Let $f(\boldsymbol{x}|\mu)$ be the joint p.d.f. of $\boldsymbol{X}$ given μ. For each set A and $i = 0, 1$,

$$\Pr(\boldsymbol{X} \in A|\mu = \mu_i) = \int \cdots_A \int f(\boldsymbol{x}|\mu_i) d\boldsymbol{x}. \quad \text{(S.9.7)}$$

It is clear that $f(\boldsymbol{x}|\mu_0) > 0$ for all $\boldsymbol{x}$ and so is $f(\boldsymbol{x}|\mu_1) > 0$ for all $\boldsymbol{x}$. Hence (S.9.7) is strictly positive for $i = 0$ if and only if it is strictly positive for $i = 1$.

(b) Both δ and δ_1 are size α_0 tests of $H_0' : \mu = \mu_0$ versus $H_1' : \mu > \mu_0$. Let

$$\begin{aligned} A &= \{\boldsymbol{x} : \delta \text{ rejects but } \delta_1 \text{ does not reject}\}, \\ B &= \{\boldsymbol{x} : \delta \text{ does not reject but } \delta_1 \text{ rejects}\}, \\ C &= \{\boldsymbol{x} : \text{ both tests reject}\}. \end{aligned}$$

Because both tests have the same size, it must be the case that

$$\Pr(\boldsymbol{X} \in A|\mu = \mu_0) + \Pr(\boldsymbol{X} \in |\mu = \mu_0) = \alpha_0 = \Pr(\boldsymbol{X} \in B|\mu = \mu_0) + \Pr(\boldsymbol{X} \in |\mu = \mu_0).$$

Hence,

$$\Pr(\boldsymbol{X} \in A|\mu = \mu_0) = Pr(\boldsymbol{X} \in B|\mu = \mu_0). \quad \text{(S.9.8)}$$

Because of the MLR and the form of the test δ_1, we know that there is a constant c such that for every $\mu > \mu_0$ and every $\boldsymbol{x} \in B$ and every $\boldsymbol{y} \in A$,

$$\frac{f(\boldsymbol{x}|\mu)}{f(\boldsymbol{x}|\mu_0)} > c > \frac{f(\boldsymbol{y}|\mu)}{f(\boldsymbol{y}|\mu_0)}. \quad \text{(S.9.9)}$$

Now,

$$\pi(\mu|\delta) = \int \cdots_A \int f(\boldsymbol{x}|\mu) d\boldsymbol{x} + \int \cdots_C \int f(\boldsymbol{x}|\mu) d\boldsymbol{x}.$$

Also,

$$\pi(\mu|\delta_1) = \int \cdots_B \int f(\boldsymbol{x}|\mu) d\boldsymbol{x} + \int \cdots_C \int f(\boldsymbol{x}|\mu) d\boldsymbol{x}.$$

It follows that, for $\mu > \mu_0$,

$$\begin{aligned}
\pi(\mu|\delta_1) - \pi(\mu|\delta) &= \int\cdots_B\int f(\boldsymbol{x}|\mu)d\boldsymbol{x} - \int\cdots_A\int f(\boldsymbol{x}|\mu)d\boldsymbol{x} \\
&= \int\cdots_B\int \frac{f(\boldsymbol{x}|\mu)}{f(\boldsymbol{x}|\mu_0)} f(\boldsymbol{x}|\mu_0)d\boldsymbol{x} - \int\cdots_A\int \frac{f(\boldsymbol{x}|\mu)}{f(\boldsymbol{x}|\mu_0)} f(\boldsymbol{x}|\mu_0)d\boldsymbol{x} \\
&> \int\cdots_B\int cf(\boldsymbol{x}|\mu_0)d\boldsymbol{x} - \int\cdots_A\int cf(\boldsymbol{x}|\mu_0)d\boldsymbol{x} = 0,
\end{aligned}$$

where the inequality follows from (S.9.9), and the final equality follows from (S.9.8).

9.4 Two-Sided Alternatives

1. If $\pi(\mu\,|\,\delta)$ is to be symmetric with respect to the point $\mu = \mu_0$, then the constants c_1 and c_2 must be chosen to be symmetric with respect to the value μ_0. Let $c_1 = \mu_0 - k$ and $c_1 = \mu_0 + k$. When $\mu = \mu_0$, the random variable $Z = n^{1/2}(\overline{X}_n - \mu_0)$ has the standard normal distribution. Therefore,

$$\begin{aligned}
\pi(\mu_0\,|\,\delta) &= \Pr(\overline{X}_n \le \mu_0 - k\,|\,\mu_0) + \Pr(\overline{X}_n \ge \mu_0 + k\,|\,\mu_0) \\
&= \Pr(Z \le -n^{1/2}k) + \Pr(Z \ge n^{1/2}k) \\
&= 2\Pr(Z \ge n^{1/2}k) = 2[1 - \Phi(n^{1/2}k)].
\end{aligned}$$

Since k must be chosen so that $\pi(\mu_0\,|\,\delta) = 0.10$, it follows that $\Phi(n^{1/2}k) = 0.95$. Therefore, $n^{1/2}k = 1.645$ and $k = 1.645n^{-1/2}$.

3. From Exercise 1, we know that if $c_1 = \mu_0 - 1.645n^{-1/2}$ and $c_2 = \mu_0 + 1.645n^{-1/2}$, then $\pi(\mu_0\,|\,\delta) = 0.10$ and, by symmetry, $\pi(\mu_0 + 1\,|\,\delta) = \pi(\mu_0 - 1\,|\,\delta)$. Also, when $\mu = \mu_0 + 1$, the random variable $n^{1/2}(\overline{X}_n - \mu_0 - 1)$ has the standard normal distribution. Therefore,

$$\begin{aligned}
\pi(\mu_0 + 1\,|\,\delta) &= \Pr(\overline{X}_n \le c_1\,|\,\mu_0 + 1) + \Pr(\overline{X}_n \ge c_2\,|\,\mu_0 + 1) \\
&= \Pr(Z \le -1.645 - n^{1/2}) + \Pr(Z \ge 1.645 - n^{1/2}) \\
&= \Phi(-1.645 - n^{1/2}) + \Phi(n^{1/2} - 1.645).
\end{aligned}$$

For $n = 9$, $\quad \pi(\mu_0 + 1\,|\,\delta) = \Phi(-4.645) + \Phi(1.355) < 0.95.$
For $n = 10$, $\quad \pi(\mu_0 + 1\,|\,\delta) = \Phi(-4.807) + \Phi(1.517) < 0.95.$
For $n = 11$, $\quad \pi(\mu_0 + 1\,|\,\delta) = \Phi(-4.962) + \Phi(1.672) > 0.95.$

5. As in Exercise 4,

$$\pi(0.1\,|\,\delta) = \Pr[Z \le 5(c_1 - 0.1)] + \Pr[Z \ge 5(c_2 - 0.1)] = \Phi(5c_1 - 0.5) + \Phi(0.5 - 5c_2).$$

Similarly,

$$\pi(0.2\,|\,\delta) = \Pr[Z \le 5(c_1 - 0.2)] + \Pr[Z \ge 5(c_2 - 0.2)] = \Phi(5c_1 - 1) + \Phi(1 - 5c_2).$$

Hence, the following two equations must be solved simultaneously:

$$\begin{aligned}
\Phi(5c_1 - 0.5) + \Phi(0.5 - 5c_2) &= 0.02, \\
\Phi(5c_1 - 1) + \Phi(1 - 5c_2) &= 0.05.
\end{aligned}$$

By trial and error, using the table of the standard normal distribution, it is found ultimately that if $5c_1 = -2.12$ and $5c_2 = 2.655$, then

$$\Phi(5c_1 - 0.5) + \Phi(0.5 - 5c_2) = \Phi(-2.62) + \Phi(-2.155) = 0.0044 + 0.0155 = 0.02.$$

and

$$\Phi(5c_1 - 0.1) + \Phi(1 - 5c_2) = \Phi(-3.12) + \Phi(-1.655) = 0.0009 + 0.0490 = 0.05.$$

7. For $\theta > 0$, the power function is $\pi(\theta \,|\, \delta) = \Pr(T \geq c \,|\, \theta)$. Hence,

$$\pi(\theta \,|\, \delta) = \begin{cases} 0 & \text{for } \theta \leq c, \\ 1 - \left(\dfrac{c}{\theta}\right)^n & \text{for } \theta > c. \end{cases}$$

The plot is in Fig. S.9.2.

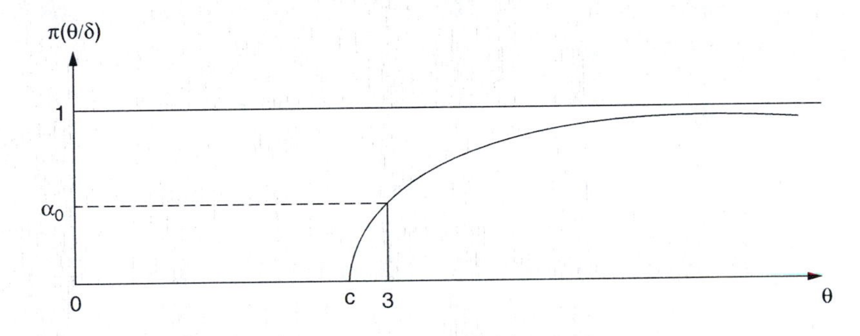

Figure S.9.2: Figure for Exercise 7 of Sec. 9.4.

9. A sketch is given in Fig. S.9.3.

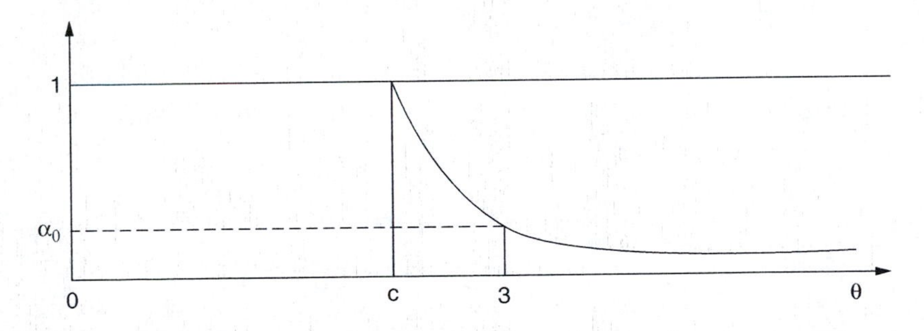

Figure S.9.3: Figure for Exercise 9 of Sec. 9.4.

11. It can be verified that if c_1 and c_2 are chosen to be symmetric with respect to the value μ_0, then the power function $\pi(\mu \,|\, \delta)$ will be symmetric with respect to the point $\mu = \mu_0$ and will attain its minimum value at $\mu = \mu_0$. Therefore, if c_1 and c_2 are chosen as in Exercise 1, the required conditions will be satisfied.

13. The first term on the right of (9.4.13) is

$$\frac{n}{\theta} \int_0^x \frac{\theta^n}{\Gamma(n)} t^{n-1} \exp(-t\theta) dt = \frac{n}{\theta} G(x; n, \theta).$$

The second term on the right of (9.4.13) is the negative of

$$\frac{n}{\theta}\int_0^x \frac{\theta^{n+1}}{\Gamma(n+1)} t^n \exp(-t\theta)dt = \frac{n}{\theta}G(x; n+1, \theta).$$

9.5 The t Test

1. We computed the summary statistics $\overline{x}_n = 1.379$ and $\sigma' = 0.3277$ in Example 8.5.4.

   (a) The test statistic is U from (9.5.2)

   $$U = 10^{1/2}\frac{1.379 - 1.2}{0.3277} = 1.727.$$

   We reject H_0 at level $\alpha_0 = 0.05$ if $U \geq 1.833$, the 0.95 quantile of the t distribution with 9 degrees of freedom. Since $1.727 \not\geq 1.833$, we do not reject H_0 at level 0.05.

   (b) We need to compute the probability that a t random variable with 9 degrees of freedom exceeds 1.727. This probability can be computed by most statistical software, and it equals 0.0591. Without a computer, one could interpolate in the table of the t distribution in the back of the book. That would yield 0.0618.

3. It must be assumed that the miles per gallon obtained from the different tankfuls are independent and identically distributed, and that each has a normal distribution. When $\mu_0 = 20$, the statistic U given by Eq. (9.5.2) has the t distribution with 8 degrees of freedom. Here, we are testing the following hypotheses:

   $$H_0: \mu \geq 20,$$
   $$H_1: \mu < 20.$$

   We would reject H_0 if $U \leq -1.860$. From the given value, it is found that $\overline{X}_n = 19$ and $S_n^2 = 22$. Hence, $U = -1.809$ and we do not reject H_0.

5. It is found from the table of the t distribution with 7 degrees of freedom that $c_1 = -2.998$ and c_2 lies between 1.415 and 1.895. Since U $= -1.979$, we do not reject H_0.

7. The random variable $T = (X - \mu)/\sigma$ will have the standard normal distribution, the random variable $Z = \sum_{i=1}^{n} Y_i^2/\sigma^2$ will have a χ^2 distribution with n degrees of freedom, and T and Z will be independent. Therefore, when $\mu = \mu_0$, the following random variable U will have the t distribution with n degrees of freedom:

   $$U = \frac{T}{[Z/n]^{1/2}} = \frac{n^{1/2}(X - \mu_0)}{\left[\sum_{i=1}^{n} Y_i^2\right]^{1/2}}.$$

   The hypothesis H_0 would be rejected if $U \geq c$.

9. When $\sigma^2 = 4$, $S_n^2/4$ has the χ^2 distribution with 9 degrees of freedom. We would reject H_0 if $S_n^2/4 \geq 16.92$. Since $S_n^2/4 = 60/4 = 15$, we do not reject H_0.

11. U_1 has the distribution of X/Y where X has a normal distribution with mean ψ and variance 1, and Y is independent of X such that mY^2 has the χ^2 distribution with m degrees of freedom. Notice that $-X$ has a normal distribution with mean $-\psi$ and variance 1 and is independent of Y. So U_2 has the distribution of $-X/Y = -U_1$. So

$$\Pr(U_2 \le -c) = \Pr(-U_1 \le -c) = \Pr(U_1 \ge c).$$

13. The test statistic is $U = 169^{1/2}(3.2 - 3)/(0.09)^{1/2} = 8.667$. The p-value can be calculated using statistical software as $1 - T_{169}(8.667) = 1.776 \times 10^{-15}$.

15. The test statistic is $U = 169^{1/2}(3.2 - 3.1)/(0.09)^{1/2} = 4.333$. The p-value can be calculated using statistical software as $2[1 - T_{169}(4.333)] = 2.512 \times 10^{-5}$.

17. The denominator of $\Lambda(\boldsymbol{x})$ is still (9.5.11). The M.L.E. $(\hat{\mu}_0, \hat{\sigma}_0^2)$ is easier to calculate in this exercise, namely $\hat{\mu}_0 = \mu_0$ (the only possible value) and

$$\hat{\sigma}_0^2 = \frac{1}{n}\sum_{i=1}^{n}(x_i - \mu_0)^2.$$

These are the same values that lead to Eq. (9.5.12) in the text. Hence, $\Lambda(\boldsymbol{x})$ has the value given in Eq. (9.5.14). For $k < 1$, $\Lambda(\boldsymbol{x}) \le k$ if and only if

$$|U| \ge ((n-1)[k^{2/n} - 1])^{1/2} = c.$$

9.6 Comparing the Means of Two Normal Distributions

1. In this example, $n = 5$, $m = 5$, $\overline{X}_m = 18.18$, $\overline{Y}_n = 17.32$, $S_X^2 = 12.61$, and $S_Y^2 = 11.01$. Then

$$U = \frac{(5+5-2)^{1/2}(18.18 - 17.32)}{\left(\frac{1}{5} + \frac{1}{5}\right)^{1/2}(11.01 + 12.61)^{1/2}} = 0.7913.$$

We see that $|U| = 0.7913$ is much smaller than the 0.975 quantile of the t distribution with 8 degrees of freedom.

3. The value $c = 1.782$ can be found from a table of the t distribution with 12 degrees of freedom. Since $U = -1.692$, H_0 is not rejected.

5. Again, H_0 should be rejected if $U < -1.356$. Since $U = -1.672$, H_0 is rejected.

7. To test the hypotheses in Exercise 6, H_0 would not be rejected if $-1.782 < U < 1.782$. The set of all values of λ for which H_0 would not be rejected will form a confidence interval for $\mu_1 - \mu_2$ with confidence coefficient 0.90. The value of U, for an arbitrary value of λ, is found to be

$$U = \frac{\sqrt{12}(-0.1558 - \lambda)}{0.3188}.$$

It is found that $-1.782 < U < 1.782$ if and only if $-0.320 < \lambda < 0.008$.

9. The p-value can be computed as the size of the test that rejects H_0 when $|U| \geq |u|$, where u is the observed value of the test statistic. Since U has the t distribution with $m+n-2$ degrees of freedom when H_0 is true, the size of the test that rejects H_0 when $|U| \geq |u|$ is the probability that a t random variable with $m+n-2$ degrees of freedom is either less than $-|u|$ or greater than $|u|$. This probability is

$$T_{m+n-2}(-|u|) + 1 - T_{m+n-2}(|u|) = 2[1 - T_{m+n-2}(|u|),$$

by the symmetry of t distributions.

11. (a) The observed value of the test statistic U is

$$u = \frac{(43+35-2)^{1/2}(8.560-5.551)}{\left(\frac{1}{43}+\frac{1}{35}\right)^{1/2}(2745.7+783.9)^{1/2}} = 1.939.$$

We would reject the null hypothesis at level $\alpha_0 = 0.01$ if $U > 2.376$, the 0.99 quantile of the t distribution with 76 degrees of freedom. Since $u < 2.376$, we do not reject H_0 at level 0.01. (The answer in the back of the book is incorrect in early printings.)

(b) For Welch's test, the approximate degrees of freedom is

$$\nu = \frac{\left(\frac{2745.7}{43\times 42}+\frac{783.9}{35\times 34}\right)^2}{\frac{1}{42^3}\left(\frac{2745.7}{43}\right)^2+\frac{1}{34^3}\left(\frac{783.9}{35}\right)^2} = 70.04.$$

The corresponding t quantile is 2.381. The test statistic is

$$\frac{8.560-5.551}{\left(\frac{2745.7}{43\times 42}+\frac{783.9}{35\times 34}\right)^{1/2}} = 2.038.$$

Once again, we do not reject H_0. (The answer in the back of the book is incorrect in early printings.) book is incorrect.)

13. The likelihood ratio statistic for this case is

$$\Lambda(\boldsymbol{x},\boldsymbol{y}) = \frac{\sup_{\{(\mu_1,\mu_2,\sigma^2):\mu_1\neq\mu_2\}} g(\boldsymbol{x},\boldsymbol{y} \mid \mu_1,\mu_2,\sigma^2)}{\sup_{\{(\mu_1,\mu_2,\sigma^2):\mu_1=\mu_2\}} g(\boldsymbol{x},\boldsymbol{y} \mid \mu_1,\mu_2,\sigma^2)}. \tag{S.9.10}$$

Maximizing the numerator of (S.9.10) is identical to maximizing the numerator of (9.6.10) when $\overline{x}_m \leq \overline{y}_n$ because we need $\mu_1 = \mu_2$ in both cases. So the M.L.E.'s are

$$\begin{aligned}\hat{\mu}_1 &= \hat{\mu}_2 = \frac{m\overline{x}_m + n\overline{y}_n}{m+n},\\ \hat{\sigma}^2 &= \frac{mn(\overline{x}_m - \overline{y}_n)^2/(m+n) + s_x^2 + s_y^2}{m+n}.\end{aligned}$$

Maximizing the denominator of (S.9.10) is identical to the maximization of the denominator of (9.6.10) when $\overline{x}_m \leq \overline{y}_n$. We use the overall M.L.E.'s

$$\hat{\mu}_1 = \overline{x}_m, \quad \hat{\mu}_2 = \overline{y}_n, \text{ and } \hat{\sigma}^2 = \frac{1}{m+n}(s_x^2 + s_y^2).$$

This makes $\Lambda(\boldsymbol{x},\boldsymbol{y})$ equal to $(1+v^2)^{-(m+n)/2}$ where v is defined in (9.6.12). So $\Lambda(\boldsymbol{x},\boldsymbol{y}) \geq k$ if and only if $v^2 \leq k'$ for some other constant k'. This translates easily to $|U| \geq c$.

9.7 The F Distributions

1. The test statistic is $V = [2745.7/42]/[783.9/34] = 2.835$. We reject the null hypothesis if V is greater than the 0.95 quantile of the F distribution with 42 and 34 degrees of freedom, which is 1.737. So, we reject the null hypothesis at level 0.05.

3. If Y has the t distribution with 8 degrees of freedom, then $X = Y^2$ will have the F distribution with 1 and 8 degrees of freedom. Also,

$$0.3 = \Pr(X > c) = \Pr(Y > \sqrt{c}) + \Pr(Y < -\sqrt{c}) = 2\Pr(Y > \sqrt{c}).$$

Therefore, $\Pr(Y > \sqrt{c}) = 0.15$. It can be found from the table given at the end of the book that $\Pr(Y > 1.108) = 0.15$. Hence, $\sqrt{c} = 1.108$ and $c = 1.228$.

5. By Eq. (9.7.1), X can be represented in the form $X = Y/Z$, where Y and Z are independent and have identical χ^2 distributions. Therefore, $\Pr(Y > Z) = \Pr(Y < Z) = 1/2$. Equivalently, $\Pr(X > 1) = \Pr(X < 1) = 1/2$. Therefore, the median of the distribution of X is 1.

7. (a) Here, $\overline{X}_m = 84/16 = 5.25$ and $\overline{Y}_n = 18/10 = 1.8$. Therefore, $S_1^2 = \sum_{i=1}^{16} X_i^2 - 16(\overline{X}_m^2) = 122$ and

$S_2^2 = \sum_{i=1}^{10} Y_i^2 - 10(\overline{Y}_n^2) = 39.6$. It follows that

$$\hat{\sigma}_1^2 = \frac{1}{16}S_1^2 = 7.625 \quad \text{and} \quad \hat{\sigma}_2^2 = \frac{1}{10}S_2^2 = 3.96.$$

If $\sigma_1^2 = \sigma_2^2$, the following statistic V will have the F distribution with 15 and 9 degrees of freedom:

$$V = \frac{S_1^2/15}{S_2^2/9}.$$

(b) If the test is to be carried out at the level of significance 0.05, then H_0 should be rejected if $V > 3.01$. It is found that $V = 1.848$. Therefore, we do not reject H_0.

9. When $\sigma_1^2 = \sigma_2^2$, V has an F distribution with 15 and 9 degrees of freedom. Therefore, $\Pr(V > 3.77) = 0.025$, which implies that $c_2 = 3.77$. Also, $1/V$ has an F distribution with 9 and 15 degrees of freedom. Therefore, $\Pr(1/V > 3.12) = 0.025$. It follows that $\Pr(V < 1/(3.12)) = 0.025$, which means that $c_1 = 1/(3.12) = 0.321$.

11. For any positive number r, the hypothesis H_0 in Exercise 9 will not be rejected if $c_1 < V/r < c_2$. The set of all values of r for which H_0 would not be rejected will form a confidence interval with confidence coefficient 0.95. But $c_1 < V/r < c_2$ if and only if $V/c_2 < r < V/c_1$. Therefore, the confidence interval will contain all values of r between $V/3.77 = 0.265V$ and $V/0.321 = 3.12V$.

13. Suppose that X has the F distribution with m and n degrees of freedom, and consider the representation of X in Eq. (9.7.1). Then $Y/m \xrightarrow{P} 1$. Therefore, as $m \to \infty$, the distribution of X will become the same as the distribution of n/Z, where Z has a χ^2 distribution with n degrees of freedom. Suppose that c is the 0.05 quantile of the χ^2 distribution with n degrees of freedom. Then $\Pr(n/Z < n/c) = 0.95$. Hence, $\Pr(X < n/c) = 0.95$, and the value n/c should be entered in the column of the F distribution with $m = \infty$.

15. The p-value will be the value of α_0 such that the observed v is exactly equal to either c_1 or c_2. The problem is deciding wheter $v = c_1$ or $v = c_2$ since, we haven't constructed a specific test. Since c_1 and c_2 are assumed to be the $\alpha_0/2$ and $1 - \alpha_0/2$ quantiles of the F distribution with $m - 1$ and $n - 1$ degrees of freedom, we must have $c_1 < c_2$ and $G_{m=1,n-1}(c_1) < 1/2$ and $G_{m-1,n-1}(c_2) > 1/2$. These inequalities allow us to choose whether $v = c_1$ or $v = c_2$. Every $v > 0$ is *some* quantile of each F distribution, indeed the $G^{-1}_{m-1,n-1}(v)$ quantile. If $G_{m-1,n-1}(v) < 1/2$, then $v = c_1$ and $\alpha_0 = 2G_{m-1,n-1}(v)$. If $G_{m-1,n-1}(v) > 1/2$, then $v = c_2$, and $\alpha_0 = 2[1 - G_{m-1,n-1}(v)]$. (There is 0 probability that $G_{m-1,n-1}(v) = 1/2$.) Hence, α_0 is the smaller of the two numbers $2G_{m-1,n-1}(v)$ and $2[1 - G_{m-1,n-1}(v)]$.

   In Example 9.7.4, $v = 0.9491$ was the observed value and $2G_{25,25}(0.9491) = 0.8971$, so this would be the p-value.

17. The test found in Exercise 9 uses the values $c_1 = 0.321$ and $c_2 = 3.77$. The likelihood ratio test rejects H_0 when $dw^8(1-w)^5) \leq k$, which is equivalent to $w^8(1-w)^5 \leq k/d$. If $V = v$, then $w = 15v/(15v+9)$. In order for a test to be a likelihood ratio test, the two values c_1 and c_2 must lead to the same value of the likelihood ratio. In particular, we must have

$$\left(\frac{15c_1}{15c_1+9}\right)^8\left(1-\frac{15c_1}{15c_1+9}\right)^5 = \left(\frac{15c_2}{15c_2+9}\right)^8\left(1-\frac{15c_2}{15c_2+9}\right)^5.$$

   Plugging the values of c_1 and c_2 from Exercise 9 into this formula we get 2.555×10^{-5} on the left and 1.497×10^{-5} on the right.

19. (a) Apply the result of Exercise 18 with $c_1 = G^{-1}_{10,20}(0.025) = 0.2952$ and $c_2 = G^{-1}_{10,20}(0.975) = 2.774$ and $\sigma_2^2/\sigma_1^2 = 1/1.01$. The result is

$$G_{10,20}(c_1/1.01) + 1 - G_{10,20}(c_2/1.01) = G_{10,20}(0.289625) + 1 - G_{10,20}(2.746209) = 0.0503.$$

   (b) Apply the result of Exercise 18 with $c_1 = G^{-1}_{10,20}(0.025) = 0.2952$ and $c_2 = G^{-1}_{10,20}(0.975) = 2.774$ and $\sigma_2^2/\sigma_1^2 = 1.01$. The result is

$$G_{10,20}(1.01 \times c_1) + 1 - G_{10,20}(1.01 \times c_2) = G_{10,20}(0.2954475) + 1 - G_{10,20}(2.80148) = 0.0498.$$

   (c) Since the answer to part (b) is less than 0.05 (the value of the power function for all parameters in the null hypothesis set), the test is not unbiased.

9.8 Bayes Test Procedures

1. In this exercise, $\xi_0 = 0.9$, $\xi_1 = 0.1$, $w_0 = 1000$, and $w_1 = 18{,}000$. Also,

$$f_0(x) = \frac{1}{(2\pi)^{1/2}}\exp\left[-\frac{1}{2}(x-50)^2\right]$$

   and

$$f_1(x) = \frac{1}{(2\pi)^{1/2}}\exp\left[-\frac{1}{2}(x-52)^2\right].$$

   By the results of this section, it should be decided that the process is out of control if

$$\frac{f_1(x)}{f_0(x)} > \frac{\xi_0 w_0}{\xi_1 w_1} = \frac{1}{2}.$$

   This inequality can be reduced to the inequality $2x - 102 > -\log\ 2$ or, equivalently, $x > 50.653$.

3. In this exercise, $\xi_0 = 0.8$, $\xi_1 = 0.2$, $w_0 = 400$, and $w_1 = 2500$. Also, if we let $y = \sum_{i=1}^{n} x_i$, then

$$f_0(\boldsymbol{X}) = \frac{\exp(-3n)\, 3^y}{\prod_{i=1}^{n}(x_i!)}$$

and

$$f_1(\boldsymbol{X}) = \frac{\exp(-7n)\, 7^y}{\prod_{i=1}^{n}(x_i!)}.$$

By the results of this section, it should be decided that the failure was caused by a major defect if

$$\frac{f_1(\boldsymbol{X})}{f_0(\boldsymbol{X})} = \exp(-4n)\left(\frac{7}{3}\right)^y > \frac{\xi_0 w_0}{\xi_1 w_1} = 0.64$$

or, equivalently, if

$$y > \frac{4n + \log(0.64)}{\log\left(\frac{7}{3}\right)}.$$

5. (a) In the notation of this section $\xi_0 = \Pr(\theta = \theta_0)$ and f_i is the p.f. or p.d.f. of $\boldsymbol{X}$ given $\theta = \theta_i$. By the law of total probability, the marginal p.f. or p.d.f. of $\boldsymbol{X}$ is $\xi_0 f_0(\boldsymbol{x}) + \xi_1 f_1(\boldsymbol{x})$. Applying Bayes' theorem for random variables gives us that

$$\Pr(\theta = \theta_0|\boldsymbol{x}) = \frac{\xi_0 f_0(\boldsymbol{x})}{\xi_0 f_0(\boldsymbol{x}) + \xi_1 f_1(\boldsymbol{x})}.$$

(b) The posterior expected value of the loss given $\boldsymbol{X} = \boldsymbol{x}$ is

$$\begin{cases} \dfrac{w_0 \xi_0 f_0(\boldsymbol{x})}{\xi_0 f_0(\boldsymbol{x}) + \xi_1 f_1(\boldsymbol{x})} & \text{if reject } H_0, \\ \dfrac{w_1 \xi_1 f_1(\boldsymbol{x})}{\xi_0 f_0(\boldsymbol{x}) + \xi_1 f_1(\boldsymbol{x})} & \text{if don't reject } H_0. \end{cases}$$

The tests δ that minimize $r(\delta)$ have the form

$$\begin{aligned} \text{Don't Reject } H_0 &\text{ if } \xi_0 w_0 f_0(\boldsymbol{x}) > \xi_1 w_1 f_1(\boldsymbol{x}), \\ \text{Reject } H_0 &\text{ if } \xi_0 w_0 f_0(\boldsymbol{x}) < \xi_1 w_1 f_1(\boldsymbol{x}), \\ \text{Do either} &\text{ if } \xi_0 w_0 f_0(\boldsymbol{x}) = \xi_1 w_1 f_1(\boldsymbol{x}). \end{aligned}$$

Notice that these tests choose the action that has smaller posterior expected loss. If neither action has smaller posterior expected loss, these tests can do either, but either would then minimize the posterior expected loss.

(c) The "reject H_0" condition in part (b) is $\xi_0 w_0 f_0(\boldsymbol{x}) < \xi_1 w_1 f_1(\boldsymbol{x})$. This is equivalent to $w_0 \Pr(\theta = \theta_0|\boldsymbol{x}) < w_1[1 - \Pr(\theta = \theta_0|\boldsymbol{x})]$. Simplifying this inequality yields $\Pr(\theta = \theta_0|\boldsymbol{x}) < w_1/(w_0 + w_1)$. Since we can do whatever we want when equality holds, and since "H_0 true" means $\theta = \theta_0$, we see that the test described in part (c) is one of the tests from part (b).

7. We shall argue indirectly. Suppose that there is $\boldsymbol{x}$ such that the p-value is not equal to the posterior probability that H_0 is true. First, suppose that the p-value is greater. Let α_0 be greater than the posterior probability and less than the p-value. Then the test that rejects H_0 when $\Pr(H_0 \text{ true}|\boldsymbol{x}) \le \alpha_0$ will reject H_0, but the level α_0 test will not reject H_0 because the p-value is greater than α_0. This contradicts the fact that the two tests are the same. The case in which the p-value is smaller is very similar.

9. (a) The null hypothesis can be rewritten as $\tau_1 \ge \tau_2$, where $\tau_i = 1/\sigma_i^2$. This can be further rewritten as $\tau_1/\tau_2 \ge 1$. Using the usual improper prior for all parameters yields the posterior distribution of τ_1 and τ_2 to be that of independent gamma random variables with τ_1 having parameters $(m-1)/2$ and $s_x^2/2$ while τ_2 has parameters $(n-1)/2$ and $s_y^2/2$. Put another way, $\tau_1 s_x^2$ has the χ^2 distribution with $m-1$ degrees of freedom independent of $\tau_2^2 s_y^2$ which has the χ^2 distribution with $n-1$ degrees of freedom. This makes the distribution of

$$W = \frac{\tau_1 s_x^2/(m-1)}{\tau_2 s_y^2/(n-1)}$$

the F distribution with $m-1$ and $n-1$ degrees of freedom. The posterior probability that H_0 is true is

$$\Pr(\tau_1/\tau_2 \ge 1) = \Pr\left(W \ge \frac{s_x^2/(m-1)}{s_y^2/(n-1)}\right) = 1 - G_{m-1,n-1}\left(\frac{s_x^2/(m-1)}{s_y^2/(n-1)}\right).$$

The posterior probability is at most α_0 if and only if

$$\frac{s_x^2/(m-1)}{s_y^2/(n-1)} \ge F_{m-1,n-1}^{-1}(1-\alpha_0).$$

This is exactly the form of the rejection region for the level α_0 F test of H_0.

(b) This is a special case of Exercise 7.

11. (a) First, let $H_0 : \theta \in \Omega'$ and $H_1 : \theta \in \Omega''$. Then $\Omega_0 = \Omega'$ and $\Omega_1 = \Omega''$. Since d_0 is the decision that H_0 is true we have $d_0 = d'$ and $d_1 = d''$. Since w_0 is the cost of type II error, and type I error is to choose $\theta \in \Omega''$ when $\theta \in \Omega'$, $w_0 = w'$, and $w_1 = w''$. It is straightforward to see that everything switches for the other case.

(b) The test procedure is to

$$\text{choose } d_1 \text{ if } \Pr(\theta \in \Omega_0|\boldsymbol{x}) < \frac{w_1}{w_0 + w_1}, \quad \text{(S.9.11)}$$

$$\text{and choose either action if the two sides are equal.} \quad \text{(S.9.12)}$$

In the first case, this translates to "choose d" if $\Pr(\theta \in \Omega'|\boldsymbol{x}) < w''/(w' + w'')$, and choose either action if the two sides are equal." This is equivalent to "choose d' if $\Pr(\theta \in \Omega'|\boldsymbol{x}) > w''/(w' + w'')$, and choose either action if the two sides are equal." This, in turn, is equivalent to "choose d' if $\Pr(\theta \in \Omega''|\boldsymbol{x}) < w'/(w' + w'')$, and choose either action if the two sides are equal." This last statement, in the second case, translates to (S.9.11). Hence, the Bayes test produces the same action (d' or d'') regardless of which hypothesis you choose to call the null and which the alternative.

9.9 Foundational Issues

1. (a) When $\mu = 0$, X has the standard normal distribution. Therefore, $c = 1.96$. Since H_0 should be rejected if $|X| > c$, then H_0 will be rejected when $X = 2$.

(b) $$\frac{f(X\,|\,\mu=0)}{f(X\,|\,\mu=5)} = \frac{\exp\left(-\frac{1}{2}X^2\right)}{\exp[-\frac{1}{2}(X-5)^2]} = \exp\left[\frac{1}{2}(25-10X)\right].$$

When $X = 2$, this likelihood ratio has the value $\exp(5/2) = 12.2$. Also,

$$\frac{f(X\,|\,\mu=0)}{f(X\,|\,\mu=-5)} = \frac{\exp\left(-\frac{1}{2}X^2\right)}{\exp\left[-\frac{1}{2}(X+5)^2\right]} = \exp[\frac{1}{2}(25+10X)].$$

When $X = 2$, this likelihood ratio has the value $\exp(45/2) = 5.9 \times 10^9$.

3. When $\mu = 0$, $100\overline{X}_n$ has the standard normal distribution. The calculated value of $100\overline{X}_n$ is $100(0.03) = 3$. The corresponding tail area is $\Pr(100\,\overline{X}_n > 3) = 0.0013$.

5. (a) We want to choose c_n so that

$$19[1-\Phi(\sqrt{n}c_n)] = \Phi(\sqrt{n}[c_n - 0.5]).$$

Solving this equation must be done numerically. For $n = 1$, the equation is solved for $c_n = 1.681$. For $n = 100$, we need $c_n = 0.3021$. For $n = 10000$, we need $c_n = 0.25$ (both sides are essentially 0).

(b) The size of the test is $1 - \Phi(c_n n^{1/2})$, which is 0.0464 for $n = 1$, 0.00126 for $n = 100$ and essentially 0 for $n = 10000$.

9.10 Supplementary Exercises

1. According to Theorem 9.2.1, we want to reject H_0 when

$$(1/2)^3 < (3/4)^x(1/4)^{3-x}.$$

We don't reject H_0 when the reverse inequality holds, and we can do either if equality holds. The inequality above can be simplified to $x > \log(8)/\log(3) = 1.892$. That is, we reject H_0 if X is 2 or 3, and we don't reject H_0 if X is 0 or 1. The probability of type I error is $3(1/2)^3 + (1/2)^3 = 1/2$ and the probability of type II error is $(1/4)^3 + 3(1/4)^2(3/4) = 5/32$.

3. It follows from Sec. 9.2 that the Bayes test procedure rejects H_0 when $f_1(x)/f_0(x) > 1$. In this problem,

$$f_1(x) = (.8)^{x-1}(.2) \quad \text{for} \quad x = 1, 2, \ldots,$$

and

$$f_0(x) = (.9)^{x-1}(.1) \quad \text{for} \quad x = 1, 2, \ldots.$$

Hence, H_0 should be rejected when $2(8/9)^{x-1} \geq 1$ or $x - 1 \leq 5.885$. Thus, H_0 should be rejected for $X \leq 6$.

5. It follows from the previous exercise and the Neyman-Pearson lemma that the optimal procedure δ specifies rejecting H_0 when $x/(1-x) > k'$ or, equivalently, when $x > k$. The constant k must be chosen so that

$$\alpha = \Pr(X > k\,|\,\theta = 2) = \int_k^1 f(x\,|\,\theta = 2)dx = (1-k)^2.$$

Hence, $k = 1 - \alpha^{1/2}$ and

$$\beta(\delta) = \Pr(X < k\,|\,\theta = 0) = k^2 = (1-\alpha^{1/2})^2.$$

7. A direct calculation shows that for $\theta_1 < \theta_2$,

$$\frac{d}{dx}\left[\frac{f(x\,|\,\theta_2)}{f(x\,|\,\theta_1)}\right] = \frac{2(\theta_1-\theta_2)}{[2(1-\theta_1)x+\theta_1]^2} < 0.$$

Hence, the ratio $f(x\,|\,\theta_2)/f(x\,|\,\theta_1)$ is a decreasing function of x or, equivalently, an increasing function of $r(x) = -x$. It follows from Theorem 9.3.1 that a UMP test of the given hypotheses will reject H_0 when $r(X) \geq c$ or , equivalently, when $X \leq k$. Hence, k must be chosen so that

$$.05 = \Pr\left(X \leq k\,|\,\theta = \frac{1}{2}\right) = \int_0^k f\left(x\,|\,\theta = \frac{1}{2}\right)dx = \frac{1}{2}(k^2+k), \quad \text{or} \quad k = \frac{1}{2}(\sqrt{1.4}-1).$$

9. Let $f_i(x)$ denote the p.d.f. of X under the hypotheses $H_i (i = 0, 1)$. Then

$$\frac{f_1(x)}{f_0(x)} = \begin{cases} \infty & \text{for } x \leq 0 \text{ or } x \geq 1, \\ \varphi(x) & \text{for } 0 < x < 1, \end{cases}$$

where $\varphi(x)$ is the standard normal p.d.f. The most powerful test δ of size 0.01 rejects H_0 when $f_1(x)/f_0(x) > k$. Since $\varphi(x)$ is strictly decreasing for $0 < x < 1$, it follows that δ will reject H_0 if $X \leq 0$, $X \geq 1$, or $0 < X < c$, where c is chosen so that $\Pr(0 < X < c\,|\,H_0) = .01$. Since X has a uniform distribution under H_0 , $c = .01$. Thus, δ specifies rejecting H_0 if $X \leq .01$ or $X \geq 1$. The power of δ under H_1 is

$$\Pr(X \leq .01\,|\,H_1) + \Pr(X \geq 1\,|\,H_1) = \Phi(.01) + [1 - \Phi(1)] = .5040 + .1587 = .6627.$$

11. It is known from Example 4 that the UMP test rejects H_0 *if* $\overline{X}_n \geq c$. Hence, c must be chosen so that

$$0.95 = \Pr(X_n \geq c\,|\,\theta = 1) = \Pr\left[Z \geq \sqrt{n}(c-1)\right],$$

where Z has the standard normal distribution. Hence, $\sqrt{n}(c-1) = -1.645$, and $c = 1 - (1.645)/n^{1/2}$. Since the power function of this test will be a strictly increasing function of θ, the size of the test will be

$$\begin{aligned} \alpha &= \sup_{\theta \leq 0} \Pr(\text{Rej. } H_0\,|\,\theta) = \Pr(\text{Rej. } H_0\,|\,\theta = 0) = \Pr\left[\overline{X}_n \geq 1 - \frac{(1.645)}{n^{1/2}}\,\middle|\,\theta = 0\right] \\ &= \Pr(Z \geq n^{1/2} - 1.645), \end{aligned}$$

where Z again has the standard normal distribution. When $n = 16$,

$$\alpha = \Pr(Z \geq 2.355) = .0093.$$

13. The χ^2 distribution with θ degrees of freedom is a gamma distribution with parameters $\alpha = \theta/2$ *and* $\beta = 1/2$. Hence, it follows from Exercise 3 of Sec. 9.3 that the joint p.d.f. of $X_i, \ldots, X_n$ has a monotone likelihood ratio in the statistic $T = \prod_{i=1}^{n} X_i$. Hence, there is a UMP test of the given hypotheses, and it specifies rejecting H_0 when $T \geq c$ or, equivalently, when $\log T = \sum_{i=1}^{n} \log X_i \geq k$.

15. It was shown in Sec. 9.7 that the F test rejects H_0 *if* $V \geq 2.20$, where V is given by (9.7.4) and 2.20 is the 0.95 quantile of the F distribution with 15 and 20 degrees of freedom. For any values of σ_1^2 *and* σ_2^2, the random variable V^* given by (9.7.5) has the F distribution with 15 and 20 degrees of freedom. When $\sigma_1^2 = 2\sigma_2^2, V^* = V/2$. Hence, the power when $\sigma_1^2 = 2\sigma_2^2$ is

$$P^*(\text{Rej. } H_0) = P^*(V \geq 2.20) = P^*\left(\frac{1}{2}V \geq 1.10\right) = \Pr(V^* \geq 1.1),$$

where P^* denotes a probability calculated under the assumption that $\sigma_1^2 = 2\sigma_2^2$.

17. (a) Carrying out a test of size α on repeated independent samples is like performing a sequence of Bernoulli trials on each of which the probability of success is α. With probability 1, a success will ultimately be obtained. Thus, sooner or later, H_0 will ultimately be rejected. Therefore, the overall size of the test is 1.

    (b) As we know from the geometric distribution, the expected number samples, or trials, until a success is obtained is $1/\alpha$.

19. At each point $\theta \in \Omega_1, \pi(\theta \mid \delta)$ must be at least as large as it is at any point in Ω_0, because δ is unbiased. But $\sup_{\theta \in \Omega_0} \pi(\theta \mid \delta) = \alpha$, at every point $\theta \in \Omega_1$.

21. Since H_0 is simple and δ has size α, then $\pi(\theta_0 \mid \delta) = \alpha$. Since δ is unbiased, $\pi(\theta \mid \delta) \geq \alpha$ for all other values of θ. Therefore, $\pi(\theta \mid \delta)$ is a minimum at $\theta = \theta_0$. Since π is assumed to be differentiable, it follows that $\pi'(\theta_0 \mid \delta) = 0$.

23. (a) The data consist of both X and Y, where X is defined in Exercise 22 and $Y = 1$ if meteorological conditions are poor and $Y = 0$ if not. The joint p.f./p.d.f. of (X, Y) given $\Theta = \theta$ is

$$\frac{1}{2(2\pi)^{1/2}10^y} \exp\left(-\frac{1-y}{2}[x-\theta]^2 - \frac{y}{200}[x-\theta]^2\right).$$

    The Bayes test will choose H_0 when

$$w_0\xi_0\frac{1}{2(2\pi)^{1/2}10^y} \exp\left(-\frac{1-y}{2}x^2 - \frac{y}{200}x^2\right)$$
$$> w_1\xi_1\frac{1}{2(2\pi)^{1/2}10^y} \exp\left(-\frac{1-y}{2}[x-10]^2 - \frac{y}{200}[x-10]^2\right).$$

    It will choose H_1 when the reverse inequality holds, and it can do either when equality holds. This inequality can be rewritten by splitting according to the value of y. That is, choose H_0 if

$$\begin{cases} x < 5 + \log(w_0\xi_0/(w_1\xi_1))/10 & \text{if } y = 0, \\ x < 5 + 10\log(w_0\xi_0/(w_1\xi_1)) & \text{if } y = 1. \end{cases}$$

    (b) In order for a test to be of the form of part (a), the two critical values c_0 and c_1 used for $y = 0$ and $y = 1$ respectively must satisfy $c_1 - 5 = 100(c_0 - 5)$. In part (a) of Exercise 22, the two critical values are $c_0 = 1.645$ and $c_1 = 16.45$. These do not even approximately satisfy $c_1 - 5 = 100(c_0 - 5)$.

    (c) In part (b) of Exercise 22, the two critical values are $c_0 = 5.069$ and $c_1 = 12.8155$. These approximately satisfy $c_1 - 5 = 100(c_0 - 5)$.

25. Let I be the random interval that corresponds to the UMP test, and let J be a random interval that corresponds to some other level α_0 test. Translating UMP into what it says about the random interval I compared to J, we have for all $\theta > c$

$$\Pr(c \in I|\theta) \leq \Pr(c \in J|\theta).$$

In other words, the observed value of I is a uniformly most accurate coefficient $1-\alpha_0$ confidence interval if, for every random interval J such that the observed value of J is a coefficient $1-\alpha_0$ confidence interval and for all $\theta_2 > \theta_1$,

$$\Pr(\theta_1 \in I|\theta = \theta_2) \leq \Pr(\theta_1 \in J|\theta = \theta_2).$$

Chapter 10

Categorical Data and Nonparametric Methods

10.1 Tests of Goodness-of-Fit

1. Let $Y = N_1$, the number of defective items, and let $\theta = p_1$, the probability that each item is defective. The level α_0 test requires us to choose c_1 and c_2 such that $\Pr(Y \le c_1|\theta = 0.1) + \Pr(Y \ge c_2|\theta = 0.1)$ is close to α_0. We can compute the probability that $Y = y$ for each $y = 0, \ldots, 100$ and arrange the numbers from smallest to largest. The smallest values correspond to large values of y down to $y = 25$, then some values corresponding to small values of y start to appear in the list. The sum of the values reaches 0.0636 when $c_1 = 4$ and $c_2 = 16$. So $\alpha_0 = 0.0636$ is the smallest α_0 for which we would reject $H_0 : \theta = 0.1$ using such a test.

3. We obtain the following frequencies:

i	0	1	2	3	4	5	6	7	8	9
N_i	25	16	19	20	20	22	24	15	14	25

Since $P_i^0 = 1/10$ for every value of i, and $n = 200$, we find from Eq. (10.1.2) that $Q = 7.4$. If Q has the χ^2 distribution with 9 degrees of freedom, $\Pr(Q \ge 7.4) = 0.6$.

5. (a) The number of successes is $n\overline{X}_n$ and the number of failures is $n(1 - \overline{X}_n)$. Therefore,

$$\begin{aligned} Q &= \frac{(n\overline{X}_n - np_0)^2}{np_0} + \frac{[n(1-\overline{X}_n) - n(1-p_0)]^2}{n(1-p_0)} \\ &= n(\overline{X}_n - p_0)^2\left(\frac{1}{p_0} + \frac{1}{1-p_0}\right) \\ &= \frac{n(\overline{X}_n - p_0)^2}{p_0(1-p_0)} \end{aligned}$$

(b) If $p = p_0$, then $E(\overline{X}_n) = p_0$ and $\mathrm{Var}(\overline{X}_n) = p_0(1-p_0)/n$. Therefore, by the central limit theorem, the c.d.f. of

$$Z = \frac{\overline{X}_n - p_0}{[p_0(1-p_0)/n]^{1/2}}$$

converges to the c.d.f. of the standard normal distribution. Since $Q = Z^2$, the c.d.f. of Q will converge to the c.d.f. of the χ^2 distribution with 1 degree of freedom.

7. We obtain the following table:

	$0 < x < 0.2$	$0.2 < x < 0.5$	$0.5 < x < 0.8$	$0.8 < x < 1.$
N_i	391	490	580	339
np_i^0	360	540	540	360

If Q has a χ^2 distribution with 3 degrees of freedom, then $\Pr(Q \geq 11.34) = 0.01$. Therefore, we should reject H_0 if $Q \geq 11.34$. It is found from Eq. (10.1.2) that Q = 11.5.

9. (a) The five intervals, each of which has probability 0.2, are as follows:

$$(-\infty, -0.842),\ (-0.842, -0.253),\ (-0.253, 0.253),\ (0.253, 0.842),\ (0.842, \infty).$$

We obtain the following table:

	N_i	np_i^0
$-\infty < x < -0.842$	15	10
$-0.842 < x < -0.253$	10	10
$-0.253 < x < 0.253$	7	10
$0.253 < x < 0.842$	12	10
$0.842 < x < \infty$	6	10

The calculated value of Q is 5.4. If Q has a χ^2 distribution with 4 degrees of freedom, then $\Pr(Q \geq 5.4) = 0.25$.

(b) The ten intervals, each of which has probability 0.1, are as given in the following table:

	N_i	np_i^0
$-\infty < x < -1.282$	8	5
$-1.282 < x < -0.842$	7	5
$-0.842 < x < -0.524$	3	5
$-0.524 < x < -0.253$	7	5
$-0.253 < x < 0$	5	5
$0 < x < 0.253$	2	5
$0.253 < x < 0.524$	5	5
$0.524 < x < 0.842$	7	5
$0.842 < x < 1.282$	2	5
$1.282 < x < \infty$	4	5

The calculated value of Q is 8.8. If Q has the χ^2 distribution with 9 degrees of freedom, then the value of $\Pr(Q \geq 8.8)$ is between 0.4 and 0.5.

10.2 Goodness-of-Fit for Composite Hypotheses

1. There are many ways to perform a χ^2 test. For example, we could divide the real numbers into the intervals $(-\infty, 15]$, $(15, 30]$, $(30, 45]$, $(45, 60]$, $(60, 75]$, $(75, 90]$, $(90, \infty)$. The numbers of observations in these intervals are 14, 14, 4, 4, 3, 0, 2

   (a) The M.L.E.'s of the parameters of a normal distribution are $\hat{\mu} = 30.05$ and $\hat{\sigma^2} = 537.51$. Using the method of Chernoff and Lehmann, we compute two different p-values with 6 and 4 degrees of freedom. The probabilities for the seven intervals are 0.2581, 0.2410, 0.2413, 0.1613, 0.0719,

0.0214, 0.0049. The expected counts are 41 times each of these numbers. This makes $Q = 24.53$. The two p-values are both smaller than 0.0005.

(b) The M.L.E.'s of the parameters of a lognormal distribution are $\hat{\mu} = 3.153$ and $\hat{\sigma^2} = 0.48111$. Using the method of Chernoff and Lehmann, we compute two different p-values with 6 and 4 degrees of freedom. The probabilities for the seven intervals are 0.2606, 0.3791, 0.1872, 0.0856, 0.0407, 0.0205, 0.0261. The expected counts are 41 times each of these numbers. This makes $Q = 5.714$. The two p-values are both larger than 0.2.

3. (a) It follows from Eqs. (10.2.2) and (10.2.6) that (aside from the multinomial coefficient)

$$\begin{aligned}\log L(\theta) &= (N_4 + N_5 + N_6)\log 2 + (2N_1 + N_4 + N_5)\log\theta_1 + (2N_2 + N_4 + N_6)\log\theta_2 \\ &\quad + (2N_3 + N_5 + N_6)\log(1 - \theta_1 - \theta_2).\end{aligned}$$

By solving the equations

$$\frac{\partial \log L(\theta)}{\partial\theta_1} = 0 \quad \text{and} \quad \frac{\partial \log L(\theta)}{\partial\theta_2} = 0,$$

we obtain the results

$$\hat{\Theta}_1 = \frac{2N_1 + N_4 + N_5}{2n} \quad \text{and} \quad \hat{\Theta}_2 = \frac{2N_2 + N_4 + N_6}{2n},$$

where $n = \sum_{i=1}^{6} N_i$.

(b) For the given values, $n = 150, \hat{\Theta}_1 = 0.2$, *and* $\hat{\Theta}_2 = 0.5$. Therefore, we obtain the following table:

i	N_i	$n\pi_i(\hat{\Theta})$
1	2	6
2	36	37.5
3	14	13.5
4	36	30
5	20	18
6	42	45

It is found from Eq. (10.2.4) that $Q = 4.37$. If Q has the χ^2 distribution with $6-1-2 = 3$ degrees of freedom, then the value of $\Pr(Q \geq 4.37)$ is approximately 0.226.

5. From the given observations, it is found that the M.L.E. of the mean Θ of the Poisson distribution is $\hat{\Theta} = \overline{X}_n = 1.5$. From the table of the Poisson distribution with $\Theta = 1.5$, we can obtain the values of $\pi_i(\hat{\Theta})$. In turn, we can then obtain the following table:

No. of tickets	N_i	$n\pi_i(\hat{\Theta})$
0	52	44.62
1	60	66.94
2	55	50.20
3	18	25.10
4	8	9.42
5 or more	7	3.70

It is found from Eq. (10.2.4) that $Q = 7.56$. Since $\hat{\Theta}$ is calculated from the original observations rather than from the grouped data, the approximate distribution of Q when H_0 is true lies between the χ^2 distribution with 4 degrees of freedom and the χ^2 distribution with 5 degrees of freedom. The two p-values for 4 and 5 degrees of freedom are 0.1091 and 0.1822.

7. There is no single correct answer to this problem. The M.L.E.'s $\hat{\mu} = \overline{X}_n$ and $\hat{\sigma}^2 = S_n^2/n$ should be calculated from the given observations. These observations should then be grouped into intervals and the observed number in each interval compared with the expected number in that interval if each of the 50 observations had the normal distribution with mean $\overline{X}_n$ and variance S_n^2/n. If the number of intervals is k, then when H_0 is true, the approximate distribution of the statistic Q will lie between the χ^2 distribution with $k-3$ degrees of freedom and the χ^2 distribution with $k-1$ degrees of freedom.

10.3 Contingency Tables

1. Table S.10.1 contains the expected counts for this example. The value of the χ^2 statistic Q calculated

Table S.10.1: Expected cell counts for Exercise 1 of Sec. 10.3.

	Good grades	Athletic ability	Popularity
Boys	117.3	42.7	67.0
Girls	129.7	47.3	74.0

from these data is $Q = 21.5$. This should be compared to the χ^2 distribution with two degrees of freedom. The tail area can be calculated using statistical software as 2.2×10^{-5}.

3. By Exercise 2,

$$Q = \sum_{i=1}^{R} \frac{N_{i1}^2}{\hat{E}_{i1}} + \sum_{i=1}^{R} \frac{N_{i2}^2}{\hat{E}_{i2}} - n.$$

But

$$\sum_{i=1}^{R} \frac{N_{i2}^2}{\hat{E}_{i2}} = \sum_{i=1}^{R} \frac{(N_{i+} - N_{i1})^2}{\hat{E}_{i2}} = \sum_{i=1}^{R} \frac{N_{i+}^2}{\hat{E}_{i2}} - 2\sum_{i=1}^{R} \frac{N_{i+}N_{i1}}{\hat{E}_{i2}} + \sum_{i=1}^{R} \frac{N_{i1}^2}{\hat{E}_{i2}}.$$

In the first two sums on the right, we let $\hat{E}_{i2} = N_{i+}N_{+2}/n$, and in the third sum we let $\hat{E}_{i2} = N_{+2}\hat{E}_{i1}/N_{+1}$. We then obtain

$$\sum_{i=1}^{R} \frac{N_{i2}^2}{\hat{E}_{i2}} = \frac{n}{N_{+2}} \sum_{i=1}^{R} N_{i+} - \frac{2n}{N_{+2}} \sum_{i=1}^{R} N_{i1} + \frac{N_{+1}}{N_{+2}} \sum_{i=1}^{R} \frac{N_{i1}^2}{\hat{E}_{i1}} = \frac{n^2}{N_{+2}} - 2n\frac{N_{+1}}{N_{+2}} + \frac{N_{+1}}{N_{+2}} \sum_{i=1}^{R} \frac{N_{i1}^2}{\hat{E}_{i1}}.$$

It follows that

$$Q = \left(1 + \frac{N_{+1}}{N_{+2}}\right) \sum_{i=1}^{R} \frac{N_{i1}^2}{\hat{E}_{i1}} + \frac{n}{N_{+2}}(n - 2N_{+1} - N_{+2}).$$

Since $n = N_{+1} + N_{+2}$,

$$Q = \frac{n}{N_{+2}} \sum_{i=1}^{R} \frac{N_{i1}^2}{\hat{E}_{i1}} - \frac{n}{N_{+2}} N_{+1}.$$

5. The values of $\hat{E}_{ij}$ are as given in the following table.

77.27	94.35	49.61	22.77
17.73	21.65	11.39	5.23

The value of Q is found from Eq. (10.3.4) to be 8.6. If Q has the χ^2 distribution with $(2-1)(4-1) = 3$ degrees of freedom, then $\Pr(Q \geq 8.6)$ lies between 0.025 and 0.05.

7. (a) The values of p_{i+} and p_{+j} are the marginal totals given in the following table:

			0.3
			0.3
			0.4
0.5	0.3	0.2	1.0

It can be verified that $p_{ij} = p_{i+}p_{+j}$ for each of the 9 entries in the table. It can be seen in advance that this relation will be satisfied for every entry in the table because it can be seen that the three rows of the table are proportional to each other or, equivalently, that the three columns are proportional to each other.

(b) Here is one example of a simulated data set

44	32	16	92
45	25	15	85
63	33	27	123
152	90	58	300

(c) The statistic Q calculated by any student from Eq. (10.3.4) will have the χ^2 distribution with $(3-1)(3-1) = 4$ degrees of freedom. For the data in part (b), the table of $\hat{E}_{ij}$ values is

46.6	27.6	17.8
43.1	25.5	16.4
62.3	36.9	23.8

The value of Q is then 2.105. The p-value 0.7165.

9. Let N_{ijk} denote the number of observations in the random sample that fall into the (i, j, k) cell, and let

$$N_{i++} = \sum_{j=1}^{C}\sum_{k=1}^{T} N_{ijk}, N_{+j+} = \sum_{i=1}^{R}\sum_{k=1}^{T} N_{ijk},$$

$$N_{++k} = \sum_{i=1}^{R}\sum_{j=1}^{C} N_{ijk}.$$

Then the M.L.E.'s are

$$\hat{p}_{i++} = \frac{N_{i++}}{n}, \hat{p}_{+j+} = \frac{N_{+j+}}{n}, \hat{p}_{++k} = \frac{N_{++k}}{n}.$$

Therefore, when H_0 is true,

$$\hat{E}_{ijk} = n\hat{p}_{i++}\hat{p}_{+j+}\hat{p}_{++k} = \frac{N_{i++}N_{+j+}N_{++k}}{n^2}.$$

Since $\sum_{i=1}^{R} \hat{p}_{i++} = \sum_{j=1}^{C} \hat{p}_{+j+} = \sum_{k=1}^{T} \hat{p}_{++k} = 1$, the number of parameters that have been estimated is $(R-1)+(C-1)+(T-1) = R+C+T-3$. Therefore, when H_0 is true, the approximate distribution of

$$Q = \sum_{i=1}^{R}\sum_{j=1}^{C}\sum_{k=1}^{T} \frac{(N_{ijk} - \hat{E}_{ijk})^2}{\hat{E}_{ijk}}$$

will be the χ^2 distribution for which the number of degrees of freedom is $RCT - 1 - (R+C+T-3) = RCT - R - C - T + 2$.

10.4 Tests of Homogeneity

1. Table S.10.2 contains the expected cell counts. The value of the χ^2 statistic is $Q = 18.8$, which should

Table S.10.2: Expected cell counts for Exercise 1 of Sec. 10.4.

	Good grades	Athletic ability	Popularity
Rural	77.0	28.0	44.0
Suburban	78.0	28.4	44.5
Urban	92.0	33.5	52.5

be compared to the χ^2 distribution with four degrees of freedom. The tail area is 8.5×10^{-4}.

3. The value of the statistic Q given by Eqs. (10.4.3) and (10.4.4) is 18.9. If Q has the χ^2 distribution with $(4-1)(5-1) = 12$ degrees of freedom, then the value of $\Pr(Q \geq 18.9)$ lies between 0.1 and 0.05.

5. The correct table to be analyzed is as follows:

Supplier	Defectives	Nondefectives
1	1	14
2	7	8
3	7	8

The value of Q found from this table is 7.2. If Q has the χ^2 distribution with $(3-1)(2-1) = 2$ degrees of freedom, then $\Pr(Q \geq 7.2) < 0.05$.

7. The proper table to be analyzed is as follows:

		After meeting		
		Favors A	Favors B	No preference
	Favors A			
Before	Favors B			
meeting	No preference			

Each person who attended the meeting can be classified in one of the nine cells of this table. If a speech was made on behalf of A at the meeting, we could evaluate the effectiveness of the speech by comparing the numbers of persons who switched from favoring B or having no preference before the meeting to favoring A after the meeting with the number who switched from favoring A before the meeting to one of the other positions after the meeting.

10.5 Simpson's Paradox

1. If population II has a relatively high proportion of men and population I has a relatively high proportion of women, then the indicated result will occur. For example, if 90 percent of population II are men and 10 percent are women, then the proportion of population II with the characteristic will be $(.9)(.6)+(.1)(.1)=.55$. If 10 percent of population I are men and 90 percent are women, then the proportion of population I with the characteristic will be only $(.1)(.8)+(.9)(.3)=.35$.

3. Assume that $\Pr(B|A)=\Pr(B|A^c)$. This means that A and B are independent. According to the law of total probability, we can write

$$\begin{aligned}\Pr(I|B) &= \Pr(I|A\cap B)\Pr(A|B)+\Pr(I|A^c\cap B)\Pr(A^c|B)\\ &= \Pr(I|A\cap B)\Pr(A)+\Pr(I|A^c\cap B)\Pr(A^c),\end{aligned}$$

where the last equality follows from the fact that A and B are independent. Similarly,

$$\Pr(I|B^c)=\Pr(I|A\cap B^c)\Pr(A)+\Pr(I|A^c\cap B^c)\Pr(A^c).$$

If the first two inequalities in (10.5.1) hold then the weighted average of the left sides of the inequalities must be larger than the same weighted average of the right sides. In particular,

$$\Pr(I|A\cap B)\Pr(A)+\Pr(I|A^c\cap B)\Pr(A^c)>\Pr(I|A\cap B^c)\Pr(A)+\Pr(I|A^c\cap B^c)\Pr(A^c).$$

But, we have just shown that this last equality is equivalent to $\Pr(I|B)>\Pr(I|B^c)$, which means that the third inequality cannot hold if the first two hold.

5. Suppose that the first two inequalities in (10.5.1) hold, and that $\Pr(A|B)=\Pr(A|B^c)$, Then

$$\begin{aligned}\Pr(I|B) &= \Pr(I\mid A\cap B)\Pr(A\mid B)+\Pr(I\mid A^c\cap B)\Pr(A^c\mid B)\\ &> \Pr(I\mid A\cap B^c)\Pr(A\mid B)+\Pr(I\mid A^c\cap B^c)\Pr(A^c\mid B)\\ &= \Pr(I\mid A\cap B^c)\Pr(A\mid B^c)+\Pr(I\mid A^c\cap B^c)\Pr(A^c\mid B^c)\\ &= \Pr(I\mid B^c).\end{aligned}$$

Hence, the final inequality in (10.5.1) must be reversed.

7. (a) Table S.10.3 shows the proportions helped by each treatment in the four categories of subjects. The proportion helped by Treatment II is higher in each category.

Table S.10.3: Table for Exercise 7a in Sec. 10.5.

	Proportion helped	
Category	Treatment I	Treatment II
Older males	.200	.667
Younger males	.750	.800
Older females	.167	.286
Younger females	.500	.640

(b) Table S.10.4 shows the proportions helped by each treatment in the two aggregated categories. Treatment I helps a larger proportion in each of the two categories

(c) When all subjects are grouped together, the proportion helped by Treatment I is $200/400=0.5$, while the proportion helped by Treatment II is $240/400=0.6$.

Table S.10.4: Table for Exercise 7b in Sec. 10.5.

Category	Proportion helped Treatment I	Treatment II
Older subjects	.433	.400
Younger subjects	.700	.667

10.6 Kolmogorov-Smirnov Tests

1. $F_n(x) = 0$ for $x < y$, and $F_n(y_1) = 0.2$. Suppose first that $F(y_1) \geq 0.1$. Since F is continuous, the values of $F(x)$ will be arbitrarily close to $F(y_1)$ for x arbitrarily close to y_1. Therefore, $\sup_{x<y_1} |F_n(x) - F(x)| = F(y_1) \geq 0.1$, and it follows that $D_n \geq 0.1$. Suppose next that $F(y_1) \leq 0.1$. Since $F_n(y_1) = 0.2$, it follows that $| F_n(y_1) - F(y_1) | \geq 0.1$. Therefore, it is again true that $D_n \geq 0.1$. We can now conclude that it must always be true that $D_n \geq 0.1$. If the values of $F(y_i)$ are as specified in the second part of the exercise, for $i = 1, \ldots, 5$, then:

$$| F_n(x) - F(x) | = \begin{cases} F(x) \leq 0.1 & \text{for } x < y_1, \\ 0.2 - 0.1 = 0.1 & \text{for } x = y_1, \\ | F(x) - 0.2 | \leq 0.1 & \text{for } y_1 < x < y_2, \\ 0.4 - 0.3 = 0.1 & \text{for } x = y_2, \\ | F(x) - 0.4 | \leq 0.1 & \text{for } y_2 < x < y_3, \\ \text{etc.} \end{cases}$$

Hence, $D_n = \sup_{-\infty < x < \infty} | F_n(x) - F(x) | = 0.1$.

3. The largest value of the difference between the sample c.d.f. and the c.d.f. of the normal distribution with mean 3.912 and variance 0.25 occurs right before $x = 4.22$, the 12th observation. For x just below 4.22, the sample c.d.f. is $F_n(x) = 0.48$, while the normal c.d.f. is $\Phi([4.22 - 3.912]/0.5) = 0.73$. The difference is $D_n^* = 0.25$. The Kolmogorov-Smirnov test statistic is $23^{1/2} \times 0.25 = 1.2$. The tail area can be found from Table 10.32 as 0.11.

5. Here,

$$F(x) = \begin{cases} \frac{3}{2}x & \text{for } 0 < x \leq 1/2, \\ \frac{1}{2}(1+x) & \text{for } \frac{1}{2} < x < 1. \end{cases}$$

Therefore, we obtain Table S.10.5. The supremum of $| F_n(x) - F(x) |$ occurs as $x \to y_{18}$ from below. Here, $F(x) \to 0.83$ while $F_n(x)$ remains at 0.68. Therefore, $D_n{}^* = 0.83 - 0.68 = 0.15$. It follows that $n^{1/2} D_n{}^* = 0.75$ and, from Table 10.32, $H(0.75) = 0.3728$. Therefore, the tail area corresponding to the observed value of $D_n{}^*$ is $1 - 0.3728 = 0.6272$.

7. We first replace each observed value x_i by the value $(x_i - 26)/2$. Then, under the null hypothesis, the transformed values will form a random sample from a standard normal distribution. When these transformed values are ordered, we obtain Table S.10.6. The maximum value of $| F_n(x) - \Phi(x) |$ is attained at $x = y_{23}$ and its value is 0.0649. Since $n = 50, n^{1/2} D_n{}^* = 0.453$. It follows from Table 10.32 that $H(0.453) = 0.02$. Therefore, the tail area corresponding to the observed value of $D_n{}^*$ is $1 - 0.02 = 0.98$.

Table S.10.5: Table for Exercise 5 in Sec. 10.6.

i	y_i	$F(y_i)$	$F_n(y_i)$	i	y_i	$F(y_i)$	$F_n(y_i)$
1	.01	.015	.04	14	.41	.615	.56
2	.06	.09	.08	15	.42	.63	.60
3	.08	.12	.12	16	.48	.72	.64
4	.09	.135	.16	17	.57	.785	.68
5	.11	.165	.20	18	.66	.83	.72
6	.16	.24	.24	19	.71	.855	.76
7	.22	.33	.28	20	.75	.875	.80
8	.23	.345	.32	21	.78	.89	.84
9	.29	.435	.36	22	.79	.895	.88
10	.30	.45	.40	23	.82	.91	.92
11	.35	.525	.44	24	.88	.94	.96
12	.38	.57	.48	25	.90	.95	1.00
13	.40	.60	.52				

Table S.10.6: Table for Exercise 7 in Sec. 10.6.

i	y_i	Φy_i	$F_n(y_i)$	i	y_i	$\Phi(y_i)$	$F_n(y_i)$
1	−2.2105	.0136	.02	26	−0.010	.4960	.52
2	−1.9265	.0270	.04	27	−0.002	.4992	.54
3	−1.492	.0675	.06	28	¼0.010	.5040	.56
4	−1.3295	.0919	.08	29	¼0.1515	.5602	.58
5	−1.309	.0953	.10	30	¼0.258	.6018	.60
6	−1.2085	.1134	.12	31	¼0.280	.6103	.62
7	−1.1995	.1152	.14	32	¼0.3075	.6208	.64
8	−1.125	.1307	.16	33	¼0.398	.6547	.66
9	−1.0775	.1417	.18	34	¼0.4005	.6556	.68
10	−1.052	.1464	.20	35	¼0.4245	.6645	.70
11	−0.961	.1682	.22	36	¼0.482	.6851	.72
12	−0.8415	.2001	.24	37	¼0.614	.7304	.74
13	−0.784	.2165	.26	38	¼0.689	.7546	.76
14	−0.767	.2215	.28	39	¼0.7165	.7631	.78
15	−0.678	.2482	.30	40	¼0.7265	.7662	.80
16	−0.6285	.2648	.32	41	¼0.9262	.8320	.82
17	−0.548	.2919	.34	42	¼1.0645	.8564	.84
18	−0.456	.3242	.36	43	¼1.120	.8686	.86
19	−0.4235	.3359	.38	44	¼1.176	.8802	.88
20	−0.340	.3669	.40	45	¼1.239	.8923	.90
21	−0.3245	.3728	.42	46	¼1.4615	.9281	.92
22	−0.309	.3787	.44	47	¼1.6315	.9487	.94
23	−0.266	.3951	.46	48	¼1.7925	.9635	.96
24	−0.078	.4689	.48	49	¼1.889	.9705	.98
25	−0.0535	.4787	.50	50	¼2.216	.9866	1.00

9. We shall denote the 25 ordered observations in the first sample by $x_1 < \cdots < x_{25}$ and shall denote the 20 ordered observations in the second sample by $y_1 < \cdots < y_{20}$. We obtain Table S.10.7. The

Table S.10.7: Table for Exercise 9 in Sec. 10.6.

x_i	y_j	$F_m(x)$	$G_n(x)$	x_i	y_j	$F_m(x)$	$G_n(x)$
−2.47		.04	0	0.51		.60	.45
−1.73		.08	0		0.52	.60	.50
−1.28		.12	0	0.59		.64	.50
−0.82		.16	0	0.61		.68	.50
−0.74		.20	0	0.64		.72	.50
	−0.71	.20	.05		0.66	.72	.55
−0.56		.24	.05		0.70	.72	.60
−0.40		.28	.05		0.96	.72	.65
−0.39		.32	.05	1.05		.76	.65
	−0.37	.32	.10	1.06		.80	.65
−0.32		.36	.10	1.09		.84	.65
	−0.30	.36	.15	1.31		.88	.65
	−0.27	.36	.20		1.38	.88	.70
−0.06		.40	.20		1.50	.88	.75
	0.00	.40	.25		1.56	.88	.80
0.05		.44	.25	1.64		.92	.80
0.06		.48	.25		1.66	.92	.85
	0.26	.48	.30	1.77		.96	.85
0.29		.52	.30		2.20	.96	.90
0.31		.56	.30		2.31	.96	.95
	0.36	.56	.35	2.36		1.00	.95
	0.38	.56	.40		3.29	1.00	1.00
	0.44	.56	.45				

maximum value of $\mid F_m(x) - G_n(x) \mid$ is attained at $x = -0.39$, where its value is $0.32 - 0.05 = 0.27$. Therefore, $D_{mn} = 0.27$ and, since $m = 25$ and $n = 20$, $(mn/[m+n])^{1/2} D_{mn} = 0.9$. From Table 10.32, $H(0.9) = 0.6073$. Hence, the tail area corresponding to the observed value of D_{mn} is $1-0.6073 = 0.3927$.

11. We shall multiply each of the observations in the second sample by 3 and then carry out the same procedure as in Exercise 9. We now obtain Table S.10.8. The maximum value of $\mid F_m(x) - G_n(x) \mid$ is attained at $x = 1.06$, where its value is $0.80 - 0.30 = 0.50$. Therefore, $D_{mn} = 0.50$ and $(mn/[m+n])^{1/2} D_{mn} = 1.667$. From Table 10.32, $H(1.667) = 0.992$. Therefore, the tail area corresponding to the observed value of D_{mn} is $1 - 0.992 = 0.008$.

10.7 Robust Estimation

1. The observed values ordered from smallest to largest are 2.1, 2.2, 21.3, 21.5, 21.7, 21.7, 21.8, 22.1, 22.1, 22.2, 22.4, 22.5, 22.9, 23.0, 63.0.

   (a) The sample mean is the average of the numbers, 22.17.

   (b) The trimmed mean for a given value of k is found by dropping k values from each end of this ordered sequence and averaging the remaining values. In this problem we get

k	1	2	3	4
kth level trimmed mean	20.57	22.02	22	22

   (c) The sample median is the middle observation, 22.1.

   (d) The median absolute deviation is 0.4. Suppose that we start iterating with the sample average 22.17. The 7th and 8th iterations are both 22.

Table S.10.8: Table for Exercise 11 in Sec. 10.6

x_i	y_j	$F_m(x)$	$G_n(x)$	x_i	y_j	$F_m(x)$	$G_n(x)$
−2.47		.04	0		0.78	.72	.30
	−2.13	.04	.05	1.05		.76	.30
−1.73		.08	.05	1.06		.80	.30
−1.28		.12	.05		1.08	.80	.35
	−1.11	.12	.10	1.09		.84	.35
	−0.90	.12	.15		1.14	.84	.40
−0.82		.16	.15	1.31		.88	.40
	−0.81	.16	.20		1.32	.88	.45
−0.74		.20	.20		1.56	.88	.50
−0.56		.24	.20	1.64		.92	.50
−0.40		.28	.20	1.77		.96	.50
−0.39		.32	.20		1.98	.96	.55
−0.32		.36	.20		2.10	.96	.60
−0.06		.40	.20	2.36		1.00	.60
	0.00	.40	.25		2.88	1.00	.65
0.05		.44	.25		4.14	1.00	.70
0.06		.48	.25		4.50	1.00	.75
0.29		.52	.25		4.68	1.00	.80
0.31		.56	.25		4.98	1.00	.85
0.51		.60	.25		6.60	1.00	.90
0.59		.64	.25		6.93	1.00	.95
0.61		.68	.25		9.87	1.00	1.00
0.64		.72	.25				

3. The distribution of $\tilde{\theta}_{.5,n}$ will be approximately normal with mean θ and standard deviation $1/[2n^{1/2}f(\theta)]$. In this exercise,

$$f(x) = \frac{1}{\sqrt{2\pi}} \quad \exp\left[-\frac{1}{2}(x-\theta)^2\right].$$

Hence, $f(\theta) = 1/\sqrt{2\pi}$. Since $n = 100$, the standard deviation of the approximate distribution of $\tilde{\theta}_{.5,n}$ is $\sqrt{2\pi}/20 = 0.1253$. It follows that the distribution of $Z = (\tilde{\theta}_{.5,n} - \theta)/0.1253$ will be approximately standard normal. Thus,

$$\Pr(|\tilde{\theta}_{.5,n} - \theta| \le 0.1) = \Pr\left(|Z| \le \frac{0.1}{0.1253}\right) = \Pr(|Z| \le 0.798) = 2\Phi(0.798) - 1 = 0.575.$$

5. Let the first density on the right side of Eq. (10.7.1) be called h. Since both h and g are symmetric with respect to μ, so also is $f(x)$. Therefore, both the sample mean $\overline{X}_n$ and the sample median $\tilde{X}_n$ are unbiased estimators of μ. It follows that the M.S.E. of $\overline{X}_n$ is equal to $\text{Var}(\overline{X}_n)$ and that the M.S.E. of $\tilde{X}_n$ is equal to $\text{Var}(\tilde{X}_n)$. The variance of a single observation X is

$$\begin{aligned}
\text{Var}(X) &= \int_{-\infty}^{\infty} (x-\mu)^2 f(x)dx \\
&= \frac{1}{2}\int_{-\infty}^{\infty} (x-\mu)^2 h(x)dx + \frac{1}{2}\int_{-\infty}^{\infty} (x-\mu)^2 g(x)dx \\
&= \frac{1}{2}(1) + \frac{1}{2}(4) = \frac{5}{2}.
\end{aligned}$$

Since $n = 100$, $\text{Var}(\overline{X}_n) = (1/100)(5/2) = 0.025$.

The variance of $\tilde{X}_n$ will be approximately $1/[4nh^2(\mu)]$. Since

$$h(x) = \frac{1}{\sqrt{2\pi}} \exp\left[-\frac{1}{2}(x-\mu)^2\right] \quad \text{and } g(x) = \frac{1}{2\sqrt{2\pi}} \exp\left[-\frac{1}{2(4)}(x-\mu)^2\right],$$

it follows that

$$f(\mu) = \frac{1}{2}h(\mu) + \frac{1}{2}g(\mu) = \frac{1}{2}\cdot\frac{1}{\sqrt{2\pi}} + \frac{1}{2}\cdot\frac{1}{2\sqrt{2\pi}} = \frac{3}{4\sqrt{2\pi}}.$$

Therefore, $\text{Var}(\tilde{X}_n)$ is approximately $2\pi/225 = 0.028$.

7. (a) The mean $\overline{X}_n$ is the mean of each X_i. Since $f(x)$ is a weighted average of two other p.d.f.'s, the $\int xf(x)dx$ is the same mixture of the means of the other two distributions. Since each of the distributions in the mixture has mean μ, so does the distribution with p.d.f. f.

   (b) The variance $\overline{X}_n$ is $1/n$ times the variance of X_i. The variance of X_i is $E(X_i^2) - \mu^2$. Since the p.d.f. of X_i is a weighted average of two other p.d.f.'s, the mean of X_i^2 is the same weighted average of the two means of X_i^2 from the two p.d.f.'s. The mean of X_i^2 from the first p.d.f. (the normal distribution with mean μ and variance σ^2) is $\mu^2+\sigma^2$. The mean of X_i^2 from the second p.d.f. (the normal distribution with mean μ and variance $100\sigma^2$) is $\mu^2 + 100\sigma^2$. The weighted average is

$$(1-\epsilon)(\mu^2+\sigma^2) + \epsilon(\mu^2+100\sigma^2) = \mu^2 + \sigma^2(1+99\epsilon).$$

   The variance of X_i is then $(1+99\epsilon)\sigma^2$, and the variance $\overline{X}_n$ is $(1+99\epsilon)\sigma^2/n$.

9. The likelihood function is

$$\frac{1}{2^n\sigma^n} \exp\left(-\frac{1}{\sigma}\sum_{i=1}^{n}|x_i-\theta|\right).$$

It is easy to see that, no matter what σ equals, the M.L.E. of θ is the number that minimizes $\sum_{i=1}|x_i-\theta|$. This is the same as the number that minimizes $\sum_{i=1}|x_i-\theta|/n$. The value $\sum_{i=1}^{n}|x_i-\theta|/n$ is the mean of $|X-\theta|$ when the c.d.f. of X is the sample c.d.f. of $X_1,\ldots,X_n$. The mean of $|X-\theta|$ is minimized by θ equal to a median of the distribution of X according to Theorem 4.5.3. The median of the sample distribution is the sample median.

11. Let x_q be the q quantile of X. The result will follow if we can prove that the q quantile of $aX+b$ is ax_q+b. Since

$$\Pr(aX+b \le ax_q+b) = \Pr(X \le x_q),$$

for all $a>0$ and b and q, it follows that ax_q+b is the q quantile of $aX+b$.

13. The Cauchy distribution is symmetric around 0, so the median is 0, and the median absolute deviation is the median of $Y=|X|$. If F is the c.d.f. of X, then the c.d.f. of Y is

$$G(y) = \Pr(Y\le y) = \Pr(|X|\le y) = \Pr(-y\le X\le y) = F(y) - F(-y),$$

because X has a continuous distribution. Because X has a symmetric distribution around 0, $F(-y) = 1-F(y)$, and $G(y) = 2F(y)-1$. The median of Y is where $G(y)=0.5$, that is $2F(y)-1=0.5$ or $F(y)=0.75$. So, the median of Y is the 0.75 quantile of X, namely $y=1$.

15. (a) The quantile function of the normal distribution with mean μ and variance σ^2 is the inverse of the c.d.f., $F(x) = \Phi([x-\mu]/\sigma)$. So,

$$F^{-1}(p) = \mu + \sigma\Phi^{-1}(p). \tag{S.10.1}$$

The IQR is

$$F^{-1}(0.75) - F^{-1}(0.25) = \sigma[\Phi^{-1}(0.75) - \Phi^{-1}(0.25)].$$

Since the standard normal distribution is symmetric around 0, $\Phi^{-1}(0.25) = -\Phi^{-1}(0.75)$, so the IQR is $2\sigma\Phi^{-1}(0.75)$.

(b) Let F be the c.d.f. of a distribution that is symmetric around its median μ. The median absolute deviation is then the value x such that $F(\mu+x) - F(\mu-x) = 0.5$. By symmetry around the median, we know that $F(\mu-x) = 1-F(\mu+x)$, so x solves $2F(\mu+x)-1 = 0.5$ or $F(\mu+x) = 0.75$. That is, $x = F^{-1}(0.75) - \mu$. For the case of normal random variables, use Eq. (S.10.1) to conclude that $x = \sigma\Phi^{-1}(0.75)$.

17. Let μ stand for the median of the distribution, and let $\mu+c$ be the 0.75 quantile. By symmetry, the 0.25 quantile is $\mu - c$. Also, $f(\mu+c) = f(\mu-c)$. The large sample joint distribution of the 0.25 and 0.75 sample quantiles is a bivariate normal distribution with means $\mu - c$ and $\mu + c$, variances both equal to $3/[16nf(\mu+c)^2]$, and covariance $1/[16nf(\mu+c)^2]$. The IQR is the difference between these two sample quantiles, so its large sample distribution is normal with mean $2c$ and variance

$$\frac{3}{16nf(\mu+c)^2} + \frac{3}{16nf(\mu+c)^2} - 2\frac{1}{16nf(\mu+c)^2} = \frac{1}{4nf(\mu+c)^2}.$$

10.8 Sign and Rank Tests

1. Let W be the number of (X_i, Y_i) pairs with $X_i \le Y_i$. Then W has a binomial distribution with parameters n and p. To test H_0, we reject H_0 if W is too large. In particular, if c is chosen so that

$$\sum_{w=c+1}^{n} \binom{n}{w}\left(\frac{1}{2}\right)^n < \alpha_0 \le \sum_{w=c}^{n} \binom{n}{w}\left(\frac{1}{2}\right)^n,$$

then we can reject H_0 if $W \ge c$ for a level α_0 test.

3. This test was performed in Example 9.6.5, and the tail area is 0.003.

5. Since there are 25 observations in the first sample, $F_m(x)$ will jump by the amount 0.04 at each observed value. Since there are 15 observations in the second sample, $G_n(x)$ will jump by the amount 0.0667 at each observed value. From the table given in the solution to Exercise 4, we obtain Table S.10.9. It can be seen from this table that the maximum value of $|F_m(x) - G_n(x)|$ occurs when x is equal to the observed value of rank 16, and its value at this point is $.60 - .0667 = .5333$. Hence, $D_{mn} = 0.5333$ and $\left(\frac{mn}{m+n}\right)^{1/2} D_{mn} = \left(\frac{375}{40}\right)^{1/2}(0.5333) = 1.633$. It is found from Table 10.32 that the corresponding tail area is almost exactly 0.01.

7. We need to show that $F(\theta + G^{-1}(p)) = p$. Compute

$$F(\theta + G^{-1}(p)) = \int_{-\infty}^{\theta+G^{-1}(p)} f(x)dx = \int_{-\infty}^{\theta+G^{-1}(p)} g(x-\theta)dx = \int_{-\infty}^{G^{-1}(p)} g(y)dy = G(G^{-1}(p)) = p,$$

where the third equality follows by making the change of variables $y = x - \theta$.

Table S.10.9: Table for Exercise 5 of Sec. 10.8.

Rank of observations	$F_m(x)$	$G_n(x)$	Rank of observations	$F_m(x)$	$G_n(x)$
1	.04	0	21	.68	.2667
2	.08	0	22	.68	.3333
3	.12	0	23	.72	.3333
4	.16	0	24	.76	.3333
5	.20	0	25	.80	.3333
6	.24	0	26	.80	.4000
7	.28	0	27	.80	.4667
8	.32	0	28	.80	.5333
9	.36	0	29	.84	.5333
10	.40	0	30	.84	.6000
11	.40	.0667	31	.88	.6000
12	.44	.0667	32	.92	.6000
13	.48	.0667	33	.92	.6667
14	.52	.0667	34	.92	.7333
15	.56	.0667	35	.96	.7333
16	.60	.0667	36	1.00	.7333
17	.60	.1333	37	1.00	.8000
18	.60	.2000	38	1.00	.8667
19	.64	.2000	39	1.00	.9333
20	.68	.2000	40	1.00	1.0000

9. To test these hypotheses, add θ_0 to each observed value y_i in the second sample and then carry out the Wilcoxon-Mann-Whitney procedure on the original values in the first sample and the new values in the second sample.

11. Let $r_1 < r_2 < \cdots < r_m$ denote the ranks of the observed values in the first sample, and let $X_{i_1} < X_{i_2} < \cdots < X_{i_m}$ denote the corresponding observed values. Then there are $r_1 - 1$ values of Y in the second sample that are smaller than X_{i_1}. Hence, there are $r_1 - 1$ pairs (X_{i_1}, Y_j) with $X_{i_1} > Y_j$. Similarly, there are $r_2 - 2$ values of Y in the second sample that are smaller than X_{i_2}. Hence, there are $r_2 - 2$ pairs (X_{i_2}, Y_j) with $X_{i_2} > Y_j$. By continuing in this way, we see that the number U is equal to

$$(r_1 - 1) + (r_2 - 2) + \cdots + (r_m - m) = \sum_{i=1}^{m} r_i - \sum_{i=1}^{m} i = S - \frac{1}{2}m(m+1).$$

13. Since S and U differ by a constant, we need to show that $\text{Var}(U)$ is given by Eq. (10.8.6). Once again, write

$$U = \sum_i \sum_j Z_{i,j},$$

where $Z_{i,j} = 1$ if $X_i \geq Y_j$ and $Z_{i,j} = 0$ otherwise. Hence,

$$\text{Var}(U) = \sum_i \sum_j \text{Var}(Z_{i,j}) + \sum_{(i',j') \neq (i,j)} \text{Cov}(Z_{i,j}, Z_{i',j'}).$$

The first sum is $mn[\Pr(X_1 \geq Y_1) - \Pr(X_1 \geq Y_1)^2]$. The second sum can be broken into three parts:

- The terms with $i' = i$ but $j' \neq j$.
- The terms with $j' = j$ but $i' \neq i$.

- The terms with both $i' \neq i$ and $j' \neq j$.

For the last set of terms $\text{Cov}(Z_{i,j}, Z_{i',j'}) = 0$ since (X_i, Y_j) is independent of $(X_{i'}, Y_{j'})$. For each term in the first set

$$E(Z_{i,j}Z_{i,j'}) = \Pr(X_1 \geq Y_1, X_1 \geq Y_2),$$

so the covariances are

$$\text{Cov}(Z_{i,j}, Z_{i,j'}) = \Pr(X_1 \geq Y_1, X_1 \geq Y_2) - \Pr(X_1 \geq Y_1)^2.$$

There are $mn(n-1)$ terms of this sort. Similarly, for the second set of terms

$$\text{Cov}(Z_{i',j}, Z_{i,j}) = \Pr(X_1 \geq Y_1, X_2 \geq Y_1) - \Pr(X_1 \geq Y_1)^2.$$

There are $nm(m-1)$ of these terms. The variance is then

$$nm\left[\Pr(X_1 \geq Y_1) + (n-1)\Pr(X_1 \geq Y_1, X_1 \geq Y_2) \right.$$
$$\left. +(m-1)\Pr(X_1 \geq Y_1, X_2 \geq Y_1) - (m+n-1)\Pr(X_1 \geq Y_1)^2\right].$$

15. (a) Arrange the observations so that $|D_1| \leq \cdots \leq |D_n|$. Then $D_i > 0$ if and only if $X_i > Y_i$ if and only if $W_i = 1$. Since rank i gets added into S_W if and only if $D_i > 0$, we see that $\sum_{i=1}^{n} iW_i$ adds just those ranks that correspond to positive D_i.

    (b) Since the distribution of each D_i is symmetric around 0, the magnitude $|D_1|, \ldots, |D_n|$ are independent of the sign indicators $W_1, \ldots, W_n$. Using the result of part (a), if we assume that the $|D_i|$ are ordered from smallest to largest, $E(S_W) = \sum_{i=1}^{n} iE(W_i)$. Since the $|D_i|$ are independent of the W_i, we have $E(W_i) = 1/2$ even after we condition on the $|D_i|$ being arranged from smallest to largest. Since $\sum_{i=1}^{n} i = n(n+1)/2$, we have $E(S_W) = n(n+1)/4$.

    (c) Since the W_i are independent before we condition on the $|D_i|$ and they are independent of the $|D_i|$, then the W_i are independent conditional on the $|D_i|$. Hence, $\text{Var}(S_W) = \sum_{i=1}^{n} i^2 \text{Var}(W_i)$. Since $\text{Var}(W_i) = 1/4$ for all i and $\sum_{i=1}^{n} i^2 = n(n+1)(2n+1)/6$, we have $\text{Var}(S_W) = n(n+1)(2n+1)/24$.

10.9 Supplementary Exercises

1. Here, $\alpha_0/2 = 0.025$. From a table of binomial probabilities we find that

$$\sum_{x=0}^{5} \binom{20}{x} 0.5^{20} = 0.021 \leq 0.025 \leq \sum_{x=0}^{6} \binom{20}{x} 0.5^{20} = 0.058.$$

So, the sign test would reject the null hypothesis that $\theta = \theta_0$ if the number W of observations with values at most θ_0 satisfies either $W \leq 5$ or $W \geq 20 - 5$. Equivalently, we would accept the null hypothesis if $6 \leq W \leq 14$. This, in turn, is true if and only if θ_0 is strictly between the sixth and fourteenth ordered values of the original data. These values are 141 and 175, so our 95 percent confidence interval is $(141, 175)$.

3. Under H_0, the proportion p_i^0 of families with i boys is as follows:

i	p_i^0	np_i^0
0	1/8	16
1	3/8	48
2	3/8	48
3	1/8	16

Hence, it follows from Eq. (10.1.2) that

$$Q = \frac{(26-16)^2}{16} + \frac{(32-48)^2}{48} + \frac{(40-48)^2}{48} + \frac{(30-16)^2}{16} = 25.1667.$$

Under H_0, Q has the χ^2 distribution with 3 degrees of freedom. Hence, the tail area corresponding to $Q = 25.1667$ is less than 0.005, the smallest probability in the table in the back of the book. It follows that H_0 should be rejected for any level of significance greater than this tail area.

5. The expected numbers of observations in each cell, as specified by Eq. (10.4.4), are presented in the following table:

	A	B	AB	O
Group 1	19.4286	11	4.8571	14.7143
Group 2	38.8571	22	9.7143	29.4286
Group 3	77.7143	44	19.4286	58.8571

It is now found from Eq. (10.4.3) that $Q = 6.9526$. Under the hypothesis H_0 that the distribution is the same in all three groups, Q will have approximately the χ^2 distribution with $3 \times 2 = 6$ degrees of freedom. It is found from the tables that the 0.9 quantile of that distribution is 10.64, so H_0 should not be rejected.

7.

$$\begin{aligned}(N_{11}-\hat{E}_{11})^2 &= \left(N_{11} - \frac{N_{1+}N_{+1}}{n}\right)^2 \\ &= \left[N_{11} - \frac{(N_{11}+N_{12})(N_{11}+N_{21})}{n}\right]^2 \\ &= \frac{1}{n^2}\left[nN_{11} - (N_{11}+N_{12})(N_{11}+N_{21})\right]^2 \\ &= \frac{1}{n^2}(N_{11}N_{22} - N_{12}N_{21})^2,\end{aligned}$$

since $n = N_{11} + N_{21} + N_{21} + N_{22}$. Exactly the same value is obtained for $(N_{12} - \hat{E}_{12})^2, (N_{21} - \hat{E}_{21})^2$, and $(N_{22} - \hat{E}_{22})^2$.

9. In this exercise, $N_{1+} = N_{2+} = N_{+1} = N_{+2} = 2n$ and $N_{11}N_{22} - N_{12}N_{21} = (n+a)^2 - (n-a)^2 = 4na$. It now follows from Exercise 8 (after we replace n by $4n$ in the expression for Q) that $Q = 4a^2/n$. Since H_0 should be rejected if $Q > 6.635$, it follows that H_0 should be rejected if $a > (6.635n)^{1/2}/2$ or $a < -(6.635n)^{1/2}/2$.

11. Results of this type are an example of Simpson's paradox. If there is a higher rate of respiratory diseases among older people than among younger people, and if city A has a higher proportion of older people than city B, then results of this type can very well occur.

13. The fundamental aspect of this exercise is that it is not possible to assess the effectiveness of the treatment without having any information about how the levels of depression of the patients would have changed over the three-month period if they had not received the treatment. In other words, without the presence of a control group of similar patients who received some other standard treatment or no treatment at all, there is little meaningful statistical analysis that can be carried out. We can compare the proportion of patients at various levels who showed improvement after the treatment with the proportion who remained the same or worsened, but without a control group we have no way of deciding whether these proportions are unusually large or small.

15. The c.d.f. of this distribution is $F(x) = x^\theta$, so the median of the distribution is the point m such that $m^\theta = 1/2$. Thus, $m = (1/2)^{1/\theta}$ and $f(m) = \theta 2^{1/\theta}/2$. It follows from Theorem 10.7.1 that the asymptotic distribution of the sample median will be normal with mean m and variance

$$\frac{1}{4nf^2(m)} = \frac{1}{n\theta^2 2^{2/\theta}}.$$

17. As shown in Exercise 5 of Sec. 10.7, $E(\overline{X}_n) = E(\tilde{X}_n) = \theta$, so the M.S.E. of each of these estimators is equal to its variance. Furthermore, $\text{Var}(\overline{X}_n) = \frac{1}{n}[\alpha \cdot 1 + (1-\alpha)\sigma^2]$ and

$$\text{Var}(\tilde{X}_n) = \frac{1}{4n[h(\theta|\theta)]^2},$$

where

$$[h(\theta|\theta)]^2 = \frac{1}{2\pi}\left(\alpha + \frac{1-\alpha}{\sigma}\right)^2.$$

(a) For $\sigma^2 = 100$, $\text{Var}(\tilde{X}_n) < \text{Var}(\overline{X}_n)$ if and only if

$$\frac{50\pi}{[10\alpha + (1-\alpha)]^2} < \alpha + 100(1-\alpha).$$

Some numerical calculations show that this inequality is satisfied for $.031 < \alpha < .994$.

(b) For $\alpha = \frac{1}{2}$, $\text{Var}(\tilde{X}_n) < \text{Var}(\overline{X}_n)$ if and only if $\sigma < .447$ or $\sigma > 1/.447 = 2.237$.

19. It follows from Exercise 18 that the joint p.d.f. $g(y_1, y_2, y_3) = 3!$, a constant, for $0 < y_1 < y_2 < y_3 < 1$. Since the required conditional p.d.f. of Y_2 is proportional to $g(y_1, y_2, y_3)$, as a function of y_2 for fixed y_1 and y_3, it follows that this conditional p.d.f. is also constant. In other words, the required conditional distribution is uniform on the interval (y_1, y_3).

21. As shown in Exercise 10 of Sec. 10.8, we add θ_0 to each observation Y_j and then carry out the Wilcoxon-Mann-Whitney test on the sum S_{θ_0} of the ranks of the X_i's among these new values $Y_1 + \theta_0, \ldots, Y_n + \theta_0$. We accept H_0 if and only if

$$\frac{|\ S_{\theta_0} - E(S)\ |}{[\text{Var}(S)]^{1/2}} < c\left(1 - \frac{\alpha}{2}\right),$$

where $E(S)$ and $\text{Var}(S)$ are given by (10.8.3) and (10.8.4). However, by Exercise 11 of Sec. 10.8,

$$S_{\theta_0} = U_{\theta_0} + \frac{1}{2}m(m+1).$$

When we make this substitution for S_{θ_0} in the above inequality, we obtain the desired result.

23. (a) We know that $\theta_p = b$ if and only if $\Pr(X \le b) = p$. So, let $Y_i = 1$ if $X_i \le b$ and $Y_i = 0$ if not. Then $Y_1, \ldots, Y_n$ are i.i.d. with a Bernoulli distribution with parameter p if and only if H_0 is true. Define $W = \sum_{i=1}^n Y_i$ To test H_0, reject H_0 if W is too big or too small. For an equal tailed level α_0 test, choose two numbers $c_1 < c_2$ such that

$$\sum_{w=0}^{c_1} \binom{n}{w} p^w (1-p)^{n-w} \le \frac{\alpha_0}{2} < \sum_{w=0}^{c_1+1} \binom{n}{w} p^w (1-p)^{n-w},$$
$$\sum_{w=c_2}^{n} \binom{n}{w} p^w (1-p)^{n-w} \le \frac{\alpha_0}{2} < \sum_{w=c_2-1}^{n} \binom{n}{w} p^w (1-p)^{n-w}.$$

Then a level α_0 test rejects H_0 if $W \le c_1$ or $W \ge c_2$.

(b) For each b, we have shown how to construct a test of $H_{0,b} : \theta_p = b$. For given observed data $X_1, \ldots, X_n$ find all values of b such that the test constructed in part (a) accepts $H_{0,b}$. The set of all such b forms our coefficient $1 - \alpha_0$ confidence interval. It is clear from the form of the test that, once we find three values $b_1 < b_2 < b_3$ such that H_{0,b_2} is accepted and H_{0,b_1} and H_{0,b_3} are rejected, we don't have to check any more values of $b < b_1$ or $b > b_3$ since all of those would be rejected also. Similarly, if we find $b_4 < b_5$ such that both H_{0,b_4} and H_{0,b_5} are accepted, the so are $H_{0,b}$ for all $b_4 < b < b_5$. This will save some time locating all of the necessary b values.

Chapter 11

Linear Statistical Models

11.1 The Method of Least Squares

1. First write $c_1x_i + c_2 = c_1(x_i - \overline{x}_n) + (c_1\overline{x}_n + c_2)$ for every i. Then

$$(c_1x_i + c_2)^2 = c_1^2(x_i - \overline{x}_n)^2 + (c_1\overline{x}_n + c_2)^2 + 2c_1(x_i - \overline{x}_n)(c_1\overline{x}_n + c_2).$$

The sum over all i from 1 to n of the first two terms on the right produce the formula we desire. The sum of the last term over all i is 0 because $c_1(c_1\overline{x}_n + c_2)$ is the same for all i and $\sum_{i=1}^{n}(x_i - \overline{x}_n) = 0$.

3. It must be shown that $\bar{y}_n = \hat{\beta}_0 + \hat{\beta}_1\bar{x}_n$. But this result follows immediately from the expression for $\hat{\beta}_0$ given in Eq. (11.1.1).

5. The least squares line will have the form $x = \gamma_0 + \gamma_1 y$, where γ_0 and γ_1 are defined similarly to $\hat{\beta}_0$ and $\hat{\beta}_1$ in Eq. (11.1.1) with the roles of x and y interchanged. Thus,

$$\gamma_1 = \frac{\sum_{i=1}^{n} x_iy_i - n\bar{x}_n\bar{y}_n}{\sum_{i=1}^{n} y_i^2 - n\bar{y}_n^2}$$

and $\hat{\gamma}_0 = \bar{x}_n - \hat{\gamma}_1\bar{y}_n$. It is found that $\hat{\gamma}_1 = 0.9394$ and $\hat{\gamma}_0 = 1.5691$. Hence, the least squares line is $x = 1.5691 + 0.9394\ y$ or, equivalently, $y = -1.6703 + 1.0645x$. This line and the line $y = -0.786 + 0.685x$ given in Fig. 11.4 can now be sketched on the same graph.

7. (a) Here, $n = 8$, $\bar{x}_n = 2.25$, $\bar{y}_n = 42.125$, $\sum_{i=1}^{n} x_iy_i = 764$, and $\sum_{i=1}^{n} x_i^2 = 51$. Therefore, by Eq. (11.1.1), $\hat{\beta}_1 = 0.548$ and $\hat{\beta}_0 = 40.893$. Hence, the least squares line is $y = 40.893 + 0.548x$.

   (b) The normal equations (11.1.8) are found to be:

$$\begin{aligned} 8\beta_0 + 18\beta_1 + 51\beta_2 &= 337, \\ 18\beta_0 + 51\beta_1 + 162\beta_2 &= 764, \\ 51\beta_0 + 162\beta_1 + 548.25\beta_2 &= 2167.5. \end{aligned}$$

   Solving these three simultaneous linear equations, we obtain the solution:

$$\hat{\beta}_0 = 38.483, \quad \hat{\beta}_1 = 3.440, \quad and \quad \hat{\beta}_2 = -0.643.$$

9. The normal equations (11.1.13) are found to be

$$\begin{aligned} 10\beta_0 + 1170\beta_1 + 18\beta_2 &= 1359, \\ 1170\beta_0 + 138,100\beta_1 + 2130\beta_2 &= 160,380, \\ 18\beta_0 + 2130\beta_1 + 38\beta_2 &= 2483. \end{aligned}$$

Solving these three simultaneous linear equations, we obtain the solution $\hat{\beta}_0 = 3.7148, \hat{\beta}_1 = 1.1013$, and $\hat{\beta}_2 = 1.8517$.

11. In Exercise 9, it is found that

$$\sum_{i=1}^{10} (y_i - \hat{\beta}_0 - \hat{\beta}_1 x_{i1} - \hat{\beta}_2 x_{i2})^2 = 102.28.$$

In Exercise 10, it is found that

$$\sum_{i=1}^{10} (y_i - \hat{\beta}_0 x_{i1} - \hat{\beta}_1 x_{i2} - \hat{\beta}_2 x_{i2}^2)^2 = 42.72.$$

Therefore, a better fit is obtained in Exercise 10.

11.2 Regression

1. After we have replaced β_0 and β_1 in (11.2.2) with $\hat{\beta}_0$ and $\hat{\beta}_1$, the maximization with respect to σ^2 is exactly the same as the maximization carried out in Example 7.5.6 in the text for finding the M.L.E. of σ^2.

3. $E(\bar{Y}_n) = \dfrac{1}{n}\sum_{i=1}^{n} E(Y_i) = \dfrac{1}{n}\sum_{i=1}^{n} (\beta_0 + \beta_1 x_i) = \beta_0 + \beta_1 \bar{x}_n.$
Hence, as shown near the end of the proof of Theorem 11.2.2,

$$E(\hat{\beta}_0) = E(\bar{Y}_n) - \bar{x}_n E(\hat{\beta}_1) = (\beta_0 + \beta_1 \bar{x}_n) - \bar{x}_n \beta_1 = \beta_0.$$

5. Since $\bar{Y}_n = \hat{\beta}_0 + \hat{\beta}_1 \bar{x}_n$, then

$$\operatorname{Var}(\bar{Y}_n) = \operatorname{Var}(\hat{\beta}_0) + \bar{x}_n^2 \operatorname{Var}(\hat{\beta}_1) + 2\bar{x}_n \operatorname{Cov}(\hat{\beta}_0, \hat{\beta}_1).$$

Therefore, if $\bar{x}_n \neq 0$,

$$\begin{aligned} \operatorname{Cov}(\hat{\beta}_0, \hat{\beta}_1) &= \frac{1}{2\bar{x}_n}\left[\operatorname{Var}(\bar{Y}_n) - \operatorname{Var}(\hat{\beta}_0) - \bar{x}_n^2 \operatorname{Var}(\hat{\beta}_1)\right] \\ &= \frac{1}{2\bar{x}_n}\left(\frac{\sigma^2}{n} - \frac{\sum_{i=1}^{n} x_i^2}{n s_x^2}\sigma^2 - \frac{\bar{x}_n^2}{s_x^2}\sigma^2 \right) \\ &= \frac{\sigma^2}{2\bar{x}_n}\left(\frac{s_x^2 - \sum_{i=1}^{n} x_i^2 - n\bar{x}_n^2}{n s_x^2} \right) \\ &= \frac{\sigma^2}{2\bar{x}_n}\left(\frac{-2n\bar{x}_n^2}{n s_x^2} \right) = \frac{-\bar{x}_n \sigma^2}{s_x^2}. \end{aligned}$$

If $\bar{x}_n = 0$, then $\hat{\beta}_0 = \bar{Y}_n = \frac{1}{n}\sum_{i=1}^{n} Y_i$, and

$$\text{Cov}(\hat{\beta}_0, \hat{\beta}_1) = \text{Cov}\left(\frac{1}{n}\sum_{i=1}^{n} Y_i, \frac{1}{s_x^2}\sum_{j=1}^{n} x_j Y_j\right) = \frac{1}{ns_x^2}\sum_{i=1}^{n}\sum_{j=1}^{n} x_j \,\text{Cov}(Y_i, Y_j),$$

by Exercise 8 of Sec. 4.6. Since $Y_1, \ldots, Y_n$ are independent and each has variance σ^2, then $\text{Cov}(Y_i, Y_j) = 0$ for $i \neq j$ and $\text{Cov}(Y_i, Y_j) = \sigma^2$ for $i = j$.

Hence,

$$\text{Cov}(\hat{\beta}_0, \hat{\beta}_1) = \frac{\sigma^2}{ns_x^2}\sum_{i=1}^{n} x_i = 0.$$

7. (a) The M.L.E.'s $\hat{\beta}_0$ and $\hat{\beta}_1$ are the same as the least squares estimates found in Sec. 11.1 for Table 11.1. The value of $\hat{\sigma}^2$ can then be found from Eq. (11.2.3).
   (b) Also, $\text{Var}(\hat{\beta}_0) = 0.2505\sigma^2$ can be determined from Eq. (11.2.5) and $\text{Var}(\hat{\beta}_1) = 0.0277\sigma^2$ from Eq. (11.2.9).
   (c) It can be found from Eq. (11.2.6) that $\text{Cov}(\hat{\beta}_0, \hat{\beta}_1) = -0.0646\sigma^2$. By using the values of $\text{Var}(\hat{\beta}_0)$ and $\text{Var}(\hat{\beta}_1)$ found in part (b), we obtain

$$\rho(\hat{\beta}_0, \hat{\beta}_1) = \frac{\text{Cov}(\hat{\beta}_0, \hat{\beta}_1)}{[\text{Var}(\hat{\beta}_0)\,\text{Var}(\hat{\beta}_1)]^{1/2}} = -0.775.$$

9. The unbiased estimator is $3\hat{\beta}_0 + c_1\hat{\beta}_1$. The M.S.E. of an unbiased estimator is its variance, and

$$\text{Var}(\hat{\theta}) = 9\,\text{Var}(\hat{\beta}_0) + 6c_1\,\text{Cov}(\hat{\beta}_0, \hat{\beta}_1) + c_1^2\,\text{Var}(\hat{\beta}_1).$$

Using the values in Exercise 7, we get

$$\text{Var}(\hat{\theta}) = \sigma^2[9 \times 0.2505 - 6c_1(0.0646) + c_1^2 0.0277].$$

We can minimize this by taking the derivative with respect to c_1 and setting the derivative equal to 0. We get $c_1 = 6.996$.

11. By Eq. (11.2.11), the M.S.E. is

$$\left[\frac{1}{ns_x^2}\sum_{i=1}^{n}(x_i - x)^2 + 1\right]\sigma^2.$$

We know that $\sum_{i=1}^{n}(x_i - x)^2$ will be a minimum (and, hence, the M.S.E. will be a minimum) when $x = \bar{x}_n$.

13. It can be found from Eq. (11.2.6) that $\text{Cov}(\hat{\beta}_0, \hat{\beta}_1) = -0.214\sigma^2$. By using the values of $\text{Var}(\hat{\beta}_0)$ and $\text{Var}(\hat{\beta}_1)$ found in Exercise 12, we obtain

$$\rho(\hat{\beta}_0, \hat{\beta}_1) = \frac{\text{Cov}(\hat{\beta}_0, \hat{\beta}_1)}{[\text{Var}(\hat{\beta}_0)\,\text{Var}(\hat{\beta}_1)]^{1/2}} = -0.891.$$

15. This exercise, is similar to Exercise 9. $\mathrm{Var}(\hat{\theta})$ attains its minimum value when $c_1 = -\bar{x}_n$.

17. It was shown in Exercise 11, that the M.S.E. of $\hat{Y}$ will be a minimum when $x = \bar{x}_n = 2.25$.

19. The formula for $E(X_2|x_1)$ is Eq. (5.10.6), which we repeat here for the case in which $\mu_1 = \mu_2 = \mu$ and $\sigma_1 = \sigma_2 = \sigma$:

$$E(X_2|x_1) = \mu + \rho\sigma\left(\frac{x_1 - \mu}{\sigma}\right) = \mu + \rho(x_1 - \mu).$$

We are asked to show that $|E(X_2|x_1) - \mu| < |x_1 - \mu|$ for all x_1. Since $0 < \rho < 1$,

$$|E(X_2|x_1) - \mu| = |\mu + \rho(x_1 - \mu) - \mu| = \rho|x_1 - \mu| < |x_1 - \mu|.$$

11.3 Statistical Inference in Simple Linear Regression

1. It is found from Table 11.9 that $\bar{x}_n = 0.42$, $\bar{y}_n = 0.33$, $\sum_{i=1}^{n} x_i^2 = 10.16$, $\sum_{i=1}^{n} x_i y_i = 5.04$, $\hat{\beta}_1 = 0.435$ and $\hat{\beta}_0 = 0.147$ by Eq. (11.1.1), and $S^2 = 0.451$ by Eq. (11.3.9). Therefore, from Eq. (11.3.19) with $n = 10$ and $\beta_0^* = 0.7$, it is found that $U_0 = -6.695$. It is found from a table of the t distribution with $n - 2 = 8$ degrees of freedom that to carry out a test at the 0.05 level of significance, H_0 should be rejected if $|U_0| > 2.306$. Therefore H_0 is rejected.

3. It follows from Eq. (11.3.22), with $\beta_1^* = 1$, that $U_1 = -6.894$. Since $|U_1| > 2.306$, the critical value found in Exercise 1, we should reject H_0.

5. The hypotheses to be tested are:

$$H_0 : 5\beta_0 - \beta_1 = 0,$$
$$H_1 : 5\beta_0 - \beta_1 \neq 0.$$

Hence, in the notation of (11.3.13), $c_0 = 5, c_1 = -1$, and $c_* = 0$. It is found that $\sum_{i=1}^{n}(c_0 x_i - c_1)^2 = 306$ and, from Eq. (11.3.14), that $U_{01} = 0.664$. It is found from a table of the t distribution with $n - 2 = 8$ degrees of freedom that to carry out a test at the 0.10 level of significance, H_0 should be rejected if $|U_{01}| > 1.860$. Therefore, H_0 is not rejected.

7.

$$\begin{aligned}
\mathrm{Cov}(\hat{\beta}_1, D) &= \mathrm{Cov}(\hat{\beta}_1, \hat{\beta}_0 + \hat{\beta}_1\bar{x}_n) \\
&= \mathrm{Cov}(\hat{\beta}_1, \hat{\beta}_0) + \bar{x}_n \,\mathrm{Cov}(\hat{\beta}_1, \hat{\beta}_1) \\
&= \mathrm{Cov}(\hat{\beta}_0, \hat{\beta}_1) + \bar{x}_n \,\mathrm{Var}(\hat{\beta}_1) \\
&= 0, \text{ by Eqs. (11.2.9) and (11.2.6).}
\end{aligned}$$

Since $\hat{\beta}_0$ and $\hat{\beta}_1$ have a bivariate normal distribution, it follows from Exercise 10 of Sec. 5.10 that D and $\hat{\beta}_1$ will also have a bivariate normal distribution. Therefore, as discussed in Sec. 5.10, since D and $\hat{\beta}_1$ are uncorrelated they are also independent.

9. Here, $\beta_0^* = 0$ and $\beta_1^* = 1$. It is found that $Q^2 = 2.759$, $S^2 = 0.451$, and $U^2 = 24.48$. It is found from a table of the F distribution with 2 and 8 degrees of freedom that to carry out a test at the 0.05 level of significance, H_0 should be rejected if $U^2 > 4.46$. Therefore, H_0 is rejected.

11. The solution here is analogous to the solution of Exercise 9. Since the confidence coefficient is again 0.95, the confidence interval will contain all values of β_1^* for which $|U_1| < 2.306$ or, equivalently, for which $-2.306 < 12.207(0.435 - \beta_1^*) < 2.306$. The interval is, therefore, found to be $0.246 < \beta_1 < 0.624$.

13. We must determine a confidence interval for $y = \beta_0 + \beta_1$ with confidence coefficient 0.99. It is found from a table of the t distribution with 8 degrees of freedom (as in Exercise 6) that this confidence interval will contain all values of c_* for which $|U_{01}| < 3.355$ or, equivalently, for which $-3.355 < 11.257(0.582 - \ b) < 3.355$. This interval reduces to $0.284 < c_* < 0.880$. This interval is, therefore, the confidence interval for y.

15. Let q be the $1 - \alpha_0/2$ quantile of the t distribution $n - 2$ degrees of freedom. A confidence interval for $\beta_0 + \beta_1 x$ contains all values of c_* for which $|U_{01}| < c$, where $c_0 = 1$ and $c_1 = x$ in Eq. (11.3.14). The inequality $|U_{01}| < q$ can be reduced to the following form

$$\hat{\beta}_0 + x\hat{\beta}_1 - q\left[\frac{S^2\sum_{i=1}^{n}(x_i - x)^2}{n(n-2)s_x^2}\right]^{1/2} < b < \hat{\beta}_0 + x\hat{\beta}_1 + q\left[\frac{S^2\sum_{i=1}^{n}(x_i - x)^2}{n(n-2)s_x^2}\right]^{1/2}$$

The length of this interval is

$$2q\left[\frac{S^2}{n(n-2)s_x^2}\quad\sum_{i=1}^{n}(x_i - x)^2\right]^{1/2}$$

The length will, therefore, be a minimum for the value of x which minimizes $\sum_{i=1}^{n}(x_i - x)^2$. We know that this quantity is a minimum when $x = \bar{x}_n$.

17. To attain a confidence coefficient of 0.95, it is found from a table of the F distribution with 2 and 8 degrees of freedom (as in Exercise 9) that the confidence ellipse for (β_0, β_1) will contain all points (β_0^*, β_1^*) for which $U^2 < 4.46$. Hence, it will contain all points for which

$$10(\beta_0^* - 0.147)^2 + 8.4(\beta_0^* - 0.147)(\beta_1^* - 0.435) + 10.16(\beta_1^* - 0.435)^2 < 0.503.$$

19. If S^2 is defined by Eq. (11.3.9), then S^2/σ^2 has a χ^2 distribution with $n - 2$ degrees of freedom. Therefore, $E(S^2/\sigma^2) = n - 2$, $E(S^2) = (n-2)\sigma^2$, and $E(S^2/[n-2]) = \sigma^2$.

21. (a) A computer is useful to perform the regressions and plots for this exercise. The two plots for parts (a) and (b) are side-by-side in Fig. S.11.1. The plot for part (a) shows residuals that are more spread out for larger values of 1970 price than they are for smaller values. This suggests that the variance of Y is not constant as X changes.

    (b) The plot for part (b) in Fig. S.11.1 has more uniform spread in the residuals as 1970 price varies. However, there appear to be two points that are not fit very well.

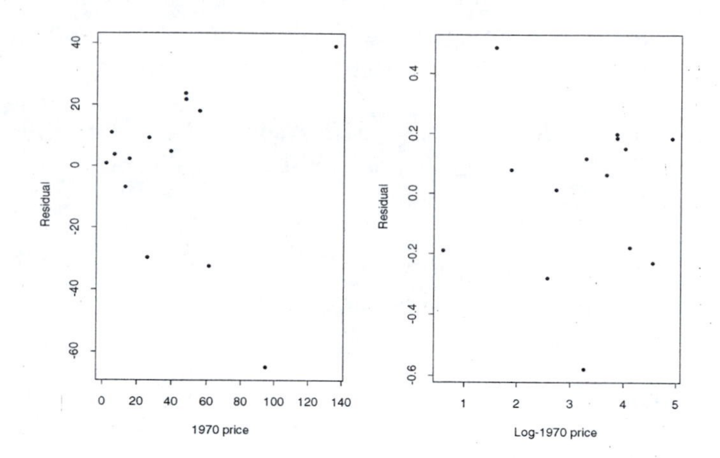

Figure S.11.1: Residual plots for Exercise 21a of Sec. 11.3. The plot on the left is for part (a) and the plot on the right is for part (b).

23. Define

$$W_{01} = \left[\frac{c_0^2}{n} + \frac{(c_0\overline{x}_n - c_1)^2}{s_x^2}\right]^{-1/2} \frac{c_0(\hat{\beta}_0 - \beta_0) + c_2(\hat{\beta}_1 - \beta_1)}{\sigma'},$$

which has the t distribution with $n-2$ degrees of freedom. Hence

$$\Pr(W_{01} \geq T_{n-2}^{-1}(1-\alpha_0)) = \alpha_0.$$

Suppose that $c_0\beta_0 + c_1\beta_1 < c_*$. Because $[(c_0^2/n) + (c_0\overline{x}_n - c_1)^2/s_x^2]/\sigma' > 0$, it follows that $W_{01} > U_{01}$. Finally, he probability of type I error is

$$\Pr(U_{01} \geq T_{n-2}^{-1}(1-\alpha_0)) \leq \Pr(W_{01} \geq T_{n-2}^{-1}(1-\alpha_0)) = \alpha_0,$$

where the first inequality follows from $W_{01} > U_{01}$.

25. (a) The simultaneous intervals are the same as (11.3.33) with $[2F_{2,n-2}^{-1}(1-\alpha_0)]^{1/2}$ replaced by $T_{n-2}^{-1}(1-\alpha_0/4)$, namely for $i = 0, 1$,

$$\beta_0 + \beta_1 x_i \pm T_{n-2}^{-1}(1-\alpha_0/4)\sigma'\left[\frac{1}{n} + \frac{(x_i - \overline{x}_n)^2}{s_x^2}\right]^{1/2}.$$

(b) Set $x = \alpha x_0 + (1-\alpha)x_1$ and solve for α. The result is, by straightforward algebra,

$$\alpha(x) = \frac{x - x_1}{x_0 - x_1}.$$

(c) First, notice that for all x,

$$\beta_0 + \beta_1 x = \alpha(x)[\beta_0 + \beta_1 x_0] + [1-\alpha(x)][\beta_0 + \beta_1 x_1]. \tag{S.11.1}$$

That is, each parameter for which we want a confidence interval is a convex combination of the parameters for which we already have confidence intervals.

Suppose that C occurs. There are three cases that depend on where $\alpha(x)$ lies relative to the interval $[0, 1]$. The first case is when $0 \leq \alpha(x) \leq 1$. In this case, the smallest of the four numbers defining

$L(x)$ and $U(x)$ is $L(x) = \alpha(x)A_0 + [1-\alpha(x)]A_1$ and the largest is $U(x) = \alpha(x)B_0 + [1-\alpha(x)]B_1$, because both $\alpha(x)$ and $1-\alpha(x)$ are nonnegative. For all such x, $A_0 < \beta_0 + \beta_1 x_0 < B_0$ and $A_1 < \beta_0 + \beta_1 x_1 < B_1$ together imply that

$$\alpha(x)A_0 + [1-\alpha(x)]A_1 < \alpha(x)[\beta_0+\beta_1 x_0] + [1-\alpha(x)][\beta_0+\beta_1 x_1] < \alpha(x)B_0 + [1-\alpha(x)]B_1.$$

Combining this with (S.11.1) and the formulas for $L(x)$ and $U(x)$ yields $L(x) < \beta_0 + \beta_1 x < U(x)$ as desired. The other two cases are similar, so we shall do one only of them. If $\alpha(x) < 0$, then $1-\alpha(x) > 0$. In this case, the smallest of the four numbers defining $L(x)$ and $U(x)$ is $L(x) = \alpha(x)B_0 + [1-\alpha(x)]A_1$, and the largest is $U(x) = \alpha(x)A_0 + [1-\alpha(x)]B_1$. For all such x, $A_0 < \beta_0 + \beta_1 x_0 < B_0$ and $A_1 < \beta_0 + \beta_1 x_1 < B_1$ together imply that

$$\alpha(x)B_0 + [1-\alpha(x)]A_1 < \alpha(x)[\beta_0+\beta_1 x_0] + [1-\alpha(x)][\beta_0+\beta_1 x_1] < \alpha(x)A_0 + [1-\alpha(x)]B_1.$$

Combining this with (S.11.1) and the formulas for $L(x)$ and $U(x)$ yields $L(x) < \beta_0 + \beta_1 x < U(x)$ as desired.

11.4 Bayesian Inference in Simple Linear Regression

1. The posterior distribution of β_1 is given as a special case of (11.4.1), namely that $U = s_x(\beta_1 - \hat{\beta}_1)/\sigma'$ has the t distribution with $n-2$ degrees of freedom. The coefficient $1-\alpha_0$ confidence interval from Sec. 11.3 has endpoints $\hat{\beta}_1 \pm T^{-1}_{n-2}(1-\alpha_0/2)\sigma'/s_x$. So, we can compute the posterior probability that β_1 is in the interval as follows:

$$\begin{aligned}&\Pr\left(\hat{\beta}_1 - T^{-1}_{n-2}(1-\alpha_0/2)\frac{\sigma'}{s_x} < \beta_1 < \hat{\beta}_1 + T^{-1}_{n-2}(1-\alpha_0/2)\frac{\sigma'}{s_x}\right)\\ &= \Pr\left(-T^{-1}_{n-2}(1-\alpha_0/2) < s_x\frac{\beta_1-\hat{\beta}_1}{\sigma'} < T^{-1}_{n-2}(1-\alpha_0/2)\right).\end{aligned} \qquad \text{(S.11.2)}$$

Since the t distributions are symmetric around 0, $-T^{-1}_{n-2}(1-\alpha_0/2) = T^{-1}_{n-2}(\alpha_0/2)$. Also, the random variable between the inequalities on the right side of (S.11.2) is U, which has the t distribution with $n-2$ degrees of freedom. Hence the right side of (S.11.2) equals

$$\Pr(U < T^{-1}(1-\alpha_0/2)) - \Pr(U \le T^{-1}(\alpha_0/2)) = 1 - \alpha_0/2 - \alpha_0/2 = 1-\alpha_0.$$

3. The joint distribution of (β_0, β_1) given τ is a bivariate normal distribution as specified in Theorem 11.4.1. Using the means, variances, and correlation given in that theorem, we compute the mean of $\beta_0 + \beta_1 x$ as $\hat{\beta}_0 + \hat{\beta}_1 x = \hat{Y}$. The variance of $\beta_0 + \beta_1 x$ given τ is

$$\frac{1}{\tau}\left[\frac{1}{n} + \frac{\overline{x}_n^2}{s_x^2} + \frac{x^2}{s_x^2} - 2xn\frac{\overline{x}_n}{\left(n\sum_{i=1}^{n} x_i^2\right)^{1/2}}\right]\left(\frac{1}{n} + \frac{\overline{x}_n^2}{s_x^2}\right)^{1/2}\frac{1}{s_x}.$$

Use the fact that $1/n + \overline{x}_n^2/s_x^2 = \sum_{i=1}^{n} x_i^2/[ns_x^2]$ to simplify the above variance to the expression

$$\frac{1}{\tau}\left[\frac{1}{n} + \frac{(x-\overline{x}_n)^2}{s_x^2}\right]^{1/2}.$$

It follows that the conditional distribution of $\tau^{1/2}(\beta_0 - \beta_1 x - \hat{Y})$ is the normal distribution with mean 0 and variance as stated in the exercise.

5. The summary data are in the solution to Exercise 4.

   (a) According to Theorem 11.4.1, the posterior distribution of β_1 is that $2.898(\beta_1 - 0.4352)/0.2374$ has the t distribution with eight degrees of freedom.

   (b) According to Theorem 11.4.1, the posterior distribution of $\beta_0 + \beta_1$ is that

$$\left(0.1 + \frac{(0.42-1)^2}{2.898^2}\right)^{-1/2} \frac{\beta_0 + \beta_1 - 0.5824}{0.2374}$$

   has the t distribution with eight degrees of freedom.

7. The conditional mean of β_0 given β_1 can be computed using results from Sec. 5.10. In particular,

$$E(\beta_0|\beta_1) = \hat{\beta}_0 - \frac{n\overline{x}_n s_x \left(\frac{1}{n} + \overline{x}_n^2/s_x^2\right)^{1/2}}{\left(n\sum_{i=1}^n x_i^2\right)^{1/2}}(\beta_1 - \hat{\beta}_1).$$

   Now, use the fact that $\sum_{i=1}^n x_i^2 = s_x^2 + n\overline{x}_n^2$. The result is

$$E(\beta_0|\beta_1) = \hat{\beta}_0 + \overline{x}_n(\beta_1 - \hat{\beta}_1).$$

11.5 The General Linear Model and Multiple Regression

1. After we have replaced $\beta_0, \dots, \beta_p$ in (11.5.4) by their M.L.E.'s $\hat{\beta}_0, \dots, \hat{\beta}_p$, the maximization with respect to σ^2 is exactly the same as the maximization carried out in Example 7.5.6 in the text for finding the M.L.E. of σ^2 or the maximization carried out in Exercise 1 of Sec. 11.2.

3. This problem is a special case of the general linear model with $p = 1$. The design matrix $\boldsymbol{Z}$ defined by Eq. (11.5.9) has dimension $n \times 1$ and is specified as follows:

$$\boldsymbol{Z} = \begin{bmatrix} x_1 \\ \vdots \\ x_n \end{bmatrix}.$$

   Therefore, $\boldsymbol{Z}'\boldsymbol{Z} = \sum_{i=1}^n x_i^2$ and $(\boldsymbol{Z}'\boldsymbol{Z})^{-1} = \dfrac{1}{\sum_{i=1}^n x_i^2}$.

   It follows from Eq. (11.5.10) that

$$\hat{\beta} = \frac{\sum_{i=1}^n x_i Y_i}{\sum_{i=1}^n x_i^2}.$$

5. It is found that $\sum_{i=1}^n x_i y_i = 342.4$ and $\sum_{i=1}^n x_i^2 = 66.8$. Therefore, from Exercises 3 and 4, $\hat{\beta} = 5.126$ and $\text{Var}(\hat{\beta}) = 0.0150\sigma^2$. Also, $S^2 = \sum_{i=1}^n (y_i - \hat{\beta}x_i)^2 = 169.94$. Therefore, by Eq. (11.5.7), $\hat{\sigma}^2 = (169.94)/10 = 16.994$.

7. The values $\hat{\beta}_0, \hat{\beta}_1$, and $\hat{\beta}_2$ were determined in Example 11.1.3.

   By Eq. (11.5.7),

$$\hat{\sigma}^2 = \frac{1}{10}S^2 = \frac{1}{10}\sum_{i=1}^{10}(y_i - \hat{\beta}_0 - \hat{\beta}_1 x_i - \hat{\beta}_2 x_i^2)^2 = \frac{1}{10}(9.37) = 0.937.$$

9. By Eq. (11.5.21), the following statistic will have the t distribution with 7 degrees of freedom when H_0 is true:

$$U_2 = \left[\frac{7}{(0.014)(9.37)}\right]^{1/2} \hat{\beta}_2 = 0.095.$$

   The corresponding two-sided tail area is greater than 0.90. The null hypothesis would not be rejected at any reasonable level of significance.

11. It is found that $\sum_{i=1}^{n}(y_i - \bar{y}_n)^2 = 26.309$. Therefore,

$$R^2 = 1 - \frac{S^2}{26.309} = 0.644.$$

13. The design matrix $\boldsymbol{Z}$ has the following form:

$$\boldsymbol{Z} = \begin{bmatrix} 1 & x_{11} & x_{12} \\ 1 & x_{21} & x_{22} \\ \cdots & \cdots & \cdots \\ 1 & x_{n1} & x_{n2} \end{bmatrix}.$$

   Therefore, $\boldsymbol{Z}'\boldsymbol{Z}$ is the 3×3 matrix of coefficients on the left side of the three equations in (11.1.14):

$$\boldsymbol{Z}'\boldsymbol{Z} = \begin{bmatrix} 10 & 23.3 & 650 \\ 23.3 & 90.37 & 1563.6 \\ 650 & 1563.6 & 42,334 \end{bmatrix}.$$

   It will now be found that

$$(\boldsymbol{Z}'\boldsymbol{Z})^{-1} = \begin{bmatrix} 222.7 & 4.832 & -3.598 \\ 4.832 & 0.1355 & -0.0792 \\ -3.598 & -0.0792 & 0.0582 \end{bmatrix}.$$

   The elements of $(\boldsymbol{Z}'\boldsymbol{Z})^{-1}$, multiplied by σ^2, are the variances and covariances of $\hat{\beta}_0, \hat{\beta}_1$, and $\hat{\beta}_2$.

15. By Eq. (11.5.21), the following statistic will have the t distribution with 7 degrees of freedom when H_0 is true:

$$U_2 = \left[\frac{7}{(0.0582)(8.865)}\right]^{1/2} (\hat{\beta}_2 + 1) = 4.319.$$

   The corresponding two-sided tail area is less than 0.01.

17.

$$\begin{aligned}
\text{Cov}(\hat{\beta}_j, A_{ij}) &= \text{Cov}\left(\hat{\beta}_j, \hat{\beta}_i - \frac{\zeta_{ij}}{\zeta_{jj}}\hat{\beta}_j\right) \\
&= \text{Cov}(\hat{\beta}_j, \hat{\beta}_i) - \frac{\zeta_{ij}}{\zeta_{jj}}\text{Cov}(\hat{\beta}_j, \hat{\beta}_j) \\
&= \text{Cov}(\hat{\beta}_i, \hat{\beta}_j) - \frac{\zeta_{ij}}{\zeta_{jj}}\text{Var}(\hat{\beta}_j) \\
&= \zeta_{ij}\sigma^2 - \frac{\zeta_{ij}}{\zeta_{jj}}\zeta_{jj}\sigma^2 = 0.
\end{aligned}$$

Just as in simple linear regression, it can be shown that the joint distribution of two estimators $\hat{\beta}_i$ and $\hat{\beta}_j$ will be a bivariate normal distribution. Since A_{ij} is a linear function of $\hat{\beta}_i$ and $\hat{\beta}_j$, the joint distribution of A_{ij} and $\hat{\beta}_j$ will also be a bivariate normal distribution. Therefore, since A_{ij} and $\hat{\beta}_j$ are uncorrelated, they are also independent.

19. (a) Since W^2 is a function only of $\hat{\beta}_i$ and $\hat{\beta}_j$, it follows that W^2 and S^2 are independent. Also, W^2 has a χ^2 distribution with 2 degrees of freedom and S^2/σ^2 has a χ^2 distribution with $n-p$ degrees of freedom. Therefore, $\dfrac{W^2/2}{S^2/[\sigma^2(n-p)]}$ has the F distribution with 2 and $n-p$ degrees of freedom.

(b) If we replace β_i and β_j in W^2 by their hypothesized values β_i^* and β_j^*, then the statistic given in part (a) will have the F distribution with 2 and $n-p$ degrees of freedom when H_0 is true and will tend to be larger when H_0 is not true. Therefore, we should reject H_0 if that statistic exceeds some constant C, where C can be chosen to obtain any specified level of significance $\alpha_0 (0 < \alpha_0 < 1)$.

21. In this problem, $i = 2, j = 3, \beta_1^* = 1, \beta_2^* = 0$ and from the values found in Exercises 12 and 13,

$$\begin{aligned}
W^2 &= \frac{(0.0582)(0.4503-1)^2 + (0.1355)(0.1725)^2 + 2(0.0792)(0.4503-1)(0.1725)}{[(0.1355)(0.0582) - (0.0792)^2]\sigma^2} \\
&= \frac{4.091}{\sigma^2}.
\end{aligned}$$

Also, $S^2 = 8.865$, as found in the solution of Exercise 12. Hence, the value of the F statistic with 2 and 7 degrees of freedom is $(7/2)(4.091/8.865) = 1.615$. The corresponding tail area is greater than 0.05.

23. We have the following relations:

$$\begin{aligned}
E(\boldsymbol{X}+\boldsymbol{Y}) &= E\begin{bmatrix} X_1+Y_1 \\ \vdots \\ X_n+Y_n \end{bmatrix} = \begin{bmatrix} E(X_1+Y_1) \\ \vdots \\ E(X_n+Y_n) \end{bmatrix} = \begin{bmatrix} E(X_1)+E(Y_1) \\ \vdots \\ E(X_n)+E(Y_n) \end{bmatrix} \\
&= \begin{bmatrix} E(X_1) \\ \vdots \\ E(X_n) \end{bmatrix} + \begin{bmatrix} E(Y_1) \\ \vdots \\ E(Y_n) \end{bmatrix} = E(\boldsymbol{X}) + E(\boldsymbol{Y}).
\end{aligned}$$

25. We know that $\text{Var}(3Y_1 + Y_2 - 2Y_3 + 8) = \text{Var}(3Y_1 + Y_2 - 2Y_3)$. By Theorem 11.5.2, with $p = 1$,

$$\text{Var}(3Y_1 + Y_2 - 2Y_3) = (3, 1, -2)\,\text{Cov}(\boldsymbol{Y})\begin{pmatrix} 3 \\ 1 \\ -2 \end{pmatrix} = 87.$$

27. In a simple linear regression, $\hat{Y}_i$ is the same linear function of X_i for all i. If $\hat{\beta}_1 > 0$, then every unit increase in X corresponds to an increase of $\hat{\beta}_1$ in $\hat{Y}$. So, a plot of residuals against $\hat{Y}$ will look the same as a plot of residuals against X except that the horizontal axis will be labeled differently. If $\hat{\beta}_1 < 0$, then a unit increase in X corresponds to a decrease of $-\hat{\beta}_1$ in $\hat{Y}$, so a plot of residuals against fitted values is a mirror image of a plot of residuals against X. (The plot is flipped horizontally around a vertical line.)

29. In Example 11.5.5, we are told that $\sigma' = 352.9$, so the residual sum of squares is 2864383. We can calculate $\sum_{i=1}^{n}(y_i - \bar{y}_n)^2$ directly from the data in Table 11.13. It equals, 26844478. It follows that

$$R^2 = 1 - \frac{2864383}{26844478} = 0.893.$$

11.6 Analysis of Variance

1. By analogy with Eq. (11.6.2),

$$\boldsymbol{Z}'\boldsymbol{Z} = \begin{bmatrix} n_1 & 0 & \cdots & 0 \\ 0 & n_2 & \cdots & 0 \\ \vdots & \vdots & \cdots & \vdots \\ 0 & 0 & \cdots & n_p \end{bmatrix} \text{ and } (\boldsymbol{Z}'\boldsymbol{Z})^{-1} = \begin{bmatrix} 1/n_1 & 0 & \cdots & 0 \\ 0 & 1/n_2 & \cdots & 0 \\ \vdots & \vdots & \cdots & \vdots \\ 0 & 0 & \cdots & 1/n_p \end{bmatrix}$$

Also,

$$\boldsymbol{Z}'\boldsymbol{Y} = \begin{bmatrix} \left(\sum_{j=1}^{n_1} Y_{1j}\right) \\ \vdots \\ \left(\sum_{j=1}^{n_p} Y_{pj}\right) \end{bmatrix} \text{ and } (\boldsymbol{Z}'\boldsymbol{Z})^{-1}\boldsymbol{Z}'\boldsymbol{Y} = \begin{bmatrix} \bar{Y}_{1+} \\ \vdots \\ \bar{Y}_{p+} \end{bmatrix}.$$

3. Use the definitions of $\bar{Y}_{i+}$ and $\bar{Y}_{++}$ in the text to compute

$$\begin{aligned} \sum_{i=1}^{p} n_i(\bar{Y}_{i+} - \bar{Y}_{++})^2 &= \sum_{i=1}^{p} n_i\bar{Y}_{i+}^2 - n\bar{Y}_{++}^2 - 2\bar{Y}_{++}\sum_{i=1}^{p} n_i\bar{Y}_{i+} \\ &= \sum_{i=1}^{p} n_i\bar{Y}_{i+}^2 + n\bar{Y}_{++}^2 - 2n\bar{Y}_{++}^2, \\ &= \sum_{i=1}^{p} n_i\bar{Y}_{i+}^2 - n\bar{Y}_{++}^2. \end{aligned}$$

5. In this problem $n_i = 10$ for $i = 1, 2, 3, 4$, and $\bar{Y}_{1+} = 105.7, \bar{Y}_{2+} = 102.0, \bar{Y}_{3+} = 93.5, \bar{Y}_{4+} = 110.8$, and $\bar{Y}_{++} = 103$.

$$\begin{aligned} \sum_{j=1}^{10}(Y_{1j} - \bar{Y}_{1+})^2 &= 303, \quad \sum_{j=1}^{10}(Y_{2j} - \bar{Y}_{2+})^2 = 544, \\ \sum_{j=1}^{10}(Y_{3j} - \bar{Y}_{3+})^2 &= 250, \quad \sum_{j=1}^{10}(Y_{4j} - \bar{Y}_{4+})^2 = 364. \end{aligned}$$

Therefore, by Eq. (11.6.9),

$$U^2 = \frac{36(1593.8)}{3(1461)} = 13.09.$$

When H_0 is true, the statistic U^2 has the F distribution with 3 and 36 degrees of freedom. The tail area corresponding to the value $U^2 = 13.09$ is found to be less than 0.025.

7. If U is defined by Eq. (9.6.3), then

$$U^2 = \frac{(m+n-2)(\bar{X}_m - \bar{Y}_n)^2}{\left(\frac{1}{m}+\frac{1}{n}\right)(S_x^2 + S_y^2)}.$$

The correspondence between the notation of Sec. 9.6 and our present notation is as follows:

Notation of Sec. 9.6	Present notation
m	n_1
n	n_2
$\bar{X}_m$	$\bar{Y}_{1+}$
$\bar{Y}_n$	$\bar{Y}_{2+}$
S_X^2	$\sum_{j=1}^{n_1}(Y_{1j} - \bar{Y}_{1+})^2$
S_Y^2	$\sum_{j=1}^{n_2}(Y_{2j} - \bar{Y}_{2+})^2$

Since $p = 2$, $\bar{Y}_{++} = n_1\bar{Y}_{1+}/n + n_2\bar{Y}_{2+}/n$. Therefore,

$$\begin{aligned} & n_1(\bar{Y}_{1+} - \bar{Y}_{++})^2 + n_2(\bar{Y}_{2+} - \bar{Y}_{++})^2 \\ = \; & n_1\left(\frac{n_2}{n}\right)^2(\bar{Y}_{1+} - \bar{Y}_{2+})^2 + n_2\left(\frac{n_1}{n}\right)^2(\bar{Y}_{1+} - \bar{Y}_{2+})^2 \\ = \; & \frac{n_1 n_2}{n}(\bar{Y}_{1+} - \bar{Y}_{2+})^2, \quad \text{since } n_1 + n_2 = n. \end{aligned}$$

Also, since $m + n$ in the notation of Sec. 9.6 is simply n in our present notation, we can now rewrite the expression for U^2 as follows:

$$U^2 = \frac{(n-2)\dfrac{n}{n_1 n_2}\displaystyle\sum_{i=1}^{2} n_i(\bar{Y}_{i+} - \bar{Y}_{++})^2}{\dfrac{n}{n_1 n_2}\displaystyle\sum_{i=1}^{2}\sum_{j=1}^{n_i}(\bar{Y}_{ij} - \bar{Y}_{i+})^2}.$$

This expression reduces to the expression for U^2 given in Eq. (11.6.9), with $p = 2$.

9. Each of the three given random variables is a linear function of the independent random variables $Y_{rs}(s = 1, \ldots, n_r$ and $r = 1, \ldots, p)$.

Let $W_1 = \sum_{r,s} a_{rs}Y_{rs}$, $W_2 = \sum_{r,s} b_{rs}Y_{rs}$, and $W_3 = \sum_{r,s} c_{rs}Y_{rs}$. We have $a_{ij} = 1 - \frac{1}{n_i}$, $a_{is} = -\frac{1}{n_i}$ for $s \neq j$, and $a_{rs} = 0$ for $r \neq i$. Also, $b_{i's} = \frac{1}{n_{i'}} - \frac{1}{n}$ and $b_{rs} = -\frac{1}{n}$ for $r \neq i'$. Finally, $c_{rs} = \frac{1}{n}$ for all values of r and s.

Now, $\text{Cov}(W_1, W_2) = \text{Cov}\left(\sum_{r,s} a_{rs}Y_{rs}, \sum_{r',s'} b_{r's'}Y_{r's'}\right) = \sum_{r,s}\sum_{r',s'} a_{rs}b_{r's'}\text{Cov}(Y_{rs}, Y_{r's'})$.

But $\text{Cov}(Y_{rs}, Y_{r's'}) = 0$ unless $r = r'$ and $s = s'$, since any two distinct Y's are independent. Also, $\text{Cov}(Y_{rs}, Y_{rs}) = \text{Var}(Y_{rs}) = \sigma^2$.

Therefore, $\text{Cov}(W_1, W_2) = \sigma^2 \sum_{r,s} a_{rs}b_{rs}$. If $i = i'$,

$$\sum_{r,s} a_{rs}b_{rs} = a_{ij}b_{ij} + \sum_{s\neq j} a_{is}b_{is} + 0 = \left(1 - \frac{1}{n_i}\right)\left(\frac{1}{n_i} - \frac{1}{n}\right) + (n_i - 1)\left(-\frac{1}{n_i}\right)\left(\frac{1}{n_i} - \frac{1}{n}\right) = 0.$$

If $i \neq i'$,

$$\sum_{r,s} a_{rs}b_{rs} = \left(1 - \frac{1}{n_i}\right)\left(-\frac{1}{n}\right) + (n_i - 1)\left(-\frac{1}{n_i}\right)\left(-\frac{1}{n}\right) = 0.$$

Similarly,

$$\text{Cov}(W_1, W_3) = \sigma^2 \sum_{r,s} a_{rs}c_{rs} = \sigma^2 \frac{1}{n}\left(a_{ij} + \sum_{s\neq j} a_{ij}\right) = 0$$

Finally,

$$\text{Cov}(W_2, W_3) = \sigma^2 \sum_{r,s} b_{rs}c_{rs} = \sigma^2 \frac{1}{n}\sum_{r,s} b_{rs} = \sigma^2 \frac{1}{n}\left[n_{i'}\left(\frac{1}{n_{i'}} - \frac{1}{n}\right) + (n - n_{i'})\left(-\frac{1}{n}\right)\right] = 0.$$

11. Write $S^2_{\text{Betw}} = \sum_{i=1}^{p} \overline{Y}^2_{i+} - n\overline{Y}^2_{++}$. Recall, that the mean of the square of a normal random variable with mean μ and variance σ^2 is $\mu^2 + \sigma^2$. The distribution of $\overline{Y}_{i+}$ is the normal distribution with mean μ_i and variance σ^2/n_i, while the distribution of $\overline{Y}_{++}$ is the normal distribution with mean $\overline{\mu}$ and variance σ^2/n. Hence the mean of S^2_{Betw} is

$$\sum_{i=1}^{p} n_i(\mu_i^2 + \sigma^2/n_i) - n(\overline{\mu} + \sigma^2/n).$$

If we collect terms in this sum, we get $(p-1)\sigma^2 + \sum_{i=1}^{p} n_i\mu_i^2 - n\overline{\mu}^2$. This simplifies to the expression stated in the exercise.

13. We know that $S^2_{\text{Tot}} = S^2_{\text{Betw}} + S^2_{\text{Resid}}$. We also know that $S^2_{\text{Betw}}/\sigma^2$ and $S^2_{\text{Resid}}/\sigma^2$ are independent. If H_0 is true then they both have χ^2 distributions, one with $p - 1$ degrees of freedom (see Exercise 2) and the other with $n - p$ degrees of freedom as in the text. The sum of two independent χ^2 random variables has χ^2 distribution with the sum of the degrees of freedom, in this case $n - 1$.

11.7 The Two-Way Layout

1. Write $S_A^2 = J\sum_{i=1}^{I}\overline{Y}_{i+}^2 - IJ\overline{Y}_{++}^2$. Recall, that the mean of the square of a normal random variable with mean μ and variance σ^2 is $\mu^2+\sigma^2$. The distribution of $\overline{Y}_{i+}$ is the normal distribution with mean μ_i and variance σ^2/J, while the distribution of $\overline{Y}_{++}$ is the normal distribution with mean μ and variance σ^2/IJ. Hence the mean of S_A^2 is $J\sum_{i=1}^{I}(\mu_i^2+\sigma^2/J) - IJ(\mu+\sigma^2/[IJ])$. If we collect terms in this sum, we get

$$(I-1)\sigma^2 + J\sum_{i=1}^{I}\mu_i^2 - IJ\mu^2 = (I-1)\sigma^2 + J\sum_{i=1}^{I}(\mu_i-\mu)^2 = (I-1)\sigma^2 + J\sum_{i=1}^{I}\alpha_i^2.$$

3. If the effects are additive, then there exist numbers Θ_i and Ψ_j such that Eq. (11.7.1) is satisfied for $i=1,\dots,I$ and $j=1,\dots,J$. Let $\bar{\Theta} = \frac{1}{I}\sum_{i=1}^{I}\Theta_i$ and $\bar{\Psi} = \frac{1}{J}\sum_{j=1}^{J}\Psi_j$, and define.

$$\begin{aligned}\mu &= \bar{\Theta}+\bar{\Psi},\\ \alpha_i &= \Theta_i - \bar{\Theta} \quad \text{for} \quad i=1,\dots,I,\\ \beta_j &= \Psi_j - \bar{\Psi} \quad \text{for} \quad j=1,\dots,J.\end{aligned}$$

Then it follows that Eqs. (11.7.2) and (11.7.3) will be satisfied.

It remains to be shown that no other set of values of μ, α_i, and β_j will satisfy Eqs. (11.7.2) and (11.7.3). Suppose that μ', α_i', and β_j' are another such set of values. Then $\mu+\alpha_i+\beta_j = \mu'+\alpha_i'+\beta_j'$ for all i and j. By summing both sides of this relation over i and j, we obtain the relation $IJ\mu = IJ\mu'$. Hence, $\mu=\mu'$. It follows, therefore, that $\alpha_i+\beta_j = \alpha_i'+\beta_j'$ for all i and j. By summing both sides of this relation over j, we obtain the result $\alpha_i=\alpha_i'$ for every value of i. Similarly, by summing both sides over i, we obtain the relation $\beta_j=\beta_j'$ for every value of j.

5. In this exercise,

$$\begin{aligned}\bar{\mu}_{++} &= \frac{1}{12}\sum_{i=1}^{3}\sum_{j=1}^{4}\mu_{ij} = \frac{1}{12}(39) = 3.25,\\ \bar{\mu}_{1+} &= \frac{5}{4} = 1.25, \bar{\mu}_{2+} = \frac{25}{4} = 6.35, \bar{\mu}_{3+} = \frac{9}{4} = 2.25,\\ \bar{\mu}_{+1} &= \frac{15}{3} = 5, \bar{\mu}_{+2} = \frac{3}{3} = 1, \bar{\mu}_{+3} = \frac{6}{3} = 2, \bar{\mu}_{+4} = \frac{15}{3} = 5.\end{aligned}$$

It follows that $\alpha_1 = 1.25-3.25 = -2$, $\alpha_2 = 6.25-3.25 = 3$, and $\alpha_3 = 2.25-3.25 = -1$. Also, $\beta_1 = 5-3.25 = 1.75$, $\beta_2 = 1-3.25 = -2.25$, $\beta_3 = 2-3.25 = -1.25$, and $\beta_4 = 5-3.25 = 1.75$.

7. $\text{Var}(\hat{\mu}) = \text{Var}\left(\frac{1}{IJ}\sum_{i=1}^{I}\sum_{j=1}^{J}Y_{ij}\right) = \frac{1}{I^2J^2}\sum_{i=1}^{I}\sum_{j=1}^{J}\text{Var}(Y_{ij}) = \frac{1}{I^2J^2}IJ\sigma^2 = \frac{1}{IJ}\sigma^2.$

The estimator $\hat{\alpha}_i$ is a linear function of the IJ independent random variables Y_{rs} ($r=1,\dots,I$ and $s=1,\dots,J$). If we let $\hat{\alpha}_i = \sum_{r=1}^{I}\sum_{s=1}^{J}a_{rs}Y_{rs}$, then it can be found that $a_{is} = \frac{1}{J} - \frac{1}{IJ}$ for $s=1,\dots,J$ and $a_{rs} = -\frac{1}{IJ}$ for $r \neq i$. Therefore,

$$\text{Var}(\hat{\alpha}_i) = \sum_{r=1}^{I}\sum_{s=1}^{J}a_{rs}^2\sigma^2 = \sigma^2\left[J\left(\frac{1}{J}-\frac{1}{IJ}\right)^2 + (I-1)J\left(\frac{1}{IJ}\right)^2\right] = \frac{I-1}{IJ}\sigma^2.$$

The value of $\text{Var}(\hat{\beta}_j)$ can be found similarly.

9. Each of the four given random variables is a linear function of the independent random variables $Y_{rs}(r = 1, \ldots, I$ and $s = 1, \ldots, J)$. Let

$$W_1 = \sum_{r=1}^{I}\sum_{s=1}^{J} a_{rs}Y_{rs},$$

$$W_2 = \sum_{r=1}^{I}\sum_{s=1}^{J} b_{rs}Y_{rs},$$

$$W_3 = \sum_{r=1}^{I}\sum_{s=1}^{J} c_{rs}Y_{rs}.$$

Then

$$a_{ij} = 1 - \frac{1}{J} - \frac{1}{I} + \frac{1}{IJ}, \; a_{rj} = -\frac{1}{I} + \frac{1}{IJ} \quad \text{for} \quad r \neq i,$$

$$a_{is} = -\frac{1}{J} + \frac{1}{IJ} \quad \text{for} \quad s \neq j, \quad \text{and} \quad a_{rs} = \frac{1}{IJ} \quad \text{for} \quad r \neq i, \quad \text{and} \quad s \neq j.$$

Also,

$$b_{i's} = \frac{1}{J} - \frac{1}{IJ} \quad \text{and} \quad b_{rs} = -\frac{1}{IJ} \quad \text{for} \quad r \neq i',$$

$$c_{rj'} = \frac{1}{I} - \frac{1}{IJ} \quad \text{and} \quad c_{rs} = -\frac{1}{IJ} \quad \text{for} \quad s \neq j'.$$

As in Exercise 12 of Sec. 11.5, $\text{Cov}(W_1, W_2) = \sigma^2 \sum_{r,s} a_{rs}b_{rs}$. If $i = i'$,

$$\begin{aligned}\sum_{r,s} a_{rs}b_{rs} &= \left(1 - \frac{1}{J} - \frac{1}{I} + \frac{1}{IJ}\right)\left(\frac{1}{J} - \frac{1}{IJ}\right) + (J-1)\left(-\frac{1}{J} + \frac{1}{IJ}\right)\left(\frac{1}{J} - \frac{1}{IJ}\right) \\ &\quad + (I-1)\left(-\frac{1}{I} + \frac{1}{IJ}\right)\left(-\frac{1}{IJ}\right) + (I-1)(J-1)\left(\frac{1}{IJ}\right)\left(-\frac{1}{IJ}\right) = 0.\end{aligned}$$

If $i \neq i'$,

$$\begin{aligned}\sum_{r,s} a_{rs}b_{rs} &= \left(1 - \frac{1}{J} - \frac{1}{I} + \frac{1}{IJ}\right)\left(-\frac{1}{IJ}\right) && (i, j \text{ term}) \\ &\quad + (J-1)\left(-\frac{1}{J} + \frac{1}{IJ}\right)\left(-\frac{1}{IJ}\right) && (i, s \text{ terms for } s \neq j) \\ &\quad + \left(-\frac{1}{I} + \frac{1}{IJ}\right)\left(\frac{1}{J} - \frac{1}{IJ}\right) && (i', j \text{ term}) \\ &\quad + (J-1)\left(\frac{1}{IJ}\right)\left(\frac{1}{J} - \frac{1}{IJ}\right) && (i', j \text{ terms for } s \neq j) \\ &\quad + (I-2)\left(-\frac{1}{I} + \frac{1}{IJ}\right)\left(-\frac{1}{IJ}\right) && (r, j \text{ terms for } r \neq i, i') \\ &\quad + (I-2)(J-1)\left(\frac{1}{IJ}\right)\left(-\frac{1}{IJ}\right) && (r, s \text{ terms for } r \neq i, i' \text{ and } s \neq j). \\ &= 0.\end{aligned}$$

Similarly, the covariance between any other pair of the four variances can be shown to be 0.

11. $$\sum_{i=1}^{I}\sum_{j=1}^{J}(Y_{ij}-\bar{Y}_{i+}-\bar{Y}_{+j}+\bar{Y}_{++})^2 = \sum_i\sum_j Y_{ij}^2 + J\sum_i \bar{Y}_{i+}^2 + I\sum_j \bar{Y}_{+j}^2 + IJ\bar{Y}_{++}^2$$
$$-2\sum_i\sum_j Y_{ij}\bar{Y}_{i+} - 2\sum_j\sum_i Y_{ij}\bar{Y}_{+j} + 2\bar{Y}_{++}\sum_i\sum_j Y_{ij}$$
$$+2\sum_i\sum_j \bar{Y}_{i+}\bar{Y}_{+j} - 2J\bar{Y}_{++}\sum_i \bar{Y}_{i+} - 2I\bar{Y}_{++}\sum_j \bar{Y}_{+j}$$
$$= \sum_i\sum_j Y_{ij}^2 + J\sum_i \bar{Y}_{i+}^2 + I\sum_i \bar{Y}_{+j}^2 + IJ\bar{Y}_{++}^2$$
$$-2J\sum_i \bar{Y}_{i+}^2 - 2I\sum_j \bar{Y}_{+j}^2 + 2IJ\bar{Y}_{++}^2$$
$$+2IJ\bar{Y}_{++}^2 - 2IJ\bar{Y}_{++}^2 - 2IJ\bar{Y}_{++}^2$$
$$= \sum_i\sum_j Y_{ij}^2 - J\sum_i \bar{Y}_{i+}^2 - I\sum_j \bar{Y}_{+j}^2 + IJ\bar{Y}_{++}^2.$$

13. The estimate of $E(Y_{ij})$ is $\hat{\mu}+\hat{\alpha}_i+\hat{\beta}_j$. From the values given in the solution of Exercise 12, we therefore obtain the following table of estimated expectations:

	1	2	3	4	5
1	15.6933	15.1933	16.1267	20.16	19.8267
2	14.2333	13.7333	14.6667	18.7	18.3667
3	15.3733	14.8733	15.8067	19.84	19.5067

Furthermore, Theorem 11.7.1 says that the M.L.E. of σ^2 is

$$\hat{\sigma}^2 = \frac{1}{15}(29.470667) = 1.9647.$$

15. It is found from Eq. (11.7.13) that

$$U_B^2 = \frac{6(22.909769)}{(29.470667)} = 4.664.$$

When the null hypothesis is true, U_B^2 will have the F distribution with $J-1=4$ and $(I-1)(J-1)=8$ degrees of freedom. The tail area corresponding to the value just calculated is found to be between 0.025 and 0.05.

11.8 The Two-Way Layout with Replications

1. Let $\mu=\bar{\Theta}_{++}, \alpha_i=\bar{\Theta}_{i+}-\bar{\Theta}_{++}, \beta_j=\bar{\Theta}_{+j}-\bar{\Theta}_{++}$, and $\gamma_{ij}=\Theta_{ij}-\bar{\Theta}_{i+}-\bar{\Theta}_{+j}+\bar{\Theta}_{++}$ for $i=1,\dots,I$ and $j=1,\dots,J$. Then it can be verified that Eqs. (11.8.4) and (11.8.5) are satisfied. It remains to be shown that no other set of values of μ, α_i, β_j, and γ_{ij} will satisfy Eqs. (11.8.4) and (11.8.5).

Suppose that $\mu', \alpha_i', \beta_j'$, and γ_{ij}' are another such set of values. Then, for all values of i and j,

$$\mu+\alpha_i+\beta_j+\gamma_{ij} = \mu'+\alpha_i'+\beta_j'+\gamma_{ij}'$$

By summing both sides of this equation over i and j, we obtain the relation $IJ\mu = IJ\mu'$. Hence, $\mu = \mu'$. It follows, therefore, that for all values of i and j,

$$\alpha_i + \beta_j + \gamma_{ij} = \alpha_i' + \beta_j + \gamma_{ij}'.$$

By summing both sides of this equation over j, we obtain the relation $J\alpha_i = J\alpha_i'$. By summing both sides of this equation over i, we obtain the relation $I\beta_j = I\beta_j'$. Hence, $\alpha_i = \alpha_i'$ and $\beta_j = \beta_j'$. It also follows, therefore, that $\gamma_{ij} = \gamma_{ij}'$.

3. The values of μ, α_i, β_j, and γ_{ij} can be determined in each part of this problem from the given values of Θ_{ij} by applying the definitions given in the solution of Exercise 1.

5. $E(\hat{\mu}) = \frac{1}{IJK}\sum_{i,j,k} E(Y_{ijk}) = \frac{1}{IJK}\sum_{i,j,k}\Theta_{ij} = \frac{1}{IJ}\sum_{i,j}\Theta_{ij} = \bar{\Theta}_{++} = \mu$, by Exercise 1;

$$\begin{aligned}
E(\hat{\alpha}_i) &= \frac{1}{JK}\sum_{j,k} E(Y_{ijk}) - E(\bar{Y}_{+++}) \\
&= \frac{1}{JK}\sum_{j,k}\Theta_{ij} - \bar{\Theta}_{++}, \text{ by the first part of this exercise} \\
&= \frac{1}{J}\sum_{j}\Theta_{ij} - \bar{\Theta}_{++} = \bar{\Theta}_{i+} - \bar{\Theta}_{++} = \alpha_i, \text{ by Exercise 1;} \\
E(\hat{\beta}_j) &= \beta_j, \text{ by a similar argument;} \\
E(\hat{\gamma}_{ij}) &= \frac{1}{K}\sum_{k} E(Y_{ijk}) - E(\hat{\mu} + \hat{\alpha}_i + \hat{\beta}_j), \text{ by Eq. (11.8.7)} \\
&= \Theta_{ij} - \bar{\Theta}_{i+} - \bar{\Theta}_{+J} + \bar{\Theta}_{++}, \text{ by the previous parts of this exercise,} \\
&= \gamma_{ij}, \text{ by Exercise 1.}
\end{aligned}$$

7. First, write S^2_{Tot} as

$$\sum_{i,j,k}[(Y_{ijk} - \bar{Y}_{ij+}) + (\bar{Y}_{ij+} - \bar{Y}_{i++} - \bar{Y}_{+j+} + \bar{Y}_{+++}) + (\bar{Y}_{i++} - \bar{Y}_{+++}) + (\bar{Y}_{+j+} - \bar{Y}_{+++})]^2. \quad \text{(S.11.3)}$$

The sums of squares of the grouped terms in the summation in (S.11.3) are S^2_{Resid}, S^2_{Int}, S^2_A, and S^2_B. Hence, to verify Eq. (11.8.9), it must be shown that the sum of each of the pairs of cross-products of the grouped terms is 0. This is verified in a manner similar to what was done in the solution to Exercise 8 in Sec. 11.7. Each of the sums of $(Y_{ijk} - \bar{Y}_{ij+})$ times one of the other terms is 0 by summing over k first. The sum of the product of the last two grouped terms is 0 because the sum factors into sums of i and j that are each 0. The other two sums are similar, and we shall illustrate this one:

$$\sum_{i,j,k}(\bar{Y}_{ij+} - \bar{Y}_{i++} - \bar{Y}_{+j+} + \bar{Y}_{+++})(\bar{Y}_{i++} - \bar{Y}_{+++}).$$

Summing over j first produces 0 in this sum. For the other one, sum over i first.

9. We shall first show that the numerators are equal:

$$\begin{aligned}
\sum_{i,j}(\bar{Y}_{ij+}-\bar{Y}_{i++}-\bar{Y}_{+j+}+\bar{Y}_{+++})^2 &= \sum_{i,j}(\bar{Y}_{ij+}^2+\bar{Y}_{i++}^2+\bar{Y}_{+j+}^2+\bar{Y}_{+++}^2-2\bar{Y}_{ij+}\bar{Y}_{i++} \\
&\quad -2\bar{Y}_{ij+}\bar{Y}_{+j+}+2\bar{Y}_{ij+}\bar{Y}_{+++}+2\bar{Y}_{i++}\bar{Y}_{+j+} \\
&\quad -2\bar{Y}_{i++}\bar{Y}_{+++}-2\bar{Y}_{+j+}\bar{Y}_{+++}) \\
&= \sum_{i,j}\bar{Y}_{ij+}^2+J\sum_i\bar{Y}_{i++}^2+I\sum_j\bar{Y}_{+j+}^2+IJ\bar{Y}_{+++}^2 \\
&\quad -2J\sum_i\bar{Y}_{i++}-2I\sum_j\bar{Y}_{+j+}^2+2IJ\bar{Y}_{+++}^2 \\
&\quad +2IJ\bar{Y}_{+++}^2-2IJ\bar{Y}_{+++}^2-2IJ\bar{Y}_{+++}^2 \\
&= \sum_{i,j}\bar{Y}_{ij+}^2-J\sum_i\bar{Y}_{i++}^2-I\sum_j\bar{Y}_{+j+}^2+IJ\bar{Y}_{+++}^2.
\end{aligned}$$

Next, we shall show that the denominators are equal:

$$\begin{aligned}
\sum_{i,j,k}(\bar{Y}_{ijk}-\bar{Y}_{ij+})^2 &= \sum_{i,j,k}(Y_{ijk}^2-2Y_{ijk}\bar{Y}_{ij+}+\bar{Y}_{ij+}^2) \\
&= \sum_{i,j}\left(\sum_k Y_{ijk}^2-2K\bar{Y}_{ij+}^2+K\bar{Y}_{ij+}^2\right) \\
&= \sum_{i,j,k}Y_{ijk}^2-K\sum_{i,j}\bar{Y}_{ij+}^2.
\end{aligned}$$

11. It is found from Eq. (11.8.12) that $U_{AB}^2 = 0.7047$. When the hypothesis is true, U_{AB}^2 has the F distribution with $(I-1)(J-1)=6$ and $IJ(K-1)=12$ degrees of freedom. The tail area corresponding to the value just calculated is found to be greater than 0.05.

13. It is found from Eq. (11.8.18) that $U_B^2 = 9.0657$. When the hypothesis is true, U_B^2 has the F distribution with $I(J-1)=9$ and 12 degrees of freedom. The tail area corresponding to the value just calculated is found to be less than 0.025.

15. The estimator $\hat{\alpha}_2$ has the normal distribution with mean α_2 and, by Exercise 6, variance $\sigma^2/12$. Hence, as in the solution of Exercise 14, when $\alpha_2 = 1$, the following statistic V will have the t distribution with 12 degrees of freedom:

$$V = \frac{\sqrt{12}(\hat{\alpha}_2-1)}{\left[\dfrac{1}{12}\sum_{i,j,k}(Y_{ijk}-\bar{Y}_{ij+})^2\right]^{1/2}}$$

The null hypothesis H_0 should be rejected if $V \geq c$, where c is an appropriate constant. It is found that

$$V = \frac{\sqrt{12}(0.7667)}{\left[\dfrac{1}{12}(10.295)\right]^{1/2}} = 2.8673$$

The corresponding tail area is between 0.005 and 0.01.

17. $\sum_{i=1}^{I} \hat{\alpha}_i = \frac{1}{J}\sum_{i=1}^{I}\sum_{j=1}^{J} \bar{Y}_{ij+} - I\hat{\mu} = I\hat{\mu} - I\hat{\mu} = 0$ and

$$\sum_{i=1}^{I} \hat{\gamma}_{ij} = \left(\sum_{i=1}^{I} \bar{Y}_{ij+} - I\hat{\mu}\right) - \sum_{i=1}^{I} \hat{\alpha}_i - I\hat{\beta}_j = I\hat{\beta}_j - 0 - I\hat{\beta}_j = 0.$$

It can be shown similarly that

$$\sum_{j=1}^{J} \hat{\beta}_j = 0 \quad \text{and} \quad \sum_{j=1}^{J} \hat{\gamma}_{ij} = 0.$$

19. Notice that we cannot reject the second null hypothesis unless we accept the first null hypothesis, since we don't even test the second hypothesis if we reject the first one. The probability that the two-stage procedure rejects at least one of the two hypotheses is then

$$\begin{aligned}&\Pr(\text{reject first null hypothesis})\\ &\quad + \Pr(\text{reject second null hypothesis and accept first null hypothesis}).\end{aligned}$$

The first term above is α_0, and the second term can be rewritten as

$$\begin{aligned}&\Pr(\text{reject second null hypothesis}|\text{accept first null hypothesis})\\ &\quad \times \Pr(\text{accept first null hypothesis}).\end{aligned}$$

This product equals $\beta_0(1-\alpha_0)$, hence the overall probability is $\alpha_0 + \beta_0(1-\alpha_0)$.

11.9 Supplementary Exercises

1. The necessary calculations were done in Example 11.3.6. The least-squares coefficients are $\hat{\beta}_0 = -0.9709$ and $\hat{\beta}_1 = 0.0206$, with $\sigma' = 8.730 \times 10^{-3}$, and $n = 17$. We also can compute $s_x^2 = 530.8$ and $\overline{x}_n = 203.0$.

   (a) A 90% confidence interval for β_1 is $\hat{\beta}_1 \pm T_{n-2}^{-1}(0.95)\sigma'/s_x$. This becomes $(0.01996, 0.02129)$.

   (b) Since 0 is not in the 90% interval in part (a), we would reject H_0 at level $\alpha_0 = 0.1$.

   (c) A 90% prediction interval for log-pressure at boiling-point equal to x is

   $$\hat{\beta}_0 + x\hat{\beta}_1 \pm T_{n-2}^{-1}(0.95)\sigma'\left(1 + \frac{1}{n} + \frac{(x-\overline{x}_n)^2}{s_x^2}\right)^{1/2}.$$

   With the data we have, this gives $[3.233, 3, 264]$. Converting this to pressure gives $(25.35, 26.16)$.

3. The conditional distribution of Y_i given $X_i = x_i$ has mean $\beta_0 + \beta_1 x_i$, where

$$\beta_0 = \mu_2 - \frac{\rho\sigma_2}{\sigma_1}\mu_1 \quad \text{and} \quad \beta_1 = \frac{\rho\sigma_2}{\sigma_1},$$

and variance $(1-\rho^2)\sigma_2^2$. Since $T = \hat{\beta}_1$, as given in Exercise 2b of Sec. 11.1, it follows that $E(T) = \beta_1 = \rho\sigma_2/\sigma_1$ and

$$\text{Var}(T) = \frac{(1-\rho^2)\sigma_2^2}{\sum_{i=1}^{n}(x_i - \bar{x}_n)^2}.$$

5. This result can be established from the formulas for the least squares line given in Sec. 11.1 or directly from the following reasoning: Let $x_1 = a$ and $x_2 = b$. The data contain one observation (a, y_1) at $x = a$ and $n-1$ observations $(b, y_2), \ldots, (b, y_n)$ at $x = b$. Let u denote the average of the $n-1$ values $y_2, \ldots, y_n$, and let h_a and h_b denote the height of the least square line at $x = a$ and $x = b$, respectively. Then the value of Q, as given by Eq. (11.1.2), is

$$Q = (y_1 - h_a)^2 + \sum_{j=2}^{n} (y_i - h_b)^2.$$

The first term is minimized by taking $h_a = y_1$ and the summation is minimized by taking $h_b = u$. Hence, Q is minimized by passing the straight line through the two points (a, y_1) and (b, u). But (a, y_1) is the point (x_1, y_1).

7. It is found from standard calculus texts that the sum of the squared distances from the points to the line is

$$Q = \frac{\sum_{i=1}^{n} (y_i - \beta_1 - \beta_2 x_i)^2}{1 + \beta_2^2}.$$

The equation $\partial Q / \partial \beta_1 = 0$ reduces to the relation $\beta_1 = \bar{y}_n - \beta_2 \bar{x}_n$. If we replace β_1 in the equation $\partial Q / \partial \beta_2 = 0$ by this quantity, we obtain the relation:

$$(1 + \beta_2^2) = \sum_{i=1}^{n} [(y_i - \bar{y}_n) - \beta_2 (x_i - \bar{x}_n)] x_i + \beta_2 \sum_{i=1}^{n} [(y_i - \bar{y}_n) - \beta_2 (x_i - \bar{x}_n)]^2 = 0.$$

Note that we can replace the factor x_i in the first summation by $x_i - \bar{x}_n$ without changing the value of the summation. If we then let $x_i' = x_i - \bar{x}_n$ and $y_i' = y_i - \bar{y}_n$, and expand the final squared term, we obtain the following relation after some algebra:

$$(\beta_2^2 - 1) \sum_{i=1}^{n} x_i' y_i' + \beta_2 \sum_{i=1}^{n} (x_i'^2 - y_i'^2) = 0.$$

Hence

$$\beta_2 = \frac{\sum_{i=1}^{n} \left(y_i'^2 - x_i'^2 \right) \pm \left[\left(\sum_{i=1}^{n} \left(y_i'^2 - x_i'^2 \right) \right)^2 + 4 \left(\sum_{i=1}^{n} x_i' y_i' \right)^2 \right]^{1/2}}{2 \sum_{i=1}^{n} x_i' y_i'}.$$

Either the plus sign or the minus sign should be used, depending on whether the optimal line has positive or negative slope.

9.

$$\begin{aligned} v^2 &= \frac{1}{n} \sum_{i=1}^{k} \sum_{j=1}^{n_i} (x_{ij} - \bar{x}_{i+} + \bar{x}_{i+} - \bar{x}_{++})^2 \\ &= \frac{1}{n} \sum_{i=1}^{k} \sum_{j=1}^{n_i} (x_{ij} - \bar{x}_{i+})^2 + \frac{1}{n} \sum_{i=1}^{k} n_i (\bar{x}_{i+} - \bar{x}_{++})^2 \\ &= \frac{1}{n} \sum_{i=1}^{k} n_i [v_i^2 + (\bar{x}_{i+} - \bar{x}_{++})^2]. \end{aligned}$$

11. It was shown in Sec. 11.8 that the quantity $S^2_{\text{Resid}}/\sigma^2$ given in Eq. (11.8.10) has a χ^2 distribution with $IJ(K-1)$ degrees of freedom. Hence, the random variable $S^2_{\text{Resid}}/[IJ(K-1)]$ is an unbiased estimator of σ^2.

13. Suppose that $\alpha_i = \beta_j = \gamma_{ij} = 0$ for all values of i and j. Then it follows from Table 11.28 that $(S^2_A + S^2_B + S^2_{\text{Int}})/\sigma^2$ will have a χ^2 distribution with $(I-1)+(J-1)+(I-1)(J-1) = IJ-1$ degrees of freedom. Furthermore, regardless of the values of α_i, β_j and γ_{ij}, $S^2_{\text{Resid}}/\sigma^2$ will have a χ^2 distribution with $IJ(K-1)$ degrees of freedom, and $S^2_A + S^2_B + S^2_{\text{Int}}$ and S^2_{Resid} will be independent. Hence, under H_0, the statistic

$$U = \frac{IJ(K-1)(S^2_A + S^2_B + S^2_{\text{Int}})}{(IJ-1)S^2_{\text{Resid}}}$$

will have the F distribution with $IJ-1$ and $IJ(K-1)$ degrees of freedom. The null hypothesis H_0 should be rejected if $U \geq c$.

15. Let $Y_1 = W_1, Y_2 = W_2 - 5$, and $Y_3 = \frac{1}{2}W_3$. Then the random vector

$$\boldsymbol{Y} = \begin{bmatrix} Y_1 \\ Y_2 \\ Y_3 \end{bmatrix}$$

satisfies the conditions of the general linear model as described in Sec. 11.5 with

$$\boldsymbol{Z} = \begin{bmatrix} 1 & 1 \\ 1 & 1 \\ 1 & -1 \end{bmatrix}, \qquad \boldsymbol{\beta} = \begin{bmatrix} \theta_1 \\ \theta_2 \end{bmatrix}.$$

Thus,

$$\boldsymbol{Z}'\boldsymbol{Z} = \begin{bmatrix} 3 & 1 \\ 1 & 3 \end{bmatrix}, \qquad (\boldsymbol{Z}'\boldsymbol{Z})^{-1} = \begin{bmatrix} 3/8 & -1/8 \\ -1/8 & 3/8 \end{bmatrix},$$

and

$$\hat{\boldsymbol{\beta}} = \begin{bmatrix} \hat{\theta}_1 \\ \hat{\theta}_2 \end{bmatrix} = (\boldsymbol{Z}'\boldsymbol{Z})^{-1}\boldsymbol{Z}'\boldsymbol{Y} = \begin{bmatrix} \frac{1}{4}Y_1 + \frac{1}{4}Y_2 + \frac{1}{2}Y_3 \\ \frac{1}{4}Y_1 + \frac{1}{4}Y_2 - \frac{1}{2}Y_3 \end{bmatrix}.$$

Also,

$$\begin{aligned} \hat{\sigma}^2 &= \frac{1}{3}(\boldsymbol{Y} - \boldsymbol{Z}\hat{\boldsymbol{\beta}})'(\boldsymbol{Y} - \boldsymbol{Z}\hat{\boldsymbol{\beta}}) \\ &= \frac{1}{3}\left[(Y_1 - \hat{\theta}_1 - \hat{\theta}_2)^2 + (Y_2 - \hat{\theta}_1 - \hat{\theta}_2)^2 + (Y_3 - \hat{\theta}_1 + \hat{\theta}_2)^2\right]. \end{aligned}$$

The following distributional properties of these M.L.E.'s are known from Sec. 11.5: $(\hat{\theta}_1, \hat{\theta}_2)$ and $\hat{\sigma}^2$ are independent; $(\hat{\theta}_1, \hat{\theta}_2)$ has a bivariate normal distribution with mean vector (θ_1, θ_2) and covariance matrix

$$\sigma^2(\boldsymbol{Z}'\boldsymbol{Z})^{-1} = \begin{bmatrix} 3/8 & -1/8 \\ -1/8 & 3/8 \end{bmatrix}\sigma^2;$$

$3\hat{\sigma}^2/\sigma^2$ has a χ^2 distribution with one degree of freedom.

17. It follows from the expressions for $\hat{\beta}_0$ and $\hat{\beta}_1$ given by Eqs. (11.1.1) and (11.2.7) that

$$\begin{aligned}
e_i &= Y_i - (\bar{Y}_n - \bar{x}_n\hat{\beta}_1) - \hat{\beta}_1 x_i \\
&= Y_i - \bar{Y}_n + \frac{(\bar{x}_n - x_i)\sum_{j=1}^{n}(x_j - \bar{x}_n)Y_j}{s_x^2} \\
&= Y_i\left[1 - \frac{1}{n} - \frac{(x_i - \bar{x}_n)^2}{s_x^2}\right] \\
&\quad - \sum_{j\neq i} Y_j\left[\frac{1}{n} + \frac{(x_i - \bar{x}_n)(x_j - \bar{x}_n)}{s_x^2}\right]
\end{aligned}$$

where $s_x^2 = \sum_{j=1}^{n}(x_j - \bar{x}_n)^2$. Since $Y_1, \ldots, Y_n$ are independent and each has variance σ^2, it follows that

$$\text{Var}(e_i) = \sigma^2\left[1 - \frac{1}{n} - \frac{(x_i - \bar{x}_n)^2}{s_x^2}\right]^2 + \sigma^2\sum_{j\neq i}\left[\frac{1}{n} + \frac{(x_i - \bar{x}_n)(x_j - \bar{x}_n)}{s_x^2}\right]^2.$$

Let $Q_i = \frac{1}{n} + \frac{(x_i - \bar{x}_n)^2}{s_x^2}$. Then

$$\begin{aligned}
\text{Var}(e_i) &= \sigma^2(1 - Q_i)^2 + \sigma^2\sum_{j=1}^{n}\left[\frac{1}{n} + \frac{(x_i - \bar{x}_n)(x_j - \bar{x}_n)}{s_x^2}\right]^2 - \sigma^2 Q_i^2 \\
&= \sigma^2[(1 - Q_i)^2 + Q_i - Q_i^2] \\
&= \sigma^2(1 - Q_i).
\end{aligned}$$

(This result could also have been obtained from the more general result to be obtained next in Exercise 18.) Since Q_i is an increasing function of $(x_i - \bar{x}_n)^2$, it follows that $\text{Var}(e_i)$ is a decreasing function of $(x_i - \bar{x}_n)^2$ and, hence, of the distance between x_i and $\bar{x}_n$.

19. Let $\bar{\theta} = \sum_{i=1}^{I} v_i\theta_i/v_+$ and $\bar{\psi} = \sum_{j=1}^{J} w_j\psi_j/w_+$, and define $\mu = \bar{\theta} + \bar{\psi}, \alpha_i = \theta_i - \bar{\theta}$, and $\beta_j = \psi_j - \bar{\psi}$. Then $E(Y_{ij}) = \theta_i + \psi_j = \mu + \alpha_i + \beta_j$ and $\sum_{i=1}^{I} v_i\alpha_i = \sum_{j=1}^{J} w_j\beta_j = 0$. To establish uniqueness, suppose that μ', α_i', and β_j' are another set of values satisfying the required conditions. Then

$$\mu + \alpha_i + \beta_j = \mu' + \alpha_i' + \beta_j' \quad \text{for all} \quad i \quad \textit{and} \quad j.$$

If we multiply both sides by $v_i w_j$ and sum over i and j, we find that $\mu = \mu'$. Hence, $\alpha_i + \beta_j = \alpha_i' + \beta_j'$. If we now multiply both sides by v_i and sum over i, we find that $\beta_j = \beta_j'$. Similarly, if we multiply both sides by w_j and sum over j, we find that $\alpha_i = \alpha_i'$.

21. As in the solution of Exercise 18 of Sec. 11.8, let

$$\begin{aligned}
\hat{\mu} &= \sum_{r,s,t} m_{rst}Y_{rst}, \\
\hat{\alpha}_i &= \sum_{r,s,t} a_{rst}Y_{rst}, \\
\hat{\beta}_j &= \sum_{r,s,t} b_{rst}Y_{rst}.
\end{aligned}$$

To show that $\text{Cov}(\hat{\mu}, \hat{\alpha}_i) = 0$, we must show that $\sum_{r,s,t} m_{rst}a_{rst} = 0$. But $m_{rst} = \frac{1}{n}$ for all r, s, t, and

$$a_{rst} = \begin{cases} \dfrac{1}{K_{i+}} - \dfrac{1}{n} & \text{for} \quad r = i, \\ -\dfrac{1}{n} & \text{for} \quad r \neq i. \end{cases}$$

Hence, it is found that $\sum m_{rst}a_{rst} = 0$. Similarly,

$$b_{rst} = \begin{cases} \dfrac{1}{K_{+j}} - \dfrac{1}{n} & \text{for} \quad s = j, \\ -\dfrac{1}{n} & \text{for} \quad s \neq j. \end{cases}$$

and $\sum m_{rst}b_{rst} = 0$, so $\text{Cov}(\hat{\mu}, \hat{\beta}_j) = 0$.

23. Consider the expression for θ_{ijk} given in this exercise. If we sum both sides of this expression over i, j, and k, then it follows from the constraints on the α's, β's, and γ's, that $\mu = \bar{\theta}_{+++}$. If we substitute this value for μ and sum both sides of the expression over j and k, we can solve the result for α_i^A. Similarly, α_j^B can be found by summing over i and k, and α_k^C by summing over i and j. After these values have been found, we can determine β_{ij}^{AB} by summing both sides over k, and determine β_{ik}^{AC} and β_{jk}^{BC} similarly. Finally, γ_{ijk} is determined by taking its value to be whatever is necessary to satisfy the required expression for θ_{ijk}. In this way, we obtain the following values:

$$\begin{aligned}
\mu &= \bar{\theta}_{+++}, \\
\alpha_i^A &= \bar{\theta}_{i++} - \bar{\theta}_{+++}, \\
\alpha_j^B &= \bar{\theta}_{+j+} - \bar{\theta}_{+++}, \\
\alpha_k^C &= \bar{\theta}_{++k} - \bar{\theta}_{+++}, \\
\beta_{ij}^{AB} &= \bar{\theta}_{ij+} - \bar{\theta}_{i++} - \bar{\theta}_{+j+} + \bar{\theta}_{+++}, \\
\beta_{ik}^{AC} &= \bar{\theta}_{i+k} - \bar{\theta}_{i++} - \bar{\theta}_{++k} + \bar{\theta}_{+++}, \\
\beta_{jk}^{BC} &= \bar{\theta}_{+jk} - \bar{\theta}_{+j+} - \bar{\theta}_{++k} + \bar{\theta}_{+++}, \\
\gamma_{ijk} &= \theta_{ijk} - \bar{\theta}_{ij+} - \bar{\theta}_{i+k} - \bar{\theta}_{+jk} + \bar{\theta}_{i++} + \bar{\theta}_{+j+} + \bar{\theta}_{++k} - \bar{\theta}_{+++}.
\end{aligned}$$

It can be verified that these quantities satisfy all the specified constraints. They are unique by the method of their construction, since they were derived as the only values that could possibly satisfy the constraints.

Chapter 12

Simulation

All exercises that involve simulation will produce different answers when run with repeatedly. Hence, one cannot expect numerical results to match perfectly with any answers given here.

For all exercises that require simulation, students will need access to software that will do some of the work for them. At a minimum they will need software to simulate uniform pseudo-random numbers on the interval $[0, 1]$. Some of the exercises require software to simulate all of the famous distributions and compute the c.d.f.'s and quantile functions of the famous distributions.

Some of the simulations require a nonnegligible programming effort. In particular, Markov chain Monte Carlo (MCMC) requires looping through all of the coordinates inside of the iteration loop. Assessing convergence and simulation standard error for a MCMC result requires running several chains in parallel. Students who do not have a lot of programming experience might need some help with these exercises.

12.1 What is Simulation?

1. Simulate a large number of exponential random variables with paremeter 1, and take their average.

3. We would expect to get a lot of very large positive observations and a lot of very large negative observations. Each time we got one, the average would either jump up (when we get a positive one) or jump down (when we get a negative one). As we sampled more and more observations, the average should bounce up and down quite a bit and never settle anywhere.

5. (a) Simulate three exponentials at a time. Call the sum of the first two X and call the third one Y. For each triple, record whether $X < Y$ or not. The proportion of times that $X < Y$ in a large sample of triples approximates $\Pr(X < Y)$.

   (b) Let Z_1, Z_2, Z_3 be i.i.d. having the exponential distribution with parameter β, and let W_1, W_2, W_3 be i.i.d. having the exponential distribution with parameter 1. Then $Z_1 + Z_2 < Z_3$ if and only if $\beta Z_1 + \beta Z_2 < \beta Z_3$. But $(\beta Z_1, \beta Z_2, \beta Z_3)$ has precisely the same joint distribution as (W_1, W_2, W_3). So, the probability that $Z_1 + Z_2 < Z_3$ is the same as the probability that $W_1 + W_2 < W_3$, and it doesn't matter which parameter we use for the exponential distribution. All simulations will approximate the same quantity as we would approximate using parameter 1.

   (c) We know that X and Y are independent and that X has the gamma distribution with parameters 2 and 0.4. The joint p.d.f. is

   $$f(x, y) = 0.4^2 x \exp(-0.4x) 0.4 \exp(-0.4y), \text{ for } x, y > 0.$$

   The integral to compute the probability is

   $$\Pr(X < Y) = \int_0^\infty \int_x^\infty 0.4^3 x \exp(-0.4[x + y]) dy dx.$$

There is also a version with the x integral on the inside.

$$\Pr(X < Y) = \int_0^\infty \int_0^y 0.4^3 x \exp(-0.4[x + y]) dx dy.$$

12.2 Why Is Simulation Useful?

1. Since $E(Z) = \mu$, the Cheybshev inequality says that $\Pr(|Z - \mu| \leq \epsilon) \geq \epsilon^2 / \operatorname{Var}(Z)$. Since Z is the average of v independent random variables with variance σ^2, $\operatorname{Var}(Z) = \sigma^2/v$. It follows that

$$\Pr(|Z - \mu| \leq \epsilon) \geq \frac{\epsilon^2 v}{\sigma^2}.$$

Now, suppose that $v \geq \sigma^2/[\epsilon^2(1 - \gamma)$, then

$$\frac{\epsilon^2 v}{\sigma^2} \geq 1 - \gamma.$$

3. We could simulate a lot (say v_0) standard normal random variables $W_1, \ldots, W_{v_0}$ and let $X_i = 7W_i + 2$. Then each X_i has the distribution of X. Let $W_i = \log(|X_i| + 1)$. We could then compute Z equal to the average of the W_i's as an estimate of $E(\log(|X| + 1))$. If we needed our estimate to be close to $E(\log(|X| + 1))$ with high probability, we could estimate the variance of W_i by the sample variance and then use (12.2.5) to choose a possibly larger simulation size.

5. (a) In my ten samples, the sample median was closest to 0 nine times, and the $k = 3$ trimmed mean was closet to 0 one time.

   (b) Although the $k = 2$ trimmed mean was never closest to 0, it was also never very far from 0, and it had the smallest average squared distance from 0. The $k = 3$ trimmed mean was a close second. Here are the six values for my first 10 simulations:

		Trimmed mean				
Estimator	Average	$k = 1$	$k = 2$	$k = 3$	$k = 4$	Median
M.S.E.	0.4425	0.1354	0.0425	0.0450	0.0509	0.0508

   These rankings were also reflected in a much larger simulation.

7. (a) The distribution of X is the contaminated normal distribution with p.d.f. given in Eq. (10.7.2) with $\sigma = 1$, $\mu = 0$.

   (b) To calculate a number in Table 10.40, we should simulate lots of samples of size 20 from the distribution in part (a) with the desired ϵ (0.05 in this case). For each sample, compute the desired estimator (the sample median in this case). Then compute the average of the squares of the estimators (since $\mu = 0$ in our samples) and multiply by 20. As an example, we did two simulations of size 10000 each and got 1.617 and 1.621.

9. The marginal p.d.f. of X is

$$\int_0^\infty \frac{\mu^3}{2} \exp(-\mu(x + 1)) d\mu = \frac{3}{(x + 1)^4},$$

for $x > 0$. The c.d.f. of X is then

$$F(x) = \int_0^x \frac{3}{(t + 1)^4} dx = 1 - \left(\frac{1}{x + 1}\right)^3,$$

for $x > 0$, and $F(x) = 0$ for $x \le 0$. The median is that x such that $F(x) = 1/2$, which is easily seen to be $2^{1/3} - 1 = 0.2599$.

11. In Example 12.2.4, μ_x and μ_y are independent with $(\mu_x - \mu_{x1})/(\beta_{x1}/[\alpha_{x1}\lambda_{x1}])^{1/2}$ having the t distribution with $2\alpha_{x1}$ degrees of freedom and $(\mu_y - \mu_{y1})/(\beta_{y1}/[\alpha_{y1}\lambda_{y1}])^{1/2}$ having the t distribution with $2\alpha_{y1}$ degrees of freedom. We should simulate lots (say v) of t random variables $T_x^{(1)}, \ldots, T_x^{(v)}$ with $2\alpha_{x1}$ degrees of freedom and just as many t random variables $T_y^{(1)}, \ldots, T_y^{(v)}$ with $2\alpha_{y1}$ degrees of freedom. Then let

$$\begin{aligned}\mu_x^{(i)} &= \mu_{x1} + T_x^{(i)}\left(\frac{\beta_{x1}}{\alpha_{x1}\lambda_{x1}}\right)^{1/2}, \\ \mu_y^{(i)} &= \mu_{y1} + T_y^{(i)}\left(\frac{\beta_{y1}}{\alpha_{y1}\lambda_{y1}}\right)^{1/2},\end{aligned}$$

for $i = 1, \ldots, v$. Then the values $\mu_x^{(i)} - \mu_y^{(i)}$ form a sample from the posterior distribution of $\mu_x - \mu_y$.

13. The function g in this exercise is $g(y, w) = w - y^2$ with partial derivatives

$$\begin{aligned}g_1(y, w) &= 2y, \\ g_2(y, w) &= 1.\end{aligned}$$

In the formula for $\mathrm{Var}(Z)$ given in Exercise 12, make the following substitutions:

Exercise 12	This exercise
$E(Y)$	$\bar{Y}$
$E(W)$	$\bar{W}$
σ_{yy}	Z/v
σ_{ww}	V/v
σ_{yw}	C/v,

where Z, V, and C are defined in Example 12.2.10. The result is $[(2\bar{Y})^2 Z + V + 4\bar{Y}C]/v$, which simplifies to (12.2.3).

15. (a) Since $S_u = S_0 \exp(\alpha u + W_u)$, we have that

$$E(S_u) = S_0 \exp(\alpha u) E\left(\exp(W_u)\right) = S_0 \exp(\alpha u)\psi(1).$$

In order for this mean to be $S_0 \exp(ru)$, it is necessary and sufficient that $\psi(1) = \exp(u[r - \alpha])$, or equivalently, $\alpha = r - \log(\psi(1))/u$.

(b) First, simulate lots (say v) of random variables $W^{(1)}, \ldots, W^{(v)}$ with the distribution of W_u. Define the function $h(s)$ as in Example 12.2.13. Define $Y^{(i)} = \exp(-ru)h(S_0 \exp[\alpha u + W^{(i)}])$, where r is the risk free interest rate and α is the number found in part (a). The sample average of the $Y^{(i)}$'s would estimate the appropriate price for the option. One should compute a simulation standard error to see if the simulation size is large enough.

12.3 Simulating Specific Distributions

1. (a) Here we are being asked to perform the simulation outlined in the solution to Exercise 10 in Sec. 12.2 with $v_0 = 2000$ simulations. Each Y_i (in the notation of that solution) can be simulated by taking a random variable U_i having uniform distribution on the interval $[0, 1]$ and setting $Y_i = -\log(1 - U_i)$. In addition to the run whose answers are in the back of the text, here are the results of two additional simulations: Approximation $= 0.0536$, sim. std. err. $= 0.0023$ and Approximation $= 0.0492$, sim. std. err. $= 0.0019$.

   (b) For the two additional simulations in part (a), the value of v to achieve the desired goal are 706 and 459.

3. The c.d.f. corresponding to g_1 is

$$G_1(x) = \int_0^x \frac{1}{2t^{1/2}} dt = x^{1/2}, \text{ for } 0 \leq x \leq 1.$$

The quantile function is then $G_1^{-1}(p) = p^2$ for $0 < p < 1$. To simulate a random variable with the p.d.f. g_1, simulate U with a uniform distribution on the interval $[0, 1]$ and let $X = U^2$. The c.d.f. corresponding to g_2 is

$$G_2(x) = \int_0^x \frac{1}{2(1-t)^{1/2}} dt = 1 - (1-x)^{1/2}, \text{ for } 0 \leq x \leq 1.$$

The quantile function is then $G_2^{-1}(p) = 1 - (1-p)^2$ for $0 < p < 1$. To simulate a random variable with the p.d.f. g_2, simulate U with a uniform distribution on the interval $[0, 1]$ and let $X = 1 - (1-U)^2$.

5. The probability of acceptance on each attempt is $1/k$. Since the attempts (trials) are independent, the number of failures X until the first acceptance is a geometric random variable with parameter $1/k$. The number of iterations until the first acceptance is $X + 1$. The mean of X is $(1 - 1/k)/(1/k) = k - 1$, so the mean of $X + 1$ is k.

7. Simulate a random sample $X_1, \ldots, X_{11}$ from the standard normal distribution. Then $\sum_{i=1}^{4} X_i^2$ has the χ^2 distribution with 4 degrees of freedom and is independent of $\sum_{i=5}^{11} X_i^2$, which has the χ^2 distribution with 7 degrees of freedom. It follows that

$$\frac{7\sum_{i=1}^{4} X_i^2}{4\sum_{i=5}^{11} X_i^2}$$

the F distribution with 4 and 7 degrees of freedom.

9. The simplest acceptance/rejection algorithm would use a uniform distribution on the interval $[0, 2]$. That is, let $g(x) = 0.5$ for $0 < x < 2$. Then $(4/3)g(x) \geq f(x)$ for all x, i.e. $k = 4/3$. We could simulate U and V both having a uniform distribution on the interval $[0, 1]$. Then let $X = 2V$ if $2f(2V) \geq (4/3)U$ and reject otherwise.

11. The χ^2 distribution with m degrees of freedom is the same as the gamma distribution with parameters $m/2$ and $1/2$. So, we should simulate $Y^{(i)}$ having the χ^2 distribution with $n-p$ degrees of freedom and set $\tau^{(i)} = Y^{(i)}/S^2_{\text{Resid}}$.

13. Let $X = F^{-1}(U)$, where F^{-1} is defined in Eq. (12.3.7) and U has a uniform distribution on the interval $[0,1]$. Let G be the c.d.f. of X. We need to show that $G = F$, where F is defined in Eq. (12.3.6). Since F^{-1} only takes the values $t_1,\ldots,t_n$, it follows that G has jumps at those values and if flat everywhere else. Since F also has jumps at $t_1,\ldots,t_n$ and is flat everywhere else, we only need to show that $F(x) = G(x)$ for $x \in \{t_1,\ldots,t_n\}$. Let $q_n = 1$. Then $X \le t_i$ if and only if $U \le q_i$ for $i = 1,\ldots,n$. Since $\Pr(U \le q_i) = q_i$, it follows that $G(t_i) = q_i$ for $i = 1,\ldots,n$. That is, $G(x) = F(x)$ for $x \in \{t_1,\ldots,t_n\}$.

15. The proof is exactly what the hint says. All joint p.d.f.'s should be considered joint p.f./p.d.f.'s and the p.d.f.'s of X and Y should be considered p.f.'s instead. The only integral over x in the proof is in the second displayed equation in the proof. The outer integral in that equation should be replaced by a sum over all possible x values. The rest of the proof is identical to the proof of Theorem 12.3.1.

17. Let $\{x_1,\ldots,x_m\}$ be the set of values that have positive probability under at least one of $g_1,\ldots,g_n$. That is, for each $j = 1,\ldots,m$ there is at least one i such that $g_i(x_j) > 0$ and for each $i = 1,\ldots,n$, $\sum_{j=1}^{m} g_i(x_j) = 1$. Then, the law of total probability says that

$$\Pr(X = x_j) = \sum_{i=1}^{n} \Pr(X = x_j | I = i)\Pr(I = i).$$

Since $\Pr(I = i) = 1/n$ for $i = 1,\ldots,n$ and $\Pr(X = x_j|I = i) = g_i(x_j)$, it follows that

$$\Pr(X = x_j) = \sum_{i=1}^{n} \frac{1}{n} g_i(x_j). \tag{S.12.1}$$

Since $x_1,\ldots,x_m$ are the only values that X can take, Eq. (S.12.1) specifies the entire p.f. of X and we see that Eq. (S.12.1) is the same as Eq. (12.3.8).

19. For $k = 1,\ldots,n$, $I = k$ if and only if $k \le nY + 1 < k+1$. Hence

$$\Pr(I = k) = \Pr\left(\frac{k-1}{n} \le Y \le \frac{k}{n}\right) = \frac{1}{n}.$$

The conditional c.d.f. of U given $I = k$ is

$$\begin{aligned}
\Pr(U \le t | I = k) &= \Pr(nY + 1 - I \le t | I = k) \\
&= \frac{\Pr(nY + 1 - k \le t, I = k)}{\Pr(I = k)} \\
&= \frac{\Pr\left(Y \le \frac{t+k-1}{n}, \frac{k-1}{n} \le Y \le \frac{k}{n}\right)}{1/n} \\
&= n \Pr\left(\frac{k-1}{n} \le Y < \frac{t+k-1}{n}\right) \\
&= t,
\end{aligned}$$

for $0 < t < 1$. So, the conditional distribution of U given $I = k$ is uniform on the interval $[0,1]$ for all k. Since the conditional distribution is the same for all k, U and I are independent and the marginal distribution of U is uniform on the interval $[0,1]$.

12.4 Importance Sampling

1. We want to approximate the integral $\int_a^b g(x)dx$. Suppose that we use importance sampling with f being the p.d.f. of the uniform distribution on the interval $[a, b]$. Then $g(x)/f(x) = (b-a)g(x)$ for $a < x < b$. Now, (12.4.1) is the same as (12.4.2).

3. (a) This is a distribution for which the quantile function is easy to compute. The c.d.f. is $F(x) = 1-(c/x)^{n/2}$ for $x > c$, so the quantile function is $F^{-1}(p) = c/(1-p)^{2/n}$. So, simulate U having a uniform distribution on the interval $[0, 1]$ and let $X = c/(1-U)^{2/n}$. Then X has the p.d.f. f.

   (b) Let

$$a = \frac{\Gamma\left[\frac{1}{2}(m+n)\right] m^{m/2}n^{n/2}}{\Gamma\left(\frac{1}{2}m\right)\Gamma\left(\frac{1}{2}n\right)}.$$

   Then the p.d.f. of Y is $g(x) = ax^{(m/2)-1}/(mx+n)^{(m+n)/2}$, for $x > 0$. Hence,

$$\Pr(Y > c) = \int_c^\infty a\frac{x^{(m/2)-1}}{(mx+n)^{(m+n)/2}}dx.$$

   We could approximate this by sampling lots of values $X^{(i)}$ with the p.d.f. f from part (a) and then averaging the values $g(X^{(i)})/f(X^{(i)})$.

   (c) The ratio $g(x)/f(x)$ is, for $x > c$,

$$\frac{g(x)}{f(x)} = \frac{ax^{(m+n)/2}}{c^{n/2}(n/2)(mx+n)^{(m+n)/2}} = \frac{a}{c^{n/2}(n/2)(m+n/x)^{(m+n)/2}}.$$

   This function is fairly flat for large x. Since we are only interested in $x > c$ in this exercise, importance sampling will have us averaging random variables $g(X^{(i)})/f(X^{(i)})$ that are nearly constant, hence the average should have small variance.

5. Let U have a uniform distribution on the interval $[0, 1]$, and let W be defined by Eq. (12.4.6). The inverse transformation is

$$u = \frac{\Phi\left(\frac{w-\mu_2}{\sigma_2}\right)}{\Phi\left(\frac{c_2-\mu_2}{\sigma_2}\right)}.$$

   The derivative of the inverse transformation is

$$\frac{1}{(2\pi)^{1/2}\sigma_2\Phi\left(\frac{c_2-\mu_2}{\sigma_2}\right)}\exp\left(-\frac{1}{2\sigma_2^2}(w-\mu_2)^2\right). \qquad \text{(S.12.2)}$$

   Since the p.d.f. of U is constant, the p.d.f. of W is (S.12.2), which is the same as (12.4.5).

7. (a) We can simulate bivariate normals by simulating one of the marginals first and then simulating the second coordinate conditional on the first one. For example, if we simulate $X_1^{(i)}$ $U^{(i)}$ as independent normal random variables with mean 0 and variance 1, we can simulate $X_2^{(i)} = 0.5X_1^{(i)} + U^{(i)}.75^{1/2}$. Three simulations of size 10000 each produced estimates of 0.8285, 0.8308, and 0.8316 with simulation standard errors of 0.0037 each time.

(b) Using the method of Example 12.4.3, we did three simulations of size 10000 each and got estimates of 0.8386, 0.8387, and 0.8386 with simulation standard errors of about 3.4×10^{-5}, about 0.01 as large as those from part (a). Also, notice how much closer the three simulations are in part (b) compared to the three in part (a).

9. The inverse transformation is $v = F(x)$ with derivative $f(x)$. So, the p.d.f. of X is $f(x)/(b-a)$ for those x that can arise as values of $F^{-1}(V)$, namely $F^{-1}(a) < x < F^{-1}(b)$.

11. Since the conditional p.d.f. of X^* given $J = j$ is f_j, the marginal p.d.f. of X^* is

$$f^*(x) = \sum_{j=1}^{k} f_j(x) \Pr(J = j) = \frac{1}{k} \sum_{j=1}^{k} f_j(x).$$

Since $f_j(x) = kf(x)$ for $q_{j-1} \le x < q_j$, for each x there is one and only one $f_j(x) > 0$. Hence, $f^*(x) = f(x)$ for all x.

13. (a) $E(Z) = E(Y^{(i)}) + kc = E(W^{(i)}) - kE(V^{(i)}) + kc$. By the usual importance sampling argument, $E(W^{(i)}) = \int g(x)dx$ and $E(V^{(i)}) = c$, so $E(Z) = \int g(x)dx$.

(b) $\text{Var}(Z) = [\sigma_W^2 + k^2\sigma_V^2 - 2k\rho\sigma_W\sigma_V]$. This is a quadratic in k that is minimized when $k = \rho\sigma_W/\sigma_V$.

15. (a) Since $U^{(i)}$ and $1 - U^{(i)}$ both have uniform distributions on the interval $[0, 1]$, $X^{(i)} = F^{-1}(U^{(i)})$ and $T^{(i)} = F^{-1}(1 - U^{(i)})$ have the same distribution.

(b) Since $X^{(i)}$ and $T^{(i)}$ have the same distribution, so do $W^{(i)}$ and $V^{(i)}$, so the means of $W^{(i)}$ and $V^{(i)}$ are both the same and they are both $\int g(x)dx$, according to the importance sampling argument.

(c) Since F^{-1} is a monotone increasing function, we know that $X^{(i)}$ and $T^{(i)}$ are decreasing functions of each other. If $g(x)/f(x)$ is monotone, then $W^{(i)}$ and $V^{(i)}$ will also be decreasing functions of each other. As such they ought to be negatively correlated since one is small when the other is large.

(d) $\text{Var}(Z) = \text{Var}(Y^{(i)})/v$, and

$$\text{Var}(Y^{(i)}) = 0.25[\text{Var}(W^{(i)}) + \text{Var}(V^{(i)}) + 2\,\text{Cov}(W^{(i)}, V^{(i)})] = 0.5(1 + \rho)\,\text{Var}(W^{(i)}).$$

Without antithetic variates, we get a variance of $\text{Var}(W^{(i)})/[2v]$. If $\rho < 0$, then $0.5(1 + \rho) < 0.5$ and $\text{Var}(Z)$ is smaller than we get without antithetic variates.

17. In Exercise 3(c), $g(x)/f(x)$ is a monotone function of x, so antithetic variates should help. In Exercise4(b), we could use control variates with $h(x) = \exp(-x)$. In Exercises 6(a) and 6(b) the ratios $g(x)/f(x)$ are monotone, so antithetic variates should help. Control variates with $h(x) = x\exp(-x^2/2)$ could also help in Exercise 6(a). Exercise 10 involves the same function, so the same methods could also be used in the stratified versions.

12.5 Markov Chain Monte Carlo

1. The conditional p.d.f. of X_2 given $X_2 = x_2$ is

$$g_1(x_1|x_2) = \frac{f(x_1, x_2)}{f_2(x_2)} = \frac{cg(x_1, x_2)}{f_2(x_2)} = \frac{c}{f_2(x_2)} h_2(x_1).$$

Let $c_2 = c/f_2(x_2)$, which does not depend on x_1.

3. Let $h(z)$ stand for the p.f. or p.d.f. of the stationary distribution and let $g(z|z')$ stand for the conditional p.d.f. or p.f. of Z_{i+1} given $Z_i = z'$, which is assumed to be the same for all i. Suppose that Z_i has the stationary distribution for some i, then (Z_i, Z_{i+1}) has the joint p.f. or p.d.f. $h(z_i)g(z_{i+1}|z_i)$. Since Z_1 does have the stationary distribution, (Z_1, Z_2) has the joint p.f. or p.d.f. $h(z_1)g(z_2|z_1)$. Hence, (Z_1, Z_2) has the same distribution as (Z_i, Z_{i+1}) whenever Z_i has the stationary distribution. The proof is complete if we can show that Z_i has the stationary distribution for every i. We shall show this by induction. We know that it is true for $i = 1$ (that is, Z_1 has the stationary distribution). Assume that each of $Z_1, \ldots, Z_k$ has the stationary distribution, and prove that Z_{k+1} has the stationary distribution. Since h is the p.d.f. or p.f. of the stationary distribution, it follows that the marginal p.d.f. or p.f. of Z_{k+1} is $\int h(z_k)g(z_{k+1}|z_k)dz_k$ or $\sum_{\text{All } z_k} h(z_k)g(z_{k+1}|z_k)$, either of which is $h(z_{k+1})$ by the definition of stationary distribution. Hence Z_{k+1} also has the stationary distribution, and the induction proof is complete.

5. The sample average of all 30 observations is 1.442, and the value of s_n^2 is 2.671. The posterior hyperparameters are then

$$\alpha_1 = 15.5, \ \lambda_1 = 31, \ \mu_1 = 1.4277, \ \text{and } \beta_1 = 1.930.$$

The method described in Example 12.5.1 says to simulate values of μ having the normal distribution with mean 1.4277 and variance $(31\tau)^{-1}$ and to simulate values of τ having the gamma distribution with parameters 16 and $1.930+0.5(\mu-1.4277)^2$. In my particular simulation, I used five Markov chains with the following starting values for μ: 0.4, 1.0, 1.4, 1.8, and 2.2. The convergence criterion was met very quickly, but we did 100 burn-in anyway. The estimated mean of $(\sqrt{\tau}\mu)^{-1}$ was 0.2542 with simulation standard error 4.71×10^{-4}.

7. There are $n_i = 6$ observations in each of $p = 3$ groups. The sample averages are 825.83, 845.0, and 775.0. The w_i values are 570.83, 200.0, and 900.0. In three separate simulations of size 10000 each, I got the following three vectors of posterior mean estimates: (826.8, 843.2, 783.3), (826.8, 843.2, 783.1), and (826.8, 843.2, 783.2).

9. In three separate simulations of size 10000 each I got posterior mean estimates for (β_0, β_1, η) of $(-0.9526, 0.02052, 1.124\times10^{-5})$, $(-0.9593, 0.02056, 1.143\times10^{-5})$, and $(-0.9491, 0.02050, 1.138\times10^{-5})$. It appears we need more than 10000 samples to get a good estimate of the posterior mean of β_0. The estimated posterior standard deviations from the three simulations were $(1.503\times10^{-2}, 7.412\times10^{-5}, 7.899\times10^{-6})$, $(2.388\times10^{-2}, 1.178\times10^{-4}, 5.799\times10^{-6})$, and $(2.287\times10^{-2}, 1.274\times10^{-4}, 6.858\times10^{-6})$.

11. For this exercise, I ran five Markov chains for 10000 iterations each. For each iteration, I obtain a vector of 10 Y_i values. Our estimated probability that X_i came from the distribution with unknown mean and variance equals the average of the 50000 Y_i values for each $i = 1, \ldots, 10$. The ten estimated probabilities for each of my three runs are listed below:

Run	Estimated Probabilities									
1	0.291	0.292	0.302	0.339	0.370	0.281	0.651	0.374	0.943	0.816
2	0.285	0.286	0.302	0.339	0.375	0.280	0.656	0.371	0.945	0.819
3	0.283	0.286	0.301	0.340	0.373	0.280	0.651	0.370	0.945	0.820

13. In part (a), the exponent in the displayed formula should have been $-1/2$.

   (a) The conditional distribution of $(\mu - \mu_0)\gamma^{1/2}$ given γ is standard normal, hence it is independent of γ. Also, the distribution of $2b_0\gamma$ is the χ^2 distribution with $2a_0$ degrees of freedom. It follows that $(\mu - \mu_0)/(b_0/a_0)^{1/2}$ has the t distribution with $2a_0$ degrees of freedom.

(b) The marginal prior distributions of τ are in the same form with the same hyperparameters in Exercise 12 and in Sec. 8.6. The marginal prior distributions of μ are in the same form also, but the hyperparameters are not identical. We need $a_0 = \alpha_0$ to make the degrees of freedom match, and we need $b_0 = \beta_0/\lambda_0$ in order to make the scale factor match.

(c) The prior p.d.f. times the likelihood equals a constant times

$$\tau^{\alpha_0+n/2-1}\gamma^{a_0+1/2-1}\exp\left(-\frac{\tau}{2}\left\{n[\bar{x}_n-\mu]^2+s_n^2+\beta_0\right\}-\frac{\gamma}{2}[\mu-\mu_0]^2+\gamma b_0\right).$$

As a function of τ this is the same as in Exercise 12. As a function of μ, it is also the same as Exercise 12 if we replace γ_0 by γ. As a function of γ, it looks like the p.d.f. of the gamma distribution with parameters $a_0 + 1/2$ and $b_0 + (\mu - \mu_0)/2$.

(d) This time, I ran 10 chains of length 10000 each for three different simulations. The three intervals are found by sorting the μ values and using the 2500th and 97500th values. The interval are $(154.4, 216.3)$, $(154.6, 215.8)$, and $(154.4, 215.9)$.

15. (a) For each censored observation X_{n+i}, we observe only that $X_{n+i} \le c$. The probability of $X_{n+i} \le c$ given θ is $1 - \exp(-c\theta)$. The likelihood times prior is a constant times

$$\theta^{n+\alpha-1}[1-\exp(-c\theta)]^m \exp\left(-\theta\sum_{i=1}^n x_i\right). \tag{S.12.3}$$

We can treat the unobserved values $X_{n+1}, \ldots, X_{n+m}$ as parameters. The conditional distribution of X_{n+i} given θ and given that $X_{n+i} \le c$ has the p.d.f.

$$g(x|\theta) = \frac{\theta\exp(-\theta x)}{1-\exp(-\theta c)}, \quad \text{for } 0 < x < c. \tag{S.12.4}$$

If we multiply the conditional p.d.f. of $(X_{n+1}, \ldots, X_{n+m})$ given θ times Eq. (S.12.3), we get

$$\theta^{n+m+\alpha-1}\exp\left(-\theta\sum_{i=1}^{n+m} x_i\right),$$

for $\theta > 0$ and $0 < x_i < c$ for $i = n+1, \ldots, n+m$. As a function of θ, this looks like the p.d.f. of the gamma distribution with parameters $n + m + \alpha$ and $\sum_{i=1}^{n+m} x_i$. As a function of X_{n+i}, it looks like the p.d.f. in Eq. (S.12.4). So, Gibbs sampling can work as follows. Pick a starting value for θ, such as one over the average of the uncensored values. Then simulate the censored observations with p.d.f. (S.12.4). This can be done using the quantile function

$$G^{-1}(p) = -\frac{\log(1-p[1-\exp(-c\theta)])}{\theta}.$$

Then, simulate a new θ from the gamma distribution mentioned above to complete one iteration.

(b) For each censored observation X_{n+i}, we observe only that $X_{n+i} \ge c$. The probability of $X_{n+i} \ge c$ given θ is $\exp(-c\theta)$. The likelihood times prior is a constant times

$$\theta^{n+\alpha-1}\exp\left(-\theta\left[mc+\sum_{i=1}^n x_i\right]\right). \tag{S.12.5}$$

We could treat the unobserved values $X_{n+1}, \ldots, X_{n+m}$ as parameters. The conditional distribution of X_{n+i} given θ and given that $X_{n+i} \ge c$ has the p.d.f.

$$g(x|\theta) = \theta\exp(-\theta[x-c]), \quad \text{for } x > c. \tag{S.12.6}$$

If we multiply the conditional p.d.f. of $(X_{n+1}, \ldots, X_{n+m})$ given θ times Eq. (S.12.5), we get

$$\theta^{n+m+\alpha-1} \exp\left(-\theta \sum_{i=1}^{n+m} x_i\right),$$

for $\theta > 0$ and $x_i > c$ for $i = n+1, \ldots, n+m$. As a function of θ, this looks like the p.d.f. of the gamma distribution with parameters $n+m+\alpha$ and $\sum_{i=1}^{n+m} x_i$. As a function of X_{n+i}, it looks like the p.d.f. in Eq. (S.12.6). So, Gibbs sampling can work as follows. Pick a starting value for θ, such as the M.L.E., $\dfrac{n+m}{mc + \sum_{i=1}^{n} x_i}$. Then simulate the censored observations with p.d.f. (S.12.6). This can be done using the quantile function

$$G^{-1}(p) = c - \frac{\log(1-p)}{\theta}.$$

Then, simulate a new θ from the gamma distribution mentioned above to complete one interaction. In this part of the exercise, Gibbs sampling is not really needed because the posterior distribution of θ is available in closed form. Notice that (S.12.5) is a constant times the p.d.f. of the gamma distribution with parameters $n+\alpha$ and $mc + \sum_{i=1}^{n} x_i$, which is then the posterior distribution of θ.

12.6 The Bootstrap

1. We could start by estimating θ by the M.L.E., $1/\overline{X}$. Then we would use the exponential distribution with parameter $1/\overline{X}$ for the distribution $\hat{F}$ in the bootstrap. The bootstrap estimate of the variance of $\overline{X}$ is the variance of a sample average $\overline{X}^*$ of a sample of size n from the distribution $\hat{F}$, i.e., the exponential distribution with parameter $1/\overline{X}$. The variance of $\overline{X}^*$ is $1/n$ times the variance of a single observation from $\hat{F}$, which equals $\overline{X}^2$. So, the bootstrap estimate is $\overline{X}^2/n$.

3. Let $n = 2k+1$. The sample median of a nonparametric bootstrap sample is the $k+1$st smallest observation in the bootstrap sample. Let x denote the smallest observation in the original sample. Assume that there are ℓ observations from the original sample that equal x. (Usually $\ell = 1$, but it is not necessary.) The sample median from the bootstrap sample equals x from the original data set if and only if at least $k+1$ observations in the bootstrap sample equal x. Since each observation in the bootstrap equals x with probability ℓ/n and the bootstrap observations are independent, the probability that at least $k+1$ of them equal x is

$$\sum_{i=k+1}^{n} \binom{n}{i} \left(\frac{\ell}{n}\right)^i \left(1 - \frac{\ell}{n}\right)^{n-i}.$$

5. This exercise is performed in a manner similar to Exercise 4.

   (a) In this case, I did three simulations of size 50000 each. The three estimates of bias were -1.684, -1.688, and -1.608.

   (b) Each time, the estimated sample size needed to achieve the desired accuracy was between 48000 and 49000.

7. (a) Each bootstrap sample consists of $\overline{X}^{*(i)}$ having a normal distribution with mean 0 and variance $31.65/11$, $\overline{Y}^{*(i)}$ having the normal distribution with mean 0 and variance $68.8/10$, $S_X^{2*(i)}$ being 31.65

times a χ^2 random variable with 10 degrees of freedom, and $S_Y^{2*(i)}$ being 68.8 times a χ^2 random variable with 9 degrees of freedom. For each sample, we compute the statistic $U^{(i)}$ displayed in Example 12.6.10 in the text. We then compute what proportion of the absolute values of the 10000 statistics exceed the 0.95 quantile of the t distribution with 19 degrees of freedom, 1.729. In three separate simulations, I got proportions of 0.1101, 0.1078, and 0.1115.

(b) To correct the level of the test, we need the 0.9 quantile of the distribution of $|U|$. For each simulation, we sort the 10000 $|U^{(i)}|$ values and select the 9000th value. In my three simulations, this value was 1.773, 1.777, and 1.788.

(c) To compute the simulation standard error of the sample quantile, I chose to split the 10000 samples into eight sets of size 1250. For each set, I sort the $|U^{(i)}|$ values and choose the 1125th one. The simulation standard error is then the the square-root of one-eighth of the the sample variance of these eight values. In my three simulations, I got the values 0.0112, 0.0136, and 0.0147.

9. (a) For each bootstrap sample, compute the sample correlation $R^{(i)}$. Then compute the sample variance of $R^{(1)}, \dots, R^{(1000)}$. This is the approximation to the bootstrap estimate of the variance of the sample correlation. I did three separate simulations and got sample variances of 4.781×10^{-4}, 4.741×10^{-4}, and 4.986×10^{-4}.

(b) The approximation to the bootstrap bias estimate is the sample average of $R^{(1)}, \dots, R^{(1000)}$ minus the original sample correlation, 0.9670. In my three simulations, I got the values -0.0030, -0.0022, and -0.0026. It looks like 1000 is not enough bootstrap samples to get a good estimate of this bias.

(c) For the simulation standard error of the variance estimate, we use the square-root of Eq. (12.2.3) where each $Y^{(i)}$ in (12.2.3) is $R^{(i)}$ in this exercise. In my three simulations, I got the values 2.231×10^{-5}, 2.734×10^{-5}, and 3.228×10^{-5}. For the simulation standard error of the bias estimate, we just note that the bias estimate is an average, so we need only calculate the square-root of 1/1000 times the sample variance of $R^{(1)}, \dots, R^{(1000)}$. In my simulations, I got 6.915×10^{-4}, 6.886×10^{-4}, and 7.061×10^{-4}.

11. (a) If X^* has the distribution F_n, then $\mu = E(X^*) = \overline{X}$,

$$\begin{aligned}
\sigma^2 &= \text{Var}(X^*) = \frac{1}{n}\sum_{i=1}^{n}(X_i - \overline{X})^2, \text{ and} \\
E([X-\mu]^3) &= \frac{1}{n}\sum_{i=1}^{n}(X_i - \overline{X})^3.
\end{aligned}$$

Plugging these values into the formula for skewness (see Definition 4.4.1) yields the formula for M_3 given in this exercise.

(b) The summary statistics of the 1970 fish price data are $\overline{X} = 41.1$, $\sum_{i=1}^{n}(X_i - \overline{X})^2/n = 1316.5$, and $\sum_{i=1}^{n}(X_i - \overline{X})^3/n = 58176$, so the sample skewness is $M_3 = 1.218$. For each bootstrap sample, we also compute the sample skewness $M_3^{*(i)}$ for $i = 1, \dots, 1000$. The bias of M_3 is estimated by the sample average of the $M_3^{*(i)}$'s minus M_3. I did three simulations and got the values -0.2537, -0.2936, and -0.2888. To estimate the standard deviation of M_3, compute the sample standard deviation of the $M_3^{*(i)}$'s. In my three simulations, I got 0.5480, 0.5590, and 0.5411.

12.7 Supplementary Exercises

1. For the random number generator that I have been using for these solutions, Fig. S.12.1 contains one such normal quantile plot. It looks fairly straight. On the horizontal axis I plotted the sorted

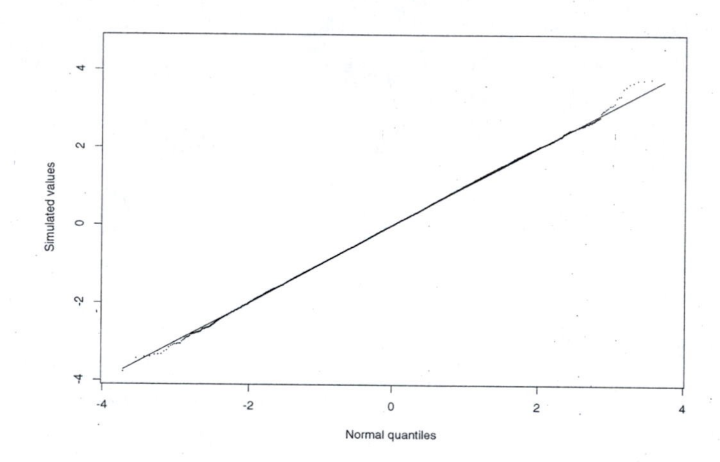

Figure S.12.1: Normal quantile plot for Exercise 1 in Sec. 12.7. A straight line has been added for reference.

pseudo-normal values and on the vertical axis, I plotted the values $\Phi^{-1}(i/10001)$ for $i = 1, \ldots, 10000$.

3. Once again, the plots are drawn in a fashion similar to Exercise 1. This time, we notice that the plot with one degree of freedom has some really serious non-linearity. This is the Cauchy distribution which has very long tails. The extreme observations from a Cauchy sample are very variable. Two of the plots are in Fig. S.12.2.

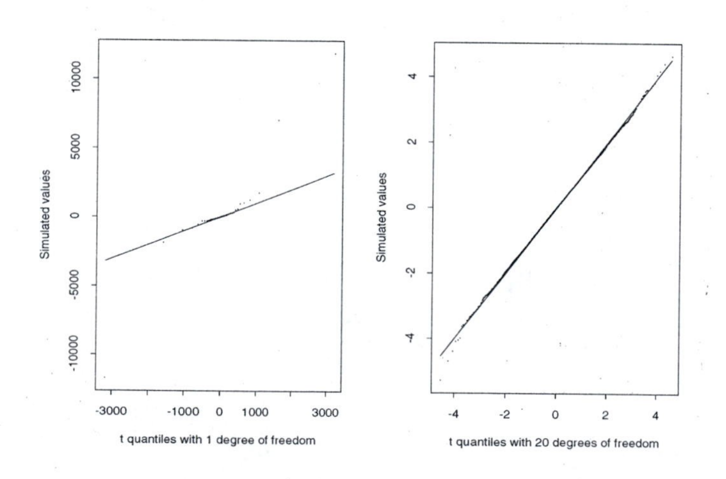

Figure S.12.2: Two t quantile plots for Exercise 3 in Sec. 12.7. The left plot has 1 degree of freedom, and the right plot has 20 degrees of freedom. Straight lines have been added for reference.

5. (a) To simulate a noncentral t random variable, we can simulate independent Z and W with Z having the normal distribution with mean 1.936 and variance 1, and W having the χ^2 distribution with 14 degrees of freedom. Then set $T = Z/(W/14)^{1/2}$.

(b) I did three separate simulations of size 1000 each and got the following three proportions with $T > 1.761$: 0.571, 0.608, 0.577. The simulation standard errors were 0.01565, 0.01544, and 0.01562.

(c) Using (12.2.5), we find that we need a bit more than 16000 simulated values.

7. (a) We need to compute the same Q statistics as in Exercise 6(b) using samples from ten different normal distributions. For each of the ten distributions, we also compute the 0.9, 0.95 and 0.99 sample quantiles of the 10000 Q statistics. Here is a table of the simulated quantiles:

		Quantile		
μ	σ^2	0.9	0.95	0.99
3.8	0.25	3.891	4.976	7.405
3.8	0.80	4.295	5.333	8.788
3.9	0.25	3.653	4.764	6.405
3.9	0.80	4.142	5.133	7.149
4.0	0.25	3.825	5.104	7.405
4.0	0.80	4.554	5.541	8.635
4.1	0.25	3.861	5.255	8.305
4.1	0.80	4.505	5.658	8.637
4.2	0.25	4.193	5.352	8.260
4.2	0.80	4.087	4.981	7.677

(b) The quantiles change a bit as the distributions change, but they are remarkably stable.

(c) Instead of starting with normal samples, we start with samples having a t distribution as described in the exercise. We compute the Q statistic for each sample and see what proportion of our 10000 Q statistics is greater than 5.2. In three simulations of this sort I got proportions of 0.12 0.118, and 01.24.

9. (a) The product of likelihood times prior is

$$\exp\left(-\frac{u_0(\psi-\psi_0)^2}{2} - \sum_{i=1}^{p}\tau_i\left[\beta + \frac{n_i(\mu_i-\overline{y}_i)^2 + w_i + \lambda_0(\mu_i-\psi)^2}{2}\right]\right)$$

$$\times\ \beta^{p\alpha_0+\epsilon_0-1}\exp(-\beta\phi_0)\prod_{i=1}^{p}\tau_i^{\alpha_0+[n_i+1]/2-1},$$

where $w_i = \sum_{j=1}^{n_i}(y_{ij}-\overline{y}_i)^2$ for $i = 1,\ldots,p$.

(b) As a function of μ_i, τ_i, or ψ this looks the same as it did in Example 12.5.6 except that β_0 needs to be replaced by β wherever it occurs. As a function of β, it looks like the p.d.f. of the gamma distribution with parameters $p\alpha_0 + \epsilon_0$ and $\phi_0 + \sum_{i=1}^{p}\tau_i$.

(c) I ran six Markov chains for 10000 iterations each, producing 60000 parameter vectors. The requested posterior means and simulation standard errors were

Parameter	μ_1	μ_2	μ_3	μ_4	$1/\tau_1$	$1/\tau_2$	$1/\tau_3$	$1/\tau_4$
Posterior mean	156.6	158.3	120.6	159.7	495.1	609.2	545.3	570.4
Sim. std. err.	0.01576	0.01836	0.02140	0.03844	0.4176	1.194	0.8968	0.7629

11. (a) We shall use the same approach as in Exercise 12 of Sec. 12.6. Let the parameter be $\theta = (\mu, \sigma_1, \sigma_2)$ (where μ is the common value of $\mu_1 = \mu_2$). Each pair of parameter values θ and θ' that have the same value of σ_2/σ_1 can be obtained from each other by multiplying μ, σ_1 and σ_2 by the same

positive constant and adding some other constant to the resulting μ. That is, there exist $a > 0$ and b such that $\theta' = (a\mu + b, a\sigma_1, a\sigma_2)$. If $X_1, \dots, X_m$ and $Y_1, \dots, Y_n$ have the distribution determined by θ, then $X_i' = aX_i + b$ for $i = 1, \dots, m$ and $Y_j' = aY_j + b$ for $j = 1, \dots, n$ have the distribution determined by θ'. We need only show that the statistic V in (9.6.13) has the same value when it is computed using the X_i's and Y_j's as when it is computed using the X_i''s and Y_j''s. It is easy to see that the numerator of V computed with the X_i''s and Y_j' equals a times the numerator of V computed using the X_i's and Y_j's. The same is true of the denominator, hence V has the same value either way and it must have the same distribution when the parameter is θ as when the parameter is θ'.

(b) By the same reasoning as in part (a), the value of ν is the same whether it is calculated with the X_i's and Y_j's or with the X_i''s and Y_j''s. Hence the distribution of ν (thought of as a random variable before observing the data) depends on the parameter only through σ_2/σ_1.

(c) For each simulation with ratio r, we can simulate $\overline{X}_m$ having the standard normal distribution and S_X^2 having the χ^2 distribution with 9 degrees of freedom. Then simulate $\overline{Y}_n$ having the normal distribution with mean 0 and variance r^2 and S_Y^2 equal to r^2 times a χ^2 random variable with 10 degrees of freedom. Make the four random variables independent when simulating. Then compute V and ν. Compute the three quantiles $T_\nu^{-1}(0.9)$, $T_\nu^{-1}(0.95)$ and $T_\nu^{-1}(0.99)$ and check whether V is greater than each quantile. Our estimates are the proportions of the 10000 simulations in which the value of V are greater than each quantile. Here are the results from one of my simulations:

	Probability		
r	0.9	0.95	0.99
1.0	0.1013	0.0474	0.0079
1.5	0.0976	0.0472	0.0088
2.0	0.0979	0.0506	0.0093
3.0	0.0973	0.0463	0.0110
5.0	0.0962	0.0476	0.0117
10.0	0.1007	0.0504	0.0113

The upper tail probabilities are very close to their nominal values.

13. (a) The fact that $E(\hat{\beta}_1) = \beta_1$ depends only on the fact that each Y_i has mean $\beta_0 + x_i\beta_1$. It does not depend on the distribution of Y_i (as long as the distribution has finite mean). Since $\hat{\beta}_1$ is a linear function of $Y_1, \dots, Y_n$, its variance depends only on the variances of the Y_i's (and the fact that they are independent). It doesn't depend on any other feature of the distribution. Indeed, we can write

$$\hat{\beta}_1 = \frac{\sum_{i=1}^{n}(x_i - \overline{x}_n)Y_i}{\sum_{j=1}^{n}(x_j - \overline{x}_n)^2} = \sum_{i=1}^{n} a_i Y_i,$$

where $a_i = (x_i - \overline{x}_n)/\sum_{j=1}^{n}(x_j - \overline{x}_n)^2$. Then $\text{Var}(\hat{\beta}_1) = \sum_{i=1}^{n} a_i^2 \text{Var}(Y_i)$. This depends only on the variances of the Y_i's, which do not depend on β_0 or β_1.

(b) Let T have the t distribution with k degrees of freedom. Then Y_i has the same distribution as $\beta_0 + \beta_1 x_i + \sigma T$, whose variance is $\sigma^2 \text{Var}(T)$. Hence, $\text{Var}(Y_i) = \sigma^2 \text{Var}(T)$. It follows that

$$\text{Var}(\hat{\beta}_1) = \sigma^2 \text{Var}(T) \sum_{i=1}^{n} a_i^2.$$

Let $v = \text{Var}(T)\sum_{i=1}^{n} a_i^2$.

(c) There are several possible simulation schemes to estimate v. The simplest might be to notice that

$$\sum_{i=1}^{n} a_i^2 = \frac{1}{\sum_{j=1}^{n}(x_i - \overline{x}_n)^2},$$

so that we only need to estimate $\text{Var}(T)$. This could be done by simulating lots of t random variables with k degrees of freedom and computing the sample variance. In fact, we can actually calculate v in closed form if we wish. According to Exercise 1 in Sec. 8.4, $\text{Var}(T) = k/(k-2)$.

15. (a) We are trying to approximate the value a that makes $\ell(a) = E[L(\theta, a)|\boldsymbol{x}]$ the smallest. We have a sample $\theta^{(1)}, \ldots, \theta^{(v)}$ from the posterior distribution of θ, so we can approximate $\ell(a)$ by $\hat{\ell}(a) = \sum_{i=1}^{v} L(\theta^{(i)}, a)/v$. We could then do a search through many values of a to find the value that minimizes $\hat{\ell}(a)$. We could use either brute force or mathematical software for minimization. Of course, we would only have the value of a that minimizes $\hat{\ell}(a)$ rather than $\ell(a)$.

(b) To compute a simulation standard error, we could draw several (say k) samples from the posterior (or split one large sample into k smaller ones) and let Z_i be the value of a that minimizes the ith version of $\hat{\ell}$. Then compute S in Eq. (12.2.2) and let the simulation standard error be $S/k^{1/2}$.